Springer Proceedings in Mathematics & Statistics

Volume 520

This book series features volumes composed of selected contributions from workshops and conferences in all areas of current research in mathematics and statistics, including data science, operations research and optimization. In addition to an overall evaluation of the interest, scientific quality, and timeliness of each proposal at the hands of the publisher, individual contributions are all refereed to the high quality standards of leading journals in the field. Thus, this series provides the research community with well-edited, authoritative reports on developments in the most exciting areas of mathematical and statistical research today.

Kusum Deep • Millie Pant •
Mohammad Khalid Pandit • Naveen Chauhan •
Ajay Kumar Mallick
Editors

Artificial Intelligence, Machine Learning and Intelligent Systems

ICAMS-2025, Hamirpur, India, February 7–8

Springer

Editors
Kusum Deep
Department of Mathematics
Indian Institute of Technology
Roorkee, Uttarakhand, India

Mohammad Khalid Pandit
Department of Computer Science and Engineering
National Institute of Technology
Hamirpur, Himachal Pradesh, India

Ajay Kumar Mallick
Department of Computer Science and Engineering
National Institute of Technology
Hamirpur, Himachal Pradesh, India

Millie Pant
Department of Applied Mathematics and Scientific Computing
Indian Institute of Technology
Roorkee, Uttarakhand, India

Naveen Chauhan
Department of Computer Science and Engineering
National Institute of Technology
Hamirpur, Himachal Pradesh, India

International Conference on Artificial Intelligence, Machine Learning and Intelligent Systems icams-1 icams-12025 ICAMS Hamirpur India 2025 02 07 2025 02 08 https://sites.google.com/nith.ac.in/icams2025/home

ISSN 2194-1009 ISSN 2194-1017 (electronic)
Springer Proceedings in Mathematics & Statistics
ISBN 978-981-95-3840-9 ISBN 978-981-95-3841-6 (eBook)
https://doi.org/10.1007/978-981-95-3841-6

Mathematics Subject Classification: 65D19, 65-XX, 49M41, 68Txx, 94A08

This Springer imprint is published by the registered company Springer Nature Singapore Pte Ltd.
The registered company address is: 152 Beach Road, #21-01/04 Gateway East, Singapore 189721, Singapore

Preface

This book presents a selection of papers presented at the International Conference on Artificial Intelligence, Machine Learning and Intelligent Systems (ICAMS 2025), held at the National Institute of Technology Hamirpur, from February 7 to 8, 2025. The conference was organized in collaboration with the Indian Institute of Technology Roorkee and the Centre for Artificial Intelligence, Islamic University of Science and Technology, Jammu and Kashmir, India. The conference focused on applications of machine learning and AI in the following areas:

1. Mathematical foundations of ML/optimization
2. Generative AI/computer vision
3. AI/ML in healthcare and agriculture
4. NLP/speech and signal processing
5. Edge AI and its applications in hardware

It brought together scholars, practitioners, and experts from around the world to discuss emerging trends and research in the field of AI and ML. These chapters reflect the vibrant discussions and innovative ideas that emerged during the event.

This book discusses various applications of computational algorithms in computer vision and pattern recognition related to engineering applications like computational biology, medicine, agriculture, etc. Authors have discussed the uses of AI-based algorithms for the detection of unconscious behaviour that can have lasting impact on detecting fatigue and drowsiness and for the detection of various diseases, including breast tumour, ovarian cancer, and paediatric pneumonia detection, all of which are very helpful from a medical point of view. Some contributions also talk about the use of AI-based techniques for detecting mental health issues and sleep apnoea. Several chapters are dedicated to a discussion of mathematical optimization of commonly used AI/ML algorithms using meta-heuristic techniques, including the infected hyena optimizer and socio-nomadic learning optimization. Several contributions also focus on the use of computational algorithms in emerging fields like electric vehicles, Indian language recognition and translation, intrusion detection, video summarization, and AI application in agriculture.

A thorough peer-review process was used to choose the papers that are included in this volume. Every contribution offers fresh perspectives on diverse AI applications, demonstrating a broad spectrum of research techniques and strategies.

I want to thank the reviewers for their helpful criticism and the authors for their diligence and hard work in preparing their papers. I also want to express my gratitude to the organizing committee for their hard work in making the conference a success. I hope that this book will not only serve as a valuable resource for researchers and practitioners but also inspire future innovations and discussions in the field of AI and ML.

Roorkee, Uttarakhand, India	Kusum Deep
Roorkee, Uttarakhand, India	Millie Pant
Hamirpur, Himachal Pradesh, India	Mohammad Khalid Pandit
Hamirpur, Himachal Pradesh, India	Naveen Chauhan
Hamirpur, Himachal Pradesh, India	Ajay Kumar Mallick

Contents

Real-Time Unconscious Human Behavior Detection Incorporating Enhanced CSPDarknet53 and YOLOv9

Md. Nazmus Sakib, Mst. Nishita Aktar, Afzal Hossain, Ahsan Ullah, Kh. Mustafizur Rahman, Shamsun Nahar, Jannatul Ferdous Mirza, and Md. Mahmud Khan

Abstract Empirical studies focus on detecting and investigating unconscious human behavior. The study proposes to improve CSPDarknet53 with spatial splits and incorporate it with the YOLOv9 object detection model.Studies assess the usefulness of the suggested model and look at the precision and temporal complexity of recognizing unconscious human behavior. Research establishes conscious human reflection time constraints 2500 milisecond $\approx$ and for unconscious behavior, currently depend on perception. Study have identified the importance of detecting unconscious human behavioral results of the significant impact on healthcare, secure vehicle accident prevention. The study proposed a sophisticated object detection model to investigate unconscious human activities using the UCF101 dataset with 13,320 videos spanning 101 action categories. Ten percent of the data is used for testing, while 90% of the data is used for training. YOLOv9's improvements in accuracy led to the development of a tool for real-time behavior study. Significant result is achieved with Precision (0. 985), Recall (0. 892), F1(0.936) and mAP50 (0.980), highlighting performance in efficient and accurate monitoring of processes of the proposed model. This research explores a novel approach to leveraging YOLOv9's advanced architecture and real-time capabilities for the specialized task of detecting unconscious human behaviors, demonstrating its practical applications and effectiveness in diverse real-world scenarios. Study associates behavioral detection with sensor capabilities and acknowledges a limited scope of comparative analysis among existing real-time object detection models.

Keywords Unconscious behavior detection · Object detection models · CSPDarknet53 · YOLOv9 · Real-time processing

M. N. Sakib (✉) · M. N. Aktar · A. Hossain · A. Ullah · K. M. Rahman · S. Nahar · J. F. Mirza · M. M. Khan
Computer Science & Engineering, World University of Bangladesh, Dhaka, Bangladesh
e-mail: afzal2@cse.wub.edu.bd; ahsan.ullah@cse.wub.edu.bd

K. Deep et al. (eds.), *Artificial Intelligence, Machine Learning and Intelligent Systems*, Springer Proceedings in Mathematics & Statistics 520,
https://doi.org/10.1007/978-981-95-3841-6_1

1 Introduction

The intricacy of subtle and unconscious behaviors makes it difficult to identify unconscious human activity, even with advances in deep learning and object detection. In real-time situations, existing models frequently fall short in detecting these behaviors, particularly when scales and environmental variables change. Additionally, the majority of research uses traditional deep learning methods or conventional machine learning, which have limitations in terms of accuracy and computing efficiency. These gaps show that a stronger and more effective framework that can handle these constraints is required. The rationale behind this work was the increasing demand for precise and instantaneous unconscious behavior detection in critical applications such as patient monitoring, secure driving systems, and emergency response.

Using YOLOv9 in conjunction with an improved CSPDarknet53 architecture for robust detection, our study fills the gap. Present techniques are ineffective in quickly and precisely identifying unconscious activities.

Investigation revealed the necessity of spatial splitting in CSPDarknet53 enhances feature extraction by dividing raw map segments, enabling a more extensive selection of spatial patterns and correlations [1]. Approach made it easier to find a wider range of spatial correlations and patterns, which was essential for the precise and effective detection of unconscious human behavior (Table 1).

Table 1 Comparison of existing techniques and research gaps

Aspect	Existing techniques	Identified gap	Proposed solution
Model accuracy	Traditional ML models (SVMs, HMMs) and older DL models (YOLOv3, YOLOv5)	Limited accuracy in detecting subtle unconscious behaviors	Enhanced CSPDarknet53 for improved feature extraction and precision
Real-time processing	Limited due to high computational complexity	Struggles to meet real-time requirements in dynamic environments	YOLOv9 integration for efficient real-time behavior detection
Scalability and adaptability	Lack of multi-scale feature aggregation and attention mechanisms	Inability to adapt to varying scales and environmental conditions	Use of FPN and PANet for robust multi-scale feature aggregation and adaptability
Behavior detection specificity	Focus on general object detection rather than specialized unconscious human behavior detection	Ineffective in detecting complex, subconscious human actions	Integration of specialized layers tailored for unconscious human behavior detection
Dataset utilization	Limited testing on general datasets, with minimal focus on behavioral datasets	Insufficient evaluation on comprehensive datasets for unconscious behaviors	Use of UCF101 dataset with extensive testing and fine-tuning

1.1 Significance of Unconscious Behavior Detection

Study identifies that conscious behavior refers to actions driven by awareness and intention. Illustrations of conscious behavior encompass developing purposeful choices, participating in problem-solving, and actively concentrating on a specific task or objectives. On the contrary, unconscious behavior denotes actions that transpire without our conscious awareness or regulation. Instances of unconscious behavior encompass routines, automatic reactions, and behaviors influenced by underlying psychological mechanisms. Additionally, the body's impact or the interactions between various mental entities might cause unconscious behavior. For instance, the unconscious mind might impact the conscious mind either directly or indirectly through a third entity, such as the body, resulting in behaviors like craving, excessive smoking, or drowsiness. Moreover, unconscious behavior can encompass processes that take place beyond conscious awareness, such as perception, memory, and sensory processing. Detection of unconscious human behavior is crucial in various contexts such as emergency response systems, patient monitoring, and vehicle safety. Conventional approaches typically depend on manual observation or less effective machine learning algorithms. Recent progress in deep learning, specifically in object detection frameworks, presents novel opportunities for instantaneous analysis.

1.2 YOLOv9 Overview

Integration of advanced methodologies in the identification of unconscious human actions can significantly improve the effectiveness of the model [1]. A key advancement involves utilizing an enhanced backbone network, such as a customized CSPDarknet53, to enhance feature extraction capabilities by capturing intricate details that are crucial for identifying subtle and subconscious behaviors. The customized CSPDarknet53 helps the model in discerning complex patterns and nuances in the data, which is essential to accurately identify subconscious behaviors that could otherwise be missed. Moreover, the integration of Feature Pyramid Networks (FPN) and Path Aggregation Networks (PANet) layers into the framework enables robust multi-scale feature aggregation. PANets and FPNs facilitate the merging of characteristics from diverse network levels, allowing the model to detect objects and behaviors of different sizes and under diverse conditions. This capability is vital for subconscious behavior detection as behaviors can manifest at varying scales and in different environments. Attention modules further enhance the model's performance by focusing on relevant features while suppressing irrelevant information. This selective attention improves the model's ability to detect subtle behaviors by reducing noise and enhancing detection accuracy. The attention mechanism ensures the model's focus remains on features most indicative of subconscious behavior, thereby enhancing overall detection precision [2]. The training process commences with pre-trained weights from YOLOv9, providing a solid foundation of existing knowledge and expediting the model's adaptation to subconscious behavior detection. Fine-tuning these weights on a specialized dataset enables the model to refine and tailor its detection capabilities to the specific attributes of subconscious behavior, thereby enhancing its performance and accuracy [3]. Several data augmentation techniques are used, such as random cropping, flipping, and color modifications, which provide diversity to the training dataset and further improve the robustness and adaptability of the model. These augmentation techniques assist the model in learning to handle different conditions and variations in the data, making it more versatile and suitable for real-world applications. Optimizing key hyperparameters such as learning rate, batch size, and optimizer configurations (e.g., Adam or SGD) is another crucial aspect of model development. By means of extensive testing, these hyperparameters are fine-tuned to achieve an equilibrium between training efficiency and model precision, guaranteeing peak performance without undue computing burden. Evaluation metrics are essential for assessing the model's effectiveness. Precision gauges the accuracy of positive detection's, indicating the number of true positive behaviors detected. Recall evaluates the model's ability to identify all relevant instances, demonstrating its efficacy in capturing all potential behaviors. A thorough evaluation of the model's performance can be obtained by comparing the average precision across several classes and Intersection over Union (IoU) requirements, while the F1 score incorporates accuracy and recall to offer a general indicator of the efficacy of detection [4].

Real-time processing capabilities are crucial for applications requiring live monitoring. The model's ability to swiftly and accurately process data in real-time is vital for its practicality in dynamic environments necessitating immediate detection of unconscious behaviors [4]. The incorporation of real-time functionality enhances the model's relevance in a variety of scenarios, ranging from security systems to health monitoring. Employing a modified CSPDarknet53 network, advanced feature aggregation methods, attention modules, pre-trained weights, data augmentation, optimized hyperparameters, and thorough evaluation metrics leads to a highly efficient model for identifying unconscious human behavior. The integration of real-time processing capabilities further guarantees the model's practicality and effectiveness in live applications [5].

1.3 Comparative Algorithms Utilized in Detecting Unconscious Behavior

An extensive analysis of an array of research indicates that a wide range of algorithmic techniques are used to identify unconscious human activity. There has been much research done on traditional machine learning methods including Support Vector Machines (SVMs), Hidden Markov Models (HMMs), and Gaussian Mixture Models (GMMs). Recurrent neural networks (RNNs) and Convolutional neural networks (CNNs), and its variations are being used for the purpose of extracting complicated patterns from complex behavioral data, thanks to recent developments in deep learning. Although these approaches have demonstrated potential, hybrid approaches that include both standard and deep learning techniques have surfaced as viable means of improving detection accuracy [6].

1.4 Partial Cross-Stage Network (CSPDarknet53)

In order to improve efficiency and performance in deep neural networks, especially for object detection tasks, CSPDarknet53, an improved version of the Darknet53 architecture, was introduced as part of the CSPNet (Cross Stage Partial Network) framework [7] (Fig. 1).

Initial Convolutional Layer A large convolutional layer is the first layer the network uses to extract basic information from the input image. This layer downsamples the spatial dimensions while capturing a wide range of contextual information by using a large kernel size, usually 7x7, with a stride of 2. After that, non-linearity is introduced using a Leaky ReLU activation function and training is stabilized using Batch Normalization (BN) [9].

Cross-Stage Partial Network (CSPNet) Modules The CSPNet module addresses two of the main issues with standard deep networks: gradient flow and processing efficiency, which is the primary innovation of CSPDarknet53.Feature mappings from the first convolutional layer are divided into two groups by the CSPNet [10]. Independently, every group goes through a number of residual blocks.Residual Block Processing: Within each group, residual blocks are used to facilitate learning by allowing gradients to flow through skip connections. Convolutional layers (often 3 $\times$ 3 kernels) make up each residual block, which are then followed by Batch Normalization and Leaky ReLU activation.Outputs from the two groups are combined after processing. In comparison to full residual connections, this partial connection approach guarantees that the network will experience better gradient flow and lower computing cost [11].

Residual Blocks An essential component of the CSPDarknet53 architecture are leftover blocks. Their purpose is to enhance network training by the acquisition

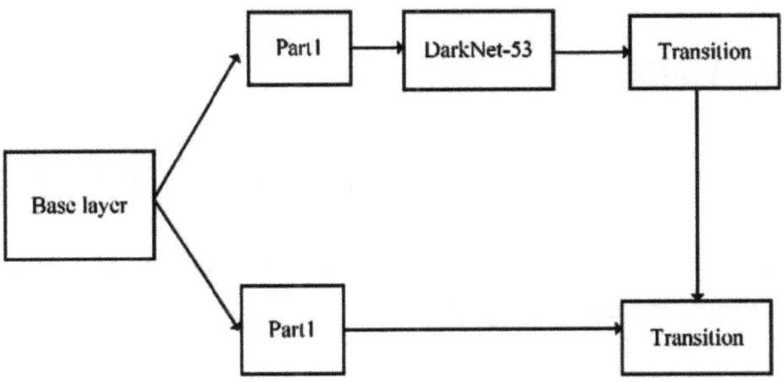

Fig. 1 Darknet architecture [8]

of residual mappings, which represent the discrepancy between the intended output and the input. Typically, every block is composed of Convolutional Layers. To capture local information, these layers make use of 3x3 convolutions.Batch Normalization Applied to each convolutional layer to normalize the feature maps and stabilize training.Leaky ReLU Activation: Contributes to the solution of the vanishing gradient problem by introducing non-linearity.

By adding each residual block's output to its input through a skip connection, gradient information is preserved and spread throughout the network [12].

Down Sampling Layers Gradient information in each residual block is dispersed throughout the network by adding its output to its input via a skip connection.Max-Pooling Layers utilized to preserve significant characteristics while shrinking the feature maps' spatial dimensionality,Strided Convolutions applied with a stride greater than 1 to achieve downs sampling and increase the receptive field of the network [13].

To strike a compromise between computational efficiency and feature resolution, down sampling is carefully carried out at several phases..

Transition Stages In CSPDarknet53, transition stages are intended to link various network segments while coordinating feature dimensions among stages. These shifts frequently entail:

Convolutional Layers: Modifying the feature map's size to make sure it works with later layers [14].

Pooling or striding: To retain feature integrity throughout transitions and match dimensions.

Final Convolutional Layers and Detection Head The last layers of CSPDarknet53 concentrate on creating the feature maps required for tasks involving detection. Among these levels are:

Convolutional Layers: Enhance the characteristics and get them ready for processing by the detecting head.

Detection Head: This part produces predictions for bounding boxes, object classes, and objectness scores in the context of object detection. To get the final detection outputs, it usually entails adding further convolutional layers and then activation functions [15].

1.5 Existing Method

Study demonstrated two categories for feature maps in CSPDarknet53 X_1 and X_2 [16].

Separate residual blocks are processed for each category. Following processing, the two feature maps are joined together:

$$X_{\text{concat}} = X1_{\text{processed}}, X2_{\text{processed}} \tag{1}$$

Here, It has been analyzed that exiting method considers linear feature maps.To address this deficiency, we advocate for an innovative methodology aimed at discerning spatial movement patterns exhibited by individuals, which seeks to mitigate the occurrence of accidents. The model, derived from the field of computer vision, possesses significant strengths in delineating the interrelations among segments in relation to temporal variables; consequently, we employ it to elucidate the interaction dynamics inherent in human motion.

1.6 Materials and Methods

Study proposes modifying CSPDarknet53 by incorporating Spatial Splits. This technique divides feature maps into groups and further splits these groups into different spatial regions. For instance, feature maps can be split into two regions based on their height or width. Utilizing leftover blocks, each spatial region is handled separately. After that, these dimensional regions yield the processed feature maps X_{input}.

Due to unconscious movement it has been observed that human behaviour contains spatial dimensions .Hence research improves detection mechanism with H×W×CH×W×C, where HH is height, WW is width, and CC is the number of channels. Perform during spatial splitting along height.

$$X_{\text{spatialSplit1}}, X_{\text{spatialSplit2}} = \text{SpatialSplit}(X_{\text{input}}, \text{axis} = H) \tag{2}$$

The modification divides the feature maps into regions based on their spatial dimensions. For splitting along the height, this results in two feature maps with dimensions $\frac{H}{2} \times W \times C$. This approach allows the network to focus on different spatial regions independently, which can be beneficial for capturing region-specific features. Each spatial region is processed through its own set of residual blocks, potentially allowing the network to learn more detailed and localized features. The processed spatial regions are then concatenated to combine the features from different regions, maintaining a comprehensive representation of the original input.

1.7 Proposed Architecture of CSPDarknet53

The UCF101 dataset, which comprises 13,320 videos categorized into 101 action classes, was employed in the research. After the data was obtained, a pipeline for preparation was established. Utilizing a Cross Stage Partial Network (CSPNet) module to fine-tune the CNN to increase its performance. Specifically, this transfer learning strategy improves the challenge of detecting unconscious behavior. The method proceeds to the spatial split stage once the model has been fine-tuned iteratively until the required condition is satisfied.Potential changes to the network's

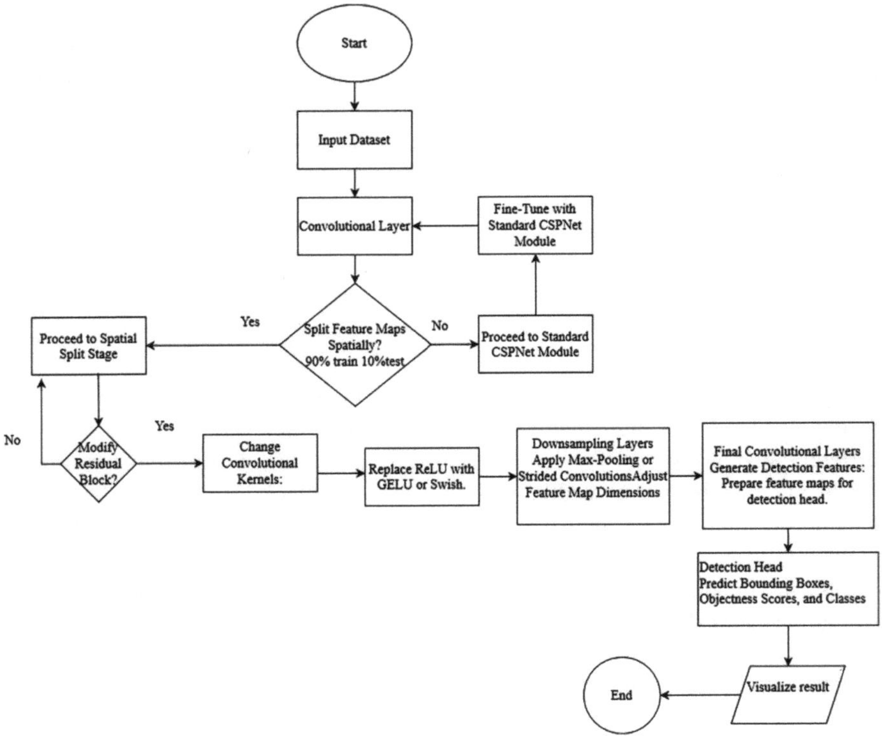

Fig. 2 Notable approach

residual block design were investigated in the study. It is well known that residual connections help to mitigate the vanishing gradient problem, which makes training deeper networks easier.In order to assess how different convolutional kernel sizes affect feature extraction, a hyperparameter sweep was performed, with smaller kernels focusing on fine-grained details and larger kernels gathering broader contextual information. In order to potentially improve model performance through the inclusion of non-linearity and mitigation of the vanishing gradient problem, the ReLU activation function was also substituted with GELU and Swish.By using common methods like max-pooling and strided convolutions to gradually lower spatial dimensions, downsampling was accomplished.Bounding box coordinates, objectness ratings, and class probabilities of observed behaviors were predicted using enhanced feature representations and finally produceo output (Fig. 2).

2 Result and Discussion

Figure 3 represents a confidence score of 75% $\approx$ 85%, indicating that the model is approximately.Due to the complexity of code optimization 25% $\approx$ 15% has been recognized as outlier.

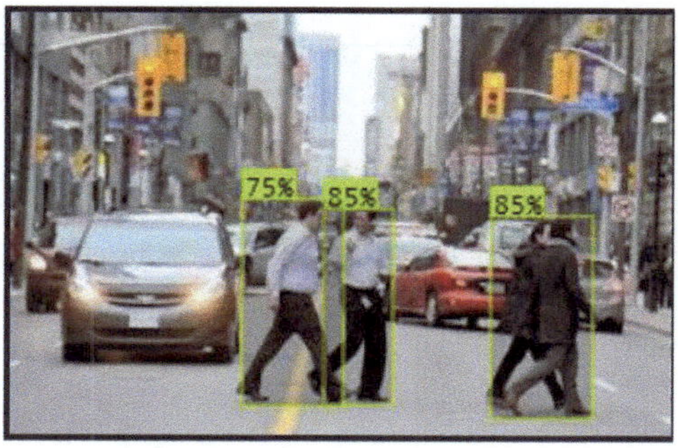

Fig. 3 Object detection

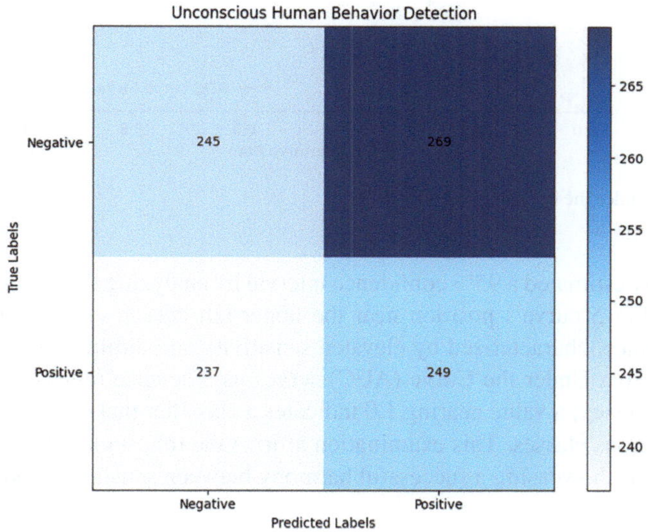

Fig. 4 Confusion matrix

Study determined a confident level for detecting object within the bounding box dimensional movement of human. Here, confidence score was set to reach a threshold value of 0.7 in order to each a certain level of human movement detection. Research found that below 0.7 threshold below leads to erroneous movement detection.

2.1 *Statistical Evaluation*

Figure 4 shows the confusion matrix that demonstrates how well a classification algorithm performs. A confusion matrix shows and summarizes how well a suggested model's classification works.Depicts a confusion matrix in which human movement is referred to as Detected Behavior and Not Detected Behavior.

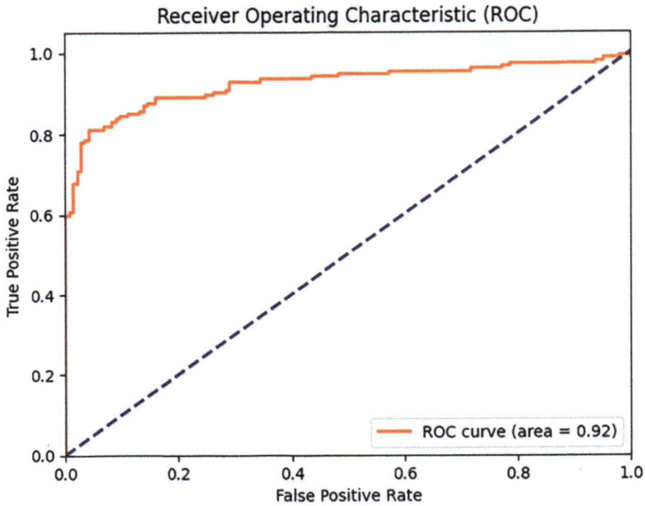

Fig. 5 Area under the curve

The study estimated a 95% confidence interval by analyzing the confusion matrix results. In Fig. 5 curve's position near the upper-left corner suggests a high level of performance, characterized by elevated sensitivity and diminished false positive rates. The Area Under the Curve (AUC) serves as a concise representation of the general efficiency; a value nearing 1.0 indicates a classifier that excels in discerning between various classes. This examination affirms the robust classification abilities of our model, showcasing a successful harmony between sensitivity and specificity.

3 Comparative Investigation

Enhanced CSPDarknet53 with Spatial Splits outperforms all other models in terms of all metrics in Fig. 6 (Table 2).

Boasting a precision of 0.985, recall of 0.892, F1 score of 0.936, and mAP50 of 0.980, showcasing its superior accuracy in both classification and object detection. YOLOv9 comes next with a precision of 0.912, recall of 0.878, F1 score of 0.895, and mAP50 of 0.925, displaying impressive performance but slightly trailing behind the Enhanced CSPDarknet53. EfficientDet demonstrates solid performance with a precision of 0.875, recall of 0.855, F1 score of 0.865, and mAP50 of 0.880, albeit falling behind the top two models. RetinaNet exhibits a precision of 0.860, recall of 0.842, F1 score of 0.851, and mAP50 of 0.865, positioning it as the least effective among these models based on the aforementioned metrics.

The time complexity of the suggested Enhanced CSPDarknet53 model with spatial splits in comparison to conventional object detection models is shown in Fig. 6. The efficiency of the suggested model in real-time applications is demonstrated by

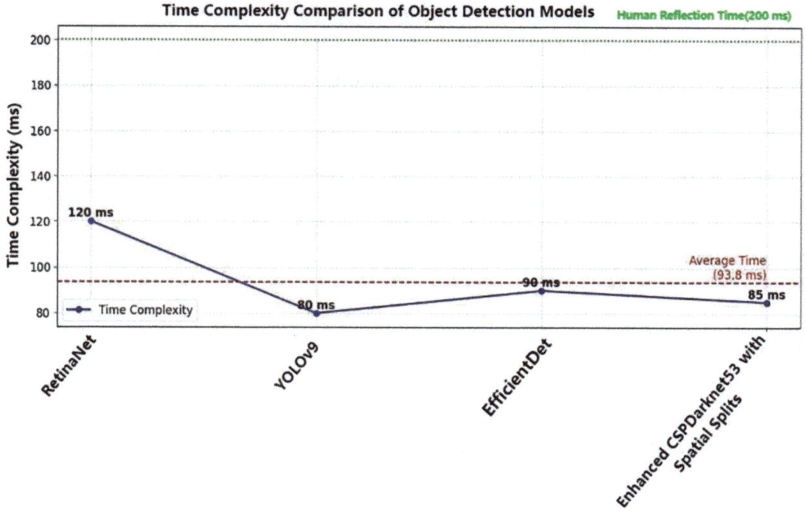

Fig. 6 Time complexity analysis of the proposed Enhanced CSPDarknet53 model, highlighting its performance improvements over traditional models

Table 2 Comparative classification report for different object detection models

Model	Precision	Recall	F1 score	mAP50
Enhanced CSPDarknet53 with spatial splits	0.985	0.892	0.936	0.980
YOLOv9	0.912	0.878	0.895	0.925
EfficientDet	0.875	0.855	0.865	0.880
RetinaNet	0.860	0.842	0.851	0.865

the fact that it achieves noticeably shorter execution times as the magnitude of the incoming data rises. Optimized spatial divides and attention techniques that lower computing overhead are responsible for this improvement.

4 Conclusion

Research presents a novel method for identifying subconscious human actions through the enhancement of CSPDarknet53 using spatial divisions and its fusion with the YOLOv9 object detection system. The model exhibits exceptional performance metrics, including a Precision of 0.985, Recall of 0.892, F1 score of 0.936, and mAP50 of 0.980, underscoring its precision and effectiveness in real-time settings. By leveraging the UCF101 dataset for training (90%) and testing (10%), the study underscores the tool's efficacy in real-time behavioral monitoring. Nevertheless, the research faces constraints such as its reliance on a specific dataset and the absence of a comparative evaluation with alternative real-time object detection models. To address these limitations, future endeavors should focus on enlarging the dataset, integrating diverse scenarios, and executing a more thorough comparison with established models to further authenticate and strengthen the method's resilience and practicality.

Acknowledgments We express profound gratitude for the invaluable input provided by the authors and co-authors during the progression of this research. Their commitment and hard work have played a crucial role in the accomplishment of this study.

References

1. C.-Y. Wang, I.-H. Yeh, and H.-Y. M. Liao, "Yolov9: Learning what you want to learn using programmable gradient information," *arXiv preprint arXiv:2402.13616*, 2024.
2. M. A. R. Alif and M. Hussain, "Yolov1 to yolov10: A comprehensive review of yolo variants and their application in the agricultural domain," *arXiv preprint arXiv:2406.10139*, 2024.
3. I. U. W. Mulyono, Y. Kusumawati, A. Susanto, C. A. Sari, H. M. M. Islam, and M. Doheir, "Hiragana character classification using convolutional neural networks methods based on adam, sgd, and rmsprops optimizer," *Scientific Journal of Informatics*, vol. 11, no. 2, pp. 467–476, 2024.
4. K. Bian and R. Priyadarshi, "Machine learning optimization techniques: a survey, classification, challenges, and future research issues," *Archives of Computational Methods in Engineering*, pp. 1–25, 2024.
5. B. F. Azevedo, A. M. A. Rocha, and A. I. Pereira, "Hybrid approaches to optimization and machine learning methods: a systematic literature review," *Machine Learning*, pp. 1–43, 2024.
6. O. M. Ogundile, A. A. Owoade, O. O. Ogundile, and O. P. Babalola, "Linear discriminant analysis based hidden markov model for detection of mysticetes' vocalisations," *Scientific African*, vol. 24, p. e02128, 2024.
7. J. Wang, Z. Zhang, B. Dai, K. Zhao, W. Shen, Y. Yin, Y. Li, *et al.*, "Cow-yolo: Automatic cow mounting detection based on non-local cspdarknet53 and multiscale neck," *International Journal of Agricultural and Biological Engineering*, vol. 17, no. 3, pp. 193–202, 2024.
8. S. Chanda, Y. N. Kumar, S. Srivastava, R. Rani, M. Shree, and A. Mohapatra, "Optimizing facial feature extraction and localization using yolov5: An empirical analysis of backbone architectures with data augmentation for precise facial region detection," *Multimedia Tools and Applications*, pp. 1–22, 2024.
9. C. J. Shallue and D. J. Eisenstein, "Reconstructing cosmological initial conditions from late-time structure with convolutional neural networks," *Monthly Notices of the Royal Astronomical Society*, vol. 520, no. 4, pp. 6256–6267, 2023.
10. M. I. M. Yusoff, "Machine learning: An overview," *Open Journal of Modelling and Simulation*, vol. 12, no. 3, pp. 89–99, 2024.
11. L. Zhou, G. Chen, L. Liu, R. Wang, and A. Knoll, "Real-time semantic segmentation in traffic scene using cross stage partial-based encoder–decoder network," *Engineering Applications of Artificial Intelligence*, vol. 126, p. 106901, 2023.
12. H. Zhang, S. Li, Q. Miao, R. Fang, S. Xue, Q. Hu, J. Hu, and S. Chan, "Surface defect detection of hot rolled steel based on multi-scale feature fusion and attention mechanism residual block," *Scientific Reports*, vol. 14, no. 1, p. 7671, 2024.
13. Y. Chen, B. Tan, J. Wu, and D. Hu, "Pjscc: A puncturing-based joint source channel coding scheme with hierarchical down-sampling layer," in *ICASSP 2024-2024 IEEE International Conference on Acoustics, Speech and Signal Processing (ICASSP)*, pp. 8976–8980, IEEE, 2024.

14. Y. Zhou, S. Zheng, and J. L. Hensen, "Machine learning-based digital district heating/cooling with renewable integrations and advanced low-carbon transition," *Renewable and Sustainable Energy Reviews*, vol. 199, p. 114466, 2024.
15. Y. Liu, Z. Zhou, Y. Wang, and C. Sun, "Head-integrated detecting method for workers under complex construction scenarios," *Buildings*, vol. 14, no. 4, p. 859, 2024.
16. "Papers with Code - CSPDarknet53 Explained — paperswithcode.com." https://paperswithcode.com/method/cspdarknet53. [Accessed 14-08-2024].

Exploring Machine Learning and IoT Technologies for Disease Detection in Cows: A Systematic Review

Devinder Kaur and Amandeep Kaur Virk

Abstract The animal farming industry has become a hotspot for research these days. Prioritizing animal welfare is essential due to the crucial role they play in providing dairy products to people. Forecasting cattle health and illnesses in advance is particularly important, considering the potential for animal-to-human disease transmission. Ensuring optimal animal health is vital because humans depend to satisfy their demand for dairy products. Today, numerous Internet of Things (IoT) based technologies are available to keep an eye on the essential health metrics of animals. Besides this, machine learning is being extensively used as a tool for supporting decisions in precise livestock farming. However, an overview of the advancements in IoT and machine learning for precision livestock farming is required, thus a thorough literature analysis has been conducted. This paper's objective is to give an assessment of the detecting illnesses in cattle using machine learning and the internet of things (IoT) for precision livestock farming and to evaluate the effectiveness of machine learning techniques. The challenges being faced by dairy farmers in application of Smart Farming techniques are also discussed. This Systematic Literature Review study will prove to be beneficial not only for researchers but for dairy farm practitioners also.

Keywords Precision livestock farming · IoT · Machine learning · Disease detection

D. Kaur (✉)
Sri Guru Granth Sahib World University, Fatehgarh Sahib, Punjab, India

Department of Computer Science, Mata Gujri College, Fatehgarh Sahib, Punjab, India
e-mail: devinderkaurcomp@matagujricollege.org

A. K. Virk
Department of Computer Science, Sri Guru Granth Sahib World University, Fatehgarh Sahib, Punjab, India
e-mail: amandeep_virk@sggswu.edu.in

K. Deep et al. (eds.), *Artificial Intelligence, Machine Learning and Intelligent Systems*, Springer Proceedings in Mathematics & Statistics 520,
https://doi.org/10.1007/978-981-95-3841-6_2

1 Introduction

The primary sources of livelihood in rural India are agriculture and practice of dairy farming. Increased food and dairy production are critically needed in the future as highlighted by the growing population. Most of India's historical livestock farming regions are following traditional practices with choices being made by farmers depending solely on their individual experience. There has, nevertheless, been a noticeable departure from these conventional methods. With the current pace, dairy farmers are being guided towards the field of precision livestock farming via precision dairy technology.

A new paradigm that combines several cutting-edge techniques with its challenges is the use of the Internet of Things in healthcare for animals. The difficulties being faced in traditional livestock farming can be overcome by focusing on the fundamental ideas, specific applications, and recognition of limitations [1]. Since human beings depend on animals to meet their demand for dairy products, maintaining the health of these animals is vital. An early prediction of the health and disease of cattle is essential given the prevalence of numerous diseases where animal to human transmission is possible [2]. Still, several elements associated with dairy agriculture such as feeding, milking, reproduction and well-being conditions are regarded as distinct entities so, it has become a very challenging task for the dairy farmers to ensure the well- being of each individual cattle. Moreover, with advancements in on-animal wearables, data from monitoring cattle behaviors has recently grown [3].

The most significant area of recent advancement is precision livestock farming. Innovative technology like big data, machine learning, artificial intelligence and sensors give farmers a choice; as opposed to responding to illnesses as they emerge, these technologies can continuously monitor critical factors employed to diagnose veterinary health, for instance movement, temperature, heart rate, and humidity. Farmers are now able to identify and prevent a big outbreak of diseases among dairy animals through use of sensors to continuously gather data, employing machine learning to anticipate deviations, and keeping an eye on the outcomes. To ensure that the best knowledge is used on the dairy operations, data-driven, and electronically controlled livestock management system are crucial. The new technology of Precision Livestock Farming enables use of livestock management techniques on farms [4]. Precision livestock farming involves managing, tracking, and modelling animal production by implementing the ideas and methods of process engineering. To aid farmers and other participants within the cycle of cattle production, precision livestock farming appears to be the only practical option [5].

The farmer may benefit from this type of system in two ways. First, it might make it possible for fewer farmers to attend a higher number of livestock, which lowers the cost of production. A similar technology can warn farmers regarding the potential for a disease even before it manifests, enabling them to take immediate action. These innovative methods leveraged for dairy farming sector can offer fresh perspectives for better decision-making in the future [6].

As improvement and maintenance of the cattle health is among the main goals of Precision Livestock Farming (PLF) [7], these technologies provide a chance for increased productivity and early detection of health problems. To extract information from a variety of data, Precision Livestock farming intends to integrate all the technology that is now accessible with intelligent software [8]. These "smart technologies," supporting dairy farmers in monitoring animal welfare is a key research topic. The transition to sustainable livestock farming methods is already generating large investments in a number of nations [9]. Recent developments in PLF technology can be broadly classified into several key areas, including individual animal health, herd-level management, data analysis and management [10]. Although these will not ever completely replace farmers' instinct and management, precise livestock monitoring technology may improve it by allowing them to make more educated choices [11]. Utilizing electronic measurement, analysis, and control systems, PLF envisions maintaining livestock health, ecological sustainability, efficiency, and economic viability [3].

Machine learning has proven to be an effective technology in dairy research for creating and enhancing farmer decision support. The capacity of machine learning algorithms to evaluate integrated data sources that define and, eventually, permit dairy cow management in accordance with all important influencing factors is provided by the multifactorial nature of the cow [12]. Machine learning techniques are capable to forecast typical behavior of an animal and create an alert when the anticipated pattern of behavior varies from a certain threshold [9].

The rapid advancement of Internet of Things (IoT) technologies and Machine Learning (ML) has revolutionized modern management of livestock, offering unprecedented opportunities for improving cattle health monitoring. Management of health effectively in livestock is crucial for ensuring productivity, reducing disease outbreaks, and enhancing animal welfare. While a wealth of research exists on these topics, there is a pressing need to consolidate the knowledge and provide a comprehensive review of the interplay between IoT and ML in this domain.

This article bridges that this gap by offering a detailed summary of IoT and machine learning tools for tracking the health of cattle. Specifically, it:

1. Summarizes the research studies, emphasizing best practices for integrating ML and IoT to track the health of livestock.
2. Highlights the most used machine learning algorithms for health monitoring and their unique applications in precision livestock farming
3. Compares these algorithms based on different assessment criteria to guide practitioners in deciding which approaches are best for their requirements
4. Discusses key challenges faced by dairy practitioners in adopting and implementing advanced technologies, including issues of scalability, cost, and data reliability.

By providing these insights, this review attempts to be a useful tool for researchers, practitioners, and industry stakeholders, opening the door for more effective and sustainable livestock management practices.

Following are the five sections that make up the review paper: The systematic literature review approach is covered in Sect. 2, along with the definition of the research questions and the criteria for selection. Section 3 outlines the findings, including the responses to the research questions. Section 4 summarizes the challenges in this field and the general conclusions are discussed in Sect. 5 of the article.

2 Methodology

2.1 Research Queries

Regarding this comprehensive review, the following research queries have been framed:

RQ1: Which applications of the Internet of Things technologies and machine learning are relevant to precision livestock farming?
RQ2: Which parameters are considered to track the health conditions of cattle?
RQ3: What algorithms are used to diagnose animal diseases using Machine Learning?
RQ4: Which evaluation parameters are used?
RQ5: Which is the most widely used algorithm?
RQ6: What issues are mentioned in the articles as challenges?

2.2 Databases

The search was performed across the following databases:

1. **ScienceDirect**: A search using the keywords **"Machine Learning" AND "Precision Livestock Farming"** was performed. Note that due to restrictions in the database, wildcards were not allowed, so the search was limited to these terms.
2. **SpringerLink**: Since SpringerLink does not allow searching within specific fields (e.g., journal title, abstract, or keywords), the search was expanded to include **"Cattle Health" AND "Machine Learning"**, capturing a broader set of studies across publication content.
3. **IEEE Xplore**: For IEEE Xplore, the search string **"Machine Learning" AND "Precision Livestock Farming"** was used to find papers that focused on the intersection of these technologies.
4. **PubMed**: A similar search strategy was used in PubMed with the search string **"Machine Learning" AND "Precision Livestock Farming"** to identify studies relevant to medical and biological aspects of livestock health.

Backward snowballing technique is conducted to identify related studies.

2.3 Search Strategy

The initial search terms included **"machine learning" AND "precision livestock farming"**, which were used to capture studies that combine these key technologies with cattle management. To narrow the focus specifically to cattle health, the search string was expanded to include **"cattle health"** as an additional keyword. The process involved examining the abstracts and summaries of the retrieved studies to identify relevant papers, after which additional relevant keywords were inserted to further refine the search.

2.3.1 Date Ranges

The literature search covered articles published between 2017 and 2022. This date range was chosen to capture both recent advancements in machine learning and IoT technologies as well as foundational research from the last decade. Research older than 2017 was excluded to guarantee the inclusion of the most current and pertinent research in the area.

2.3.2 Search Strings

Each database's search string required to be customized. Since it is not allowed to use wildcards in ScienceDirect, so, the search keywords "Machine Learning" AND "precision Livestock farming" was used for ScienceDirect. Because SpringerLink does not allow searching in the journal title, abstract, or keywords, "Cattle Health" AND "Machine Learning" was used to search across the publication. For IEEE Xplore and PubMed "Machine Learning" AND "Precision Livestock Farming" was used as search string.

2.3.3 Criteria for Selection

To ensure that the systematic review comprises only relevant papers, both **inclusion** and **exclusion** criteria were applied:

Criteria for Exclusion:

1. Research that was not centered on the application of technology in **dairy production** were excluded.
2. Non-English language papers were excluded to maintain consistency in the review.
3. Papers that were reprints or duplicates of previous studies were excluded.
4. Studies that were not available as **open access** were excluded to ensure accessibility of the research.

Inclusion Criteria:

1. Studies that discussed the use of **IoT** or **Machine Learning** for health-related forecasting in cattle were included.
2. Both **primary research studies** and **review articles** that addressed the key themes of the research questions were included.

Study Selection Process: The studies retrieved from each database were carefully screened based on the inclusion and exclusion criteria. After filtering for relevance, the selected studies were compared against the research questions to ensure they provided insightful observations regarding the use of machine learning and IoT in cattle health management. The final list of included publications was expanded to ensure extensive coverage of the topic.

2.4 Data Extraction

The vital information was taken from the studies and then synthesized to respond to the research questions. Table 1 provides a summary of the studies on machine learning for cattle illness prediction and Table 2 presents the studies which focus upon applying IoT for cattle health monitoring. All the studies that were chosen were read in their entirety. The extracted data was summarized using a spreadsheet. To tackle the research inquiries, data from all the research studies was obtained, that included aims, machine learning algorithms, publication year, performance, evaluation criteria, and obstacles. To respond to the study questions, all the gathered data was categorized and synthesized into different classifications. The findings of this SLR are presented in the following section.

The following 16 publications which focus on use of machine learning systems to detect health status of cows have been considered for the systematic review. Out of which ten studies have been selected from ScienceDirect, one from SpringerLink and five studies from PubMed have been chosen.

The following ten research studies focus on the application of IoT in livestock farming. Among this list four studies have been selected from ScienceDirect, two from SpringerLink, three from IEEE Xplore and one from PubMed.

Tables 1 and 2 show that there is broad usage of machine learning techniques and the Internet of Things (IoT) in the livestock farming field especially for the monitoring and predicting cattle health.

3 Results

This paper aims to give a general summary of the precision cattle farming using machine learning and the Internet of Things for disease diagnosis and animal health monitoring. The results attained from the study can be outlined as follows:

Table 1 List of the selected studies using machine learning in livestock farming for a systematic literature review

Source	References
ScienceDirect	[13]
	[14]
	[15]
	[16]
	[17]
	[18]
	[19]
	[20]
	[21]
	[22]
PubMed	[12]
	[23]
	[24]
	[25]
	[26]
SpringerLink	[27]

Table 2 List of research employing IoT in cattle farming that have been chosen for a comprehensive literature review

Source	References list
ScienceDirect	[28]
	[29]
	[30]
	[31]
SpringerLink	[32]
	[33]
IEEE xplore	[34]
	[35]
PubMed	[36]

In the initial search, 206 studies were found from diverse electronic databases. Out of these, 26 research articles that fulfil the criteria for selection were then further evaluated to help uncover pertinent data that would help respond to the research inquiries.

The key applications of IoT and machine learning for cattle health monitoring were recognized. About 38% of the studies were devoted to cattle health monitoring using some algorithms for machine learning for disease detection.

As evident from Fig. 1, 14 machine learning algorithms generally used to detect diseases in cattle have been identified. Random Forest (RF) is the most widely used method, followed by Support vector machines, decision trees, and K nearest neighbours. From our selection, five common health parameters identified in prediction of health status of cattle are body temperature, heart rate, motion, sound, rumination time, which were used in most of the studies.

Figure 2 displays the journal—wise distribution of relevant studies selected for the systematic review and Fig. 3 reflects the distribution of the research findings

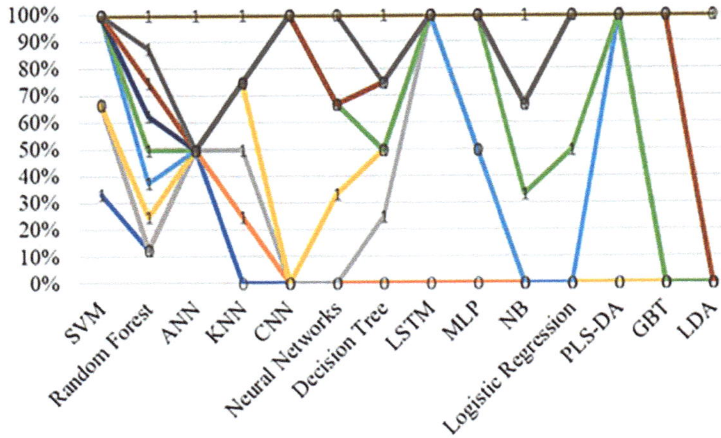

Fig. 1 Machine learning methodologies employed in selected studies

Fig. 2 Publications involving the use of IoT technologies and machine learning for precision livestock management distributed by journal

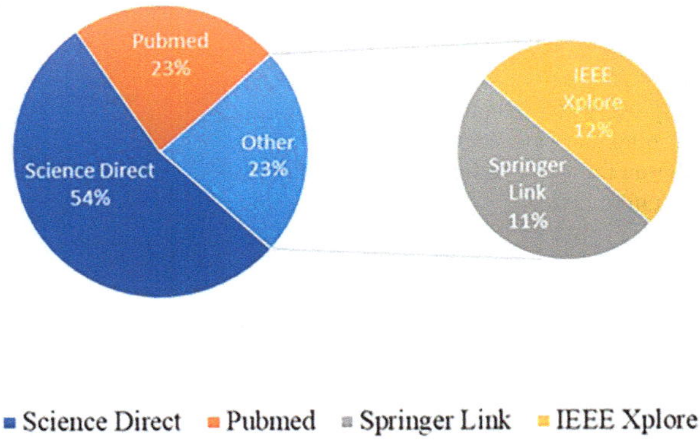

Table 3 Database distribution of the publications that were culled for the systematic literature review

Source	Initial search papers: number	Number of papers eligible
ScienceDirect	134	14
PubMed	27	6
IEEE xplore	34	3
SpringerLink	11	3
Total	**206**	**26**

considered by year. Table 3 presents the distribution of searched papers according to the corresponding database used.

As evident from Table 3, most of the papers that were taken into consideration for the review were found on ScienceDirect and PubMed.

From Fig. 3, it can be clearly realized that most of the publications considered in this review were published between the year 2017 and 2021 with the maximum

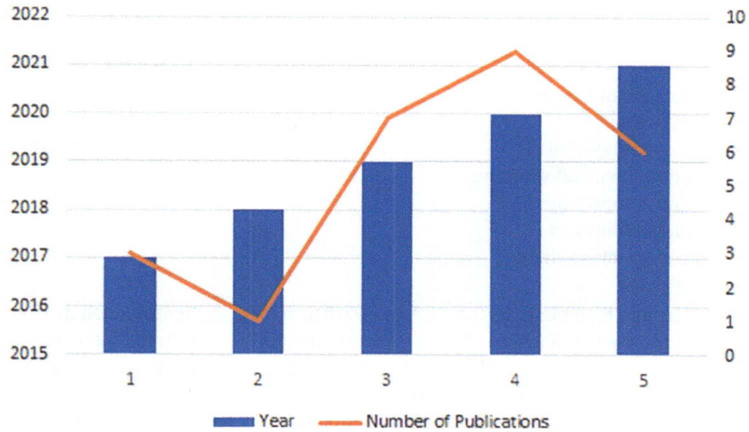

Fig. 3 Year-wise distribution of publications considered for the systematic review

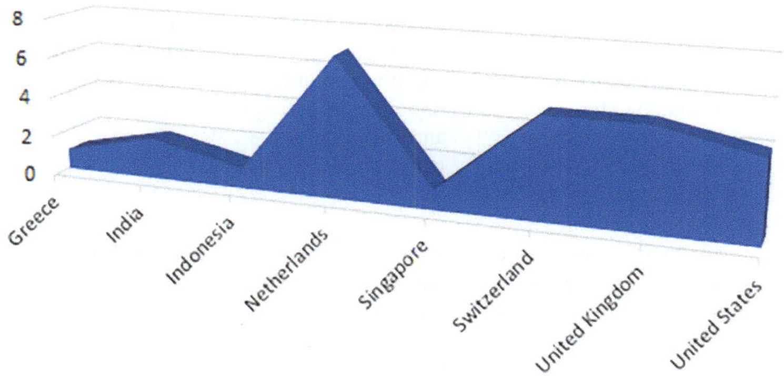

Fig. 4 Country-wise number of publications on precision livestock farming

number of publications from year 2021. Figure 4, provides information on the country-specific distribution of the articles that have been published using machine learning and IoT based applications for cattle farming. Netherlands tops the list with the most articles published in this discipline (seven research publications), followed by the UK and Switzerland with five articles apiece. India is significantly lagging in this area.

4 Livestock Management with Machine Learning and the Internet of Things

From results of the study, it has been found that IoT and machine learning are commonly used in livestock production, especially for monitoring animal health. Table 4 summarizes the studies related with prediction of animal/cattle health status

Table 4 Summary of studies on IoT applications for tracking the health of cattle

Citations	Objective of the study	Observed parameters	Technology applied
[2]	Review of the presently available technology-based equipment and solutions, and compare the capabilities and constraints of these systems	Skin temperature, rumination duration and heart rate	IoT based Sensor to measure numerous factors pertaining to health
[17]	Explore how big data, sensors and machine learning can boost cattle care, increase production efficiency, and cut costs	Animal voice signals, movement signals, and other similar behavioural information	Sensors include cameras, infrared thermal imaging, temperature sensors, motion detectors, accelerometers, RFID, pedometers, facial recognition systems, and microphones
[28]	Examination of the Current Status of Precision Dairy Wearable Technology for identifying Lameness, Metabolic Abnormalities Mastitis and Metritis in Dairy Cows	Monitoring feed bunk usage, feeding duration, rumination time, activity levels, and rumen pH and temperature	Precision wearable technology including devices like neck collars leg tags and torso bands
[30]	Forecasting cow disease utilizing a system for monitoring the health of cattle using fuzzy IoT	Heart Rate, Body temperature, details on breathing, and information about rumination	Sensors, WiFi module, Arduino
[31]	Detecting mastitis and foot and mouth disease (FMD) in cows using Internet of Things	Sound, motion, temperature, etc.	Raspberry Pi3 controller, temperature sensor, motion sensor, and rumination sensor
[32]	Description of sensors integrated to a wireless sensor network for tracking cattle health	Motion, temperature, heart rate	Sensor modules, GSM, and an android application
[33]	Demonstration of a method for automatically tracking dairy cows' health using the Internet of Things	Feeding, ruminating, resting and movement	Sensors, WiFi access points, the Hub and routers, as well as a database and application server, make up the CowDevice

(continued)

Table 4 (continued)

Citations	Objective of the study	Observed parameters	Technology applied
[34]	Classify the health status of cows as normal, less-normal, or abnormal health based on sensor data	Motion, Body temperature and heart rate	A web application for health analysis and data visualisation, a Raspberry Pi base station for managing a local server, and a collar with many sensors for body temperature, heart rate, and movement information
[35]	Introduction of a gadget to track the cattle's health metrics and identify any declines	Humidity, rumination, heart rate, and temperature	A combination of sensors, an Arduino NANO, and an Xbee module
[36]	Investigation of relevant sensor literature and machine learning methods, and to show what improvements in swine welfare can be made using these applications	Motion, sound, and heat	Remote sensors, such as accelerometers, cameras, microphones, and thermometers

using any of the IoT based techniques. As an answer to RQ2, the parameters which are primarily used to predict the cattle health status are body temperature, motion, rumination time, heartbeat, pulse rate, sound, and humidity. Several sensors are available which can measure these parameters [37]. These sensors can be combined as a single unit to be used as a collar device or leg tag. The selected research studies focus mainly on detection of three diseases mastitis, lameness, and estrus.

The analysis of the selected papers revealed that the management of livestock has benefited from the deployment of machine learning, for instance, to forecast the probability of conception, raised up somatic cell counts, and the start of calving. Farm animal medicine is experiencing an increase in "big data," which has the capacity to be transformed to "smart data" and fully utilize the information already being gathered. To facilitate the clinical diagnosis, the professionals frequently need a significant amount of time to examine and interpret the data. However, the time and training needed to make such a diagnosis can be considerably decreased with the use of automated diagnostic support technologies. In a matter of seconds, a machine learning system can provide a diagnosis, but a human would need to spend 30–60 min analyzing all the data to identify and detect a condition like mastitis. A routine use of such a quick, accessible diagnostic test might result in a more reliable diagnostic report [25]. A completely automated system, running round-the-clock, costs the same as four professional farm visits every day of the week. This make automated systems more affordable than expert farm visits [8]. Using such equipment, advancements in the field of PLF field will lead the farmer recognize the true value of technology [6]. Technology advancements, which frequently produce enormous volumes of data, are being increasingly embraced by dairy farmers. To enhance farm wide management and decision support, integrating and interpreting this data has proven difficult [38].

Machine learning algorithms can hold and analyze large categorical data sets, which is frequently challenging to assess using conventional statistical techniques.

Therefore, it has been advised by the statisticians whereas traditional methods of data analysis are frequently skewed according to the researcher's hypothesis, machine learning algorithms learn from the data, they usually yield superior outcomes[12]. Therefore, machine learning may turn out to be highly valuable for precision livestock farming and provide support to the livestock farmers in safeguarding the welfare of the animals.

4.1 Machine Learning Algorithms for Cattle Health Prediction

It was discovered that to be able to evaluate the cattle's well-being, 14 algorithms for machine learning have been employed which utilizes the data acquired by IoT based techniques from the 16 relevant papers that were retrieved using the selection criteria. Table 5 summarizes the observations made from the selected studies. To predict animal health, machine learning algorithms such as Partial Least Squares Discriminant Analysis (PLS-DA), Random Forest, Support Vector Machine (SVM), Neural Network (NN), Convolutional Neural Network, Decision Tree, K-Nearest Neighbour, Multilayer Perceptron (MLP), Long Short-Term Memory (LSTM), Naive Bayes, Artificial Neural Networks, Linear Discriminant analysis, Logistic Regression, and Gradient Boosted Trees are commonly employed (RQ3).

4.2 Evaluation Parameters for Comparison of the ML Algorithms

To address RQ4, the parameters utilized to determine the precision of application of machine learning methods to forecast cow health have been identified. The parameters which are most used for this purpose are sensitivity, specificity, accuracy, and number of true positives. In two studies, the effectiveness of algorithms for machine learning was contrasted based on specificity and specificity, Four have employed accuracy as a measure for comparison, two have used error rate and area under curve respectively as evaluation parameters whereas two studies have used number of true positives for the same purpose.

An extensive description of the algorithms used in the study is provided in Table 5, along with a comparison of how well they performed in terms of specificity, accuracy, and sensitivity, true positives, area under the curve, and error rate.

Table 5 Assessment of performance of Machine Learning Techniques to forecast cattle health

References	Objective of the study	Algorithms employed	Performance	Best algorithm
[27]	Reviewing current technologies and resolving their shortcomings to provide a novel approach for identifying clinical mastitis	K-Nearest Neighbor and the Support Vector Machine	SVM's accuracy is 86.66%, but KNN's is 73.33%	Support Vector Machine
[13]	To forecast calving in dairy cows using automated activity, laying, and rumination monitors	Neural Network, Random Forest, and Linear Discriminant Analysis	Random forest generates a sensitivity of 25% and specificity 89% using a combination of HR tag and IceQube whereas Linear discriminant analysis provides a sensitivity of 75% and specificity 93.4%, Neural Network produces a sensitivity of 100% and 86.8% specificity	Neural Network
[22]	To determine how machine learning techniques might enhance the processing of on-farm sensor data	K-nearest neighbor neural networks, decision trees, and random forest	Models using decision trees and random forests can discriminate between milk fever, KNN can diagnose subacute ruminal acidosis with 83% true positives, however because of the 66% false positive rate, it is not really effective; The accuracy of neural networks and random forests in predicting heart rate, skin temperature, and vaginal temperature is the highest	Random Forest and Neural Networks

(continued)

Table 5 (continued)

References	Objective of the study	Algorithms employed	Performance	Best algorithm
[23]	To detect the health problems in dairy cattle	Convolutional neural network, Random Forest, Neural Network (NN), Partial Least Squares Discriminant Analysis (PLS-DA), Support Vector Machine	PLS-DA showed Sensitivity (65.60%) and Specificity (79.59%); Random Forest generated Sensitivity (46.22%) Specificity (79.26%) whereas SVM produced Sensitivity (66.39%) Specificity (76.39%); Neural Networks generated Sensitivity (61.74%) Specificity (97%) and Convolutional Neural Network resulted in Sensitivity (57.02%) Specificity (92.5%)	Neural Network
[14]	To identify the best models for sub-clinical mastitis prediction	Logistic Regression, Naive Bayes, Deep Learning (DL), Random Forest, Gradient-Boosted Tree (GBT), and Decision Tree	The best reliable model for predicting subclinical bovine mastitis was found to be the gradient-boosted tree (with an accuracy of 84.9%)	Gradient Boosted Tree
[25]	To ascertain whether mastitis may be predicted using machine learning techniques by simulating a specialist clinical judgement	Random Forest	Mastitis recognised with a high degree of precision (98%) using the random forest method	Random Forest
[26]	To present an innovative unified computational analysis based on machine learning techniques to forecast lameness	Artificial neural networks, Random Forest, and support vector machines	Support Vector Machines Random Forest able to discriminate between samples that are positive and negative with a greater degree of clarity than Artificial Neural Networks, but only Random Forest can do it with 100% accuracy	Random Forest

(continued)

Table 5 (continued)

[18]	To measure the efficiency of machine learning methods for predicting cow lameness	Naive Bayes (NB), Random Forest (RF), Multilayer Perceptron (MLP), and Logistic Regression (LR)	LR, NB, MLP, and RF AOCs (0.67, 0.66, 0.62, and 0.62 respectively) (0.67). (0.61). Nevertheless, the F1-score in NB (27%) is lower than in LR (1%)	Naïve Bayes (NB)
[21]	To explore the utilization of machine learning (ML) for detecting abnormal behavior in cows through continuous remote monitoring	Long Short-Term Memory (LSTM), Multilayer Perceptron (MLP), Decision Tree, and K-Nearest Neighbors for Regression	KNNR performed most efficiently by detecting 83% of cases (true-positives) of subacute ruminal acidosis (SARA)	KNNR
[22]	To investigate the ability of machine learning techniques to predict metabolic status using on-farm data and to determine whether the metabolic status of individual cows in early lactation can be classified based on their plasma values	ANN, Random Forest, Naive Bayes, K-Nearest Neighbour, and Decision Tree	With error rates of 12.4–22.6% and 12.4–20.9%, respectively, Random Forest and Support Vector Machine algorithms can forecast metabolic state using on-farm cow data	Random forests and support vector machines

For finding out the answer to RQ5, Column 3 of Table 5 was analyzed and it has been found that the most often utilized machine learning method for assessing cow health is Random Forest. Out of the selected studies, eight have used Random Forest. Neural Networks has been identified as best algorithm in three studies, whereas in four studies, on the other hand two studies have concluded Support Vector Machine as best algorithm and lastly two publications have recognized Gradient Boost Tree and Naïve Bayes as best performing algorithms respectively.

5 Research Gaps

Numerous study gaps have appeared as IoT and machine learning technologies are increasingly integrated in cow health monitoring. These gaps indicate areas where existing studies are inadequate. Among these gaps are:

IoT Systems' Inability to Process Data in Real Time: IoT technology has advanced, but many current systems still have trouble processing data in real time. Timely interventions and herd management decision-making are hampered by delays in data gathering and processing. To guarantee immediate data processing and meaningful insights, research is required to create more effective algorithms and structures.

Machine learning models for some diseases have limited accuracy: Although machine learning methods have demonstrated potential in the identification of subclinical disorders and diseases like mastitis, the precision of these models can fluctuate greatly amongst various illnesses. The resilience of machine learning models need to be improved, especially in terms of correctly identifying fewer common illnesses or ailments with comparable symptoms.

Difficulties in Expanding IoT Systems to More Herds: There is a knowledge gap about how IoT systems may be efficiently scaled to handle larger herds because many studies concentrate on small-scale deployments. To guarantee that these technologies can be implemented more widely in commercial dairy operations, issues including data management, network dependability, and system integration need to be resolved.

Farmers' Cost-Effectiveness and Practicality of Implementation: Many farmers continue to have concerns about the viability of integrating modern technologies, especially when it comes to cost-effectiveness. Research is necessary to find low-cost solutions, assess farmers' return on investment, and create intuitive systems that do not require a lot of technical know-how to utilise.

Long-Term Effect of Constant Monitoring: Most of the current research concentrates on short-term health indicators, with few research studies assessing the long-term impacts of continuous observation on cow health and productivity.

Examining the ways in which long-term data collection affects herd management procedures and results may offer farmers insightful information.

6 Challenges

The studies that were taken into consideration for this evaluation acknowledged the numerous difficulties faced while conducting the research. In response to RQ6, some of the obstacles encountered with deployment of IoT with machine learning for cattle management, have been identified. Numerous studies have found that it is challenging to handle the enormous amount of data that is gathered by technology, which is typically stored on distant cloud servers and frequently used for commercial gains. There are several situations when technology cannot be applied efficiently because farmers might not be able to deploy cutting-edge technology or they may be unwilling to do so because of different physical, environment, and contextual constraints [17]. The effectiveness of the algorithms utilized in precision livestock farming is yet additional formidable challenge [11]. The difficulty in gathering data and conducting insightful analysis has limited the use of machine learning techniques [36]. Furthermore, machine learning models' predictions are only applicable to data from population that the model was trained on because these models are data-based [39].

Few studies reported that IoT tools should be designed according to the region, farm sizes and technological level etc. Another finding from the studies is the necessity of continuing to create reliable sensors and Internet of Things (IoT) systems that can be applied as PLF instruments to enable secure and private data transfer and storage [40]. For all this to be possible, it takes committed experts within the fields of computer science, biology, and engineering who are knowledgeable about farm production systems and cattle biology. To cope with multidimensional datasets, meticulous data editing, quality control procedures, sophisticated machine learning techniques, and statistical models are required.

6.1 Potential Solutions and Recommendations for the Challenges Identified

While the challenges identified in the implementation of IoT and machine learning for cattle management are substantial, several potential solutions and recommendations can help mitigate these obstacles and enhance the efficiency of these technologies.

Technological Adoption by Farmers
- Tailored training programs and user-friendly interfaces help farmers adopt new technologies.
- Financial incentives and community-based knowledge sharing can increase adoption, especially in under-resourced *areas.*

Enhancing Algorithm Effectiveness
- Data preprocessing, feature selection, and ensemble learning methods can improve algorithm performance.
- Collaborating with domain experts can help design more effective models tailored to livestock management.

Improving Data Collection and Analysis
- Enhance sensor calibration and adopt multi-source data fusion for more accurate data.

- Develop open-source platforms for shared data analysis to foster collaboration and innovation.

Advanced Data Handling

- Use cloud-based platforms and big data frameworks to manage large datasets and perform advanced analytics.
- Develop automated data pipelines and real-time analytics tools for better data management and decision-making.

By adopting these solutions, the challenges in IoT and machine learning implementation for cattle management can be addressed, leading to improved technology adoption, data handling, and ultimately better livestock management practices.

7 Discussion

This study emphasis the immense potential of combining Machine Learning (ML) and Internet of Things (IoT) technologies for cattle health monitoring. For researchers, the review provides a concise summary of developments in these fields and offers a foundation for future work in algorithm development, sensor integration, and data analytics. It emphasizes the importance of refining predictive models for more accurate disease detection, which is crucial for advancing the field. Additionally, the findings help identify gaps in current research, directing future studies toward areas such as multi-modal data fusion, real-time decision-making, and context-specific algorithm optimization.

For practitioners, the integration of IoT and ML presents opportunities to improve animal welfare, health management, and overall farm efficiency. By implementing automated health monitoring systems, farmers can detect health issues earlier, resulting in cost savings from reduced veterinary care, fewer deaths, and more efficient use of resources. The ability to continuously monitor cattle health provides valuable data that can inform breeding decisions, improve livestock productivity, and enhance overall farm sustainability. Moreover, the real-time nature of these systems allows for quicker interventions, ultimately improving herd health outcomes.

The adoption of machine learning and IoT technologies in cattle health monitoring has substantial economic implications for the livestock industry. On the one hand, the initial investment in technology—such as sensors, cloud infrastructure, and training for personnel—may be seen as a financial barrier, particularly for smaller farms. However, in the long term, these technologies can generate substantial cost savings by enabling early disease detection, reducing unnecessary veterinary interventions, improving herd productivity, and optimizing resource usage. The precision offered by these systems can result in better-informed decisions about animal health, thus leading to healthier livestock, higher yields, and fewer losses due to undiagnosed diseases.

Furthermore, the economic benefits extend beyond the individual farm level to the entire livestock industry. Precision livestock farming could increase the sustainability of cattle production and improve the industry's capacity to more efficiently meet the world's food needs if it is widely adopted. For example, improved cattle health translates into better meat and milk quality, potentially enhancing the marketability of livestock products. Government subsidies, industry partnerships, and public-private collaborations can help mitigate the initial costs and facilitate broader adoption of these technologies, especially among smaller-scale farms.

There are also broader economic implications for the environment, as precision farming technologies contribute to reducing waste and environmental impact by optimizing the use of inputs. The technology's role in fostering more sustainable farming practices can help ensure the continued viability of the cattle industry, while also aligning with global sustainability goals and increasing demand for ethically produced agricultural products.

8 Conclusion

The Internet of Things in conjunction with machine learning technologies to identify diseases in cattle for precision livestock farming are discussed in this Systematic Literature Review. The Internet of Things has gradually been used to automate several tasks in livestock farming. Approaches to machine learning have proven to be beneficial and quite efficient for autonomous monitoring of cattle health in recent years. The outcomes of this investigation reveal that models for machine learning learn independently, have attracted a lot of attention However, there are several difficulties still to be resolved, the most important of which is Livestock health tracking in a farm environment. Although it is impossible to determine which machine learning algorithm is the most effective in the current setting, several ML methods including Neural Networks, Random Forest, and Support Vector Machine are used more frequently than others.

These findings will help to direct future research into creating a stand-alone system to check the cattle's health. A customized, applied technology can be created at a lower cost because the livestock industry uses many animals and operations. The fact that multiple phases and procedures are involved in the creation and maintenance of PLF applications is a crucial factor that needs to be further researched to ensure the precision and relevance of the generated systems in commercial settings.

9 Future Research Directions

Internet of Things (IoT) and machine learning (ML) integration for disease detection and health monitoring in cattle presents significant opportunities for advancing precision livestock farming. However, several challenges remain that warrant focused research efforts. Future studies should explore the development of autonomous and adaptive health monitoring systems that can operate in real-time under the dynamic conditions of a farm environment.

Given the variety of machine learning models in use, further comparative studies are needed to optimize the choice of algorithms specific to cattle health monitoring. While techniques like Random Forest, Neural Networks, and Support Vector Machines have shown promising results, research is required to determine the most effective algorithms in different farming scenarios, accounting for factors such as farm size, environment, and cattle breed.

The continued development of sensor technologies, including cameras, microphones, and wearable devices, is essential for enhancing health data collection's precision and dependability. Researchers should also explore cost-effective and scalable IoT solutions that cater to the diverse needs of the livestock industry. These advancements would enable more widespread adoption of precision farming technologies, making them accessible even to smaller-scale farms.

Additionally, there is a need for research into the integration of multi-modal data (e.g., combining sensor data with environmental factors) to enhance the precision of predictive models. Finally, understanding the challenges in the real-world using these technologies is critical. Future studies ought to concentrate on ensuring that these solutions are not only technically robust but also commercially viable. This includes addressing the scalability of systems and ensuring their long-term sustainability in diverse farming environments.

References

1. Karthick, G.S., Sridhar, M., Pankajavalli, P.B.: Internet of things in animal healthcare (IoTAH): Review of recent advancements in architecture, sensing technologies, and real-time monitoring. *SN Computer Science* 1(5), 1–16 (2020).
2. Chaudhry, A.A., Mumtaz, R., Zaidi, S.M.H., Tahir, M.A., School, S.H.M.: Internet of Things (IoT) and machine learning (ML) enabled livestock monitoring. In: *2020 IEEE 17th International Conference on Smart Communities: Improving Quality of Life Using ICT, IoT and AI (HONET)*, pp. 151–155. IEEE (2020).
3. Banhazi, T.M., Lehr, H., Black, J.L., Crabtree, H., Schofield, P., Tscharke, M., Berckmans, D.: Precision livestock farming: An international review of scientific and commercial aspects. *International Journal of Agricultural and Biological Engineering* 5(3), 1–9 (2012).
4. Banhazi, T.M., Black, J.L.: Precision livestock farming: A suite of electronic systems to ensure the application of best practice management on livestock farms. *Australian Journal of Multi-disciplinary Engineering* 7(1), 1–14 (2009).
5. Vranken, E., Berckmans, D.: Precision livestock farming for pigs. *Animal Frontiers* 7(1), 32–37 (2017).

6. Norton, T., Chen, C., Larsen, M.L.V., Berckmans, D.: Precision livestock farming: Building digital representations to bring the animals closer to the farmer. *Animal* 13(12), 3009–3017 (2019).
7. Pomar, J., López, V., Pomar, C.: Agent-based simulation framework for virtual prototyping of advanced livestock precision feeding systems. *Computers and Electronics in Agriculture* 78(1), 88–97 (2011).
8. Berckmans, D.: Precision livestock farming technologies for welfare management in intensive livestock systems. *Rev. Sci. Tech* 33(1), 189–196 (2014).
9. Kaur, D., Virk, A.K.: IoT and machine learning-based systems for predicting cattle health status for precision livestock farming. In: *2022 International Conference on Smart Generation Computing, Communication and Networking (SMART GENCON)*, pp. 1–6. IEEE (2022).
10. Makinde, A., Islam, M.M., Scott, S.D.: Opportunities for ACI in PLF: Applying animal- and user-centred design to precision livestock farming. In: *Proceedings of the Sixth International Conference on Animal-Computer Interaction*, pp. 1–6 (2019).
11. Halachmi, I., Guarino, M., Bewley, J., Pastell, M.: Smart animal agriculture: Application of real-time sensors to improve animal well-being and production. *Annual Review of Animal Biosciences* 7(1), 403–425 (2019).
12. Cockburn, M.: Application and prospective discussion of machine learning for the management of dairy farms. *Animals* 10(9), 1690 (2020).
13. Borchers, M.R., Chang, Y.M., Proudfoot, K.L., Wadsworth, B.A., Stone, A.E., Bewley, J.M.: Machine-learning-based calving prediction from activity, lying, and ruminating behaviors in dairy cattle. *Journal of Dairy Science* 100(7), 5664–5674 (2017).
14. Ebrahimi, M., Mohammadi-Dehcheshmeh, M., Ebrahimie, E., Petrovski, K.R.: Comprehensive analysis of machine learning models for prediction of sub-clinical mastitis: Deep Learning and Gradient-Boosted Trees outperform other models. *Computers in Biology and Medicine* 114, 103456 (2019).
15. Garcia, R., Aguilar, J., Toro, M., Pinto, A., Rodriguez, P.: A systematic literature review on the use of machine learning in precision livestock farming. *Computers and Electronics in Agriculture* 179, 105826 (2020).
16. Mahmud, M.S., Zahid, A., Das, A.K., Muzammil, M., Khan, M.U.: A systematic literature review on deep learning applications for precision cattle farming. *Computers and Electronics in Agriculture* 187, 106313 (2021).
17. Neethirajan, S.: The role of sensors, big data and machine learning in modern animal farming. *Sensing and Bio-Sensing Research* 100367 (2020).
18. Schillings, J., Bennett, R., Rose, D.C.: Exploring the potential of Precision Livestock Farming technologies to help address farm animal welfare. *Frontiers in Animal Science* 2, 1–7 (2021).
19. Shahinfar, S., Khansefid, M., Haile-Mariam, M., Pryce, J.E.: Machine learning approaches for the prediction of lameness in dairy cows. *Animal* 15(11), 100391 (2021).
20. Slob, N., Catal, C., Kassahun, A.: Application of machine learning to improve dairy farm management: A systematic literature review. *Preventive Veterinary Medicine* 187, 105237 (2021).
21. Wagner, N., Antoine, V., Mialon, M.M., Lardy, R., Silberberg, M., Koko, J., Veissier, I.: Machine learning to detect behavioral anomalies in dairy cows under subacute ruminal acidosis. *Computers and Electronics in Agriculture* 170, 105233 (2020).
22. Xu, W., van Knegsel, A.T., Vervoort, J.J., Bruckmaier, R.M., van Hoeij, R.J., Kemp, B., Saccenti, E.: Prediction of metabolic status of dairy cows in early lactation with on-farm cow data and machine learning algorithms. *Journal of Dairy Science* 102(11), 10186–10201 (2019).
23. Contla Hernández, B., Lopez-Villalobos, N., Vignes, M.: Identifying health status in grazing dairy cows from milk mid-infrared spectroscopy by using machine learning methods. *Animals* 11(8), 2154 (2021).
24. Cowton, J., Kyriazakis, I., Plötz, T., Bacardit, J.: A combined deep learning GRU-autoencoder for the early detection of respiratory disease in pigs using multiple environmental sensors. *Sensors* 18(8), 2521 (2018).

25. Hyde, R.M., Down, P.M., Bradley, A.J., Breen, J.E., Hudson, C., Leach, K.A., Green, M.J.: Automated prediction of mastitis infection patterns in dairy herds using machine learning. *Scientific Reports* 10(1), 1–8 (2020).
26. Liakos, K., Moustakidis, S.P., Tsiotra, G., Bartzanas, T., Bochtis, D., Parisses, C.: Machine learning-based computational analysis method for cattle lameness prediction. In: HAICTA, pp. 128–139 (2017).
27. Ankitha, K., Manjaiah, D.H.: Comparison of KNN and SVM algorithms to detect clinical mastitis in cows using Internet of Animal Health Things. In: Intelligent Data Engineering and Analytics, pp. 51–60 (2021).
28. Eckelkamp, E.A.: Invited review: Current state of wearable precision dairy technologies in disease detection. Applied Animal Science 35(2), 209–220 (2019).
29. Eckelkamp, E.A., Bewley, J.M.: On-farm use of disease alerts generated by precision dairy technology. Journal of Dairy Science 103(2), 1566–1582 (2020).
30. Suseendran, G., Balaganesh, D.: Smart cattle health monitoring system using IoT sensors. Materials Today: Proceedings (2021).
31. Vyas, S., Shukla, V., Doshi, N.: FMD and mastitis disease detection in cows using Internet of Things (IoT). Procedia Computer Science 160, 728–733 (2019).
32. Priya, M.K., Jayaram, B.G.: WSN-based electronic livestock of dairy cattle and physical parameters monitoring. In: Emerging Research in Electronics, Computer Science and Technology, pp. 37–45. Springer, Singapore (2019).
33. Unold, O., Nikodem, M., Piasecki, M., Szyc, K., Maciejewski, H., Bawiec, M., Grzywacz, M., Jedrzejewski, A., Zdunek, M.: IoT-based cow health monitoring system. In: International Conference on Computational Science, pp. 344–356. Springer, Cham (2020).
34. Pratama, Y.P., Basuki, D.K., Sukaridhoto, S., Yusuf, A.A., Yulianus, H., Faruq, F., Putra, F.B.: Designing of a smart collar for dairy cow behavior monitoring with application monitoring in microservices and Internet of Things-based systems. In: 2019 International Electronics Symposium (IES), pp. 527–533. IEEE (2019).
35. Swain, K.B., Mahato, S., Patro, M., Pattnayak, S.K.: Cattle health monitoring system using Arduino and LabVIEW for early detection of diseases. In: 2017 Third International Conference on Sensing, Signal Processing and Security (ICSSS), pp. 79–82. IEEE (2017).
36. Benjamin, M., Yik, S.: Precision livestock farming in swine welfare: A review for swine practitioners. Animals 9(4), 133 (2019).
37. Suparwito, H., Wong, K.W., Xie, H., Rai, S., Thomas, D.A.: A hierarchical classification method used to classify livestock behavior from sensor data. In: International Conference on Multi-disciplinary Trends in Artificial Intelligence, pp. 204–215. Springer, Cham (2019).
38. Williams, M.L., Mac Parthaláin, N., Brewer, P., James, W.P.J., Rose, M.T.: A novel behavioral model of the pasture-based dairy cow from GPS data using data mining and machine learning techniques. Journal of Dairy Science 99(3), 2063–2075 (2016).
39. Ferris, M.C., Christensen, A., Wangen, S.R.: Symposium review: Dairy Brain—Informing decisions on dairy farms using data analytics. Journal of Dairy Science 103(4), 3874–3881 (2020).
40. Milan, H.F., Perano, K.M., Gebremedhin, K.G.: Survey and future prospects in precision dairy farming. In: 10th International Livestock Environment Symposium (ILES X), p. 1. American Society of Agricultural and Biological Engineers (2018).

Hybrid ButterFlower Algorithm

Aayushi Padia, Virinchi M., Prerana Bhat, Sharan Thummagunti, and Utkarsh Mahadeo Khaire

Abstract This research introduces an innovative Hybrid ButterFlower Algorithm (HBFA) combining the Butterfly Optimization Algorithm (BOA) and the Sunflower Optimization Algorithm (SFO) to enhance search efficiency and convergence accuracy. BOA's fragrance-based global and local search mechanisms are integrated with SFO's attraction-repulsion dynamics to balance exploration and exploitation effectively. An adaptive control mechanism dynamically tunes the influence of each algorithm, preventing premature convergence and improving search diversity. Simulation results on 10 benchmark functions give insight into the hybrid algorithm's ability to outperform both BOA and SFO in convergence speed, solution accuracy, and robustness across multimodal optimization problems.

Keywords Butterfly optimization algorithm · Sunflower optimization algorithm · Metaheuristic optimization

1 Introduction

Metaheuristic optimization algorithms play an important role in solving intricate, nonlinear complexities that traditional optimization methods frequently encounter difficulties with due to extensive search space dimensionality or being stuck in local optimal solutions. The Butterfly Optimization Algorithm (BOA) [2] and the Sunflower Optimization Algorithm (SFO) [5] are recognized for their efficacy in tackling multi-modal optimization difficulties. These algorithms, inspired by nature, mimic biological behaviors and provide clear benefits in both global exploration and local exploitation.

BOA mimics the foraging behavior of butterflies using a scent-based search method, with each butterfly's position indicating a potential solution and its

A. Padia (✉) · Virinchi M. · P. Bhat · S. Thummagunti · U. M. Khaire
Data Science and Artificial Intelligence, Indian Institute of Information Technology, Dharwad, India

K. Deep et al. (eds.), *Artificial Intelligence, Machine Learning and Intelligent Systems*, Springer Proceedings in Mathematics & Statistics 520,
https://doi.org/10.1007/978-981-95-3841-6_3

scent directing the search toward better solutions. Although BOA is efficient at quickly converging, it may encounter issues with premature convergence in intricate landscapes. Conversely, SFO emulates the heliotropic movement of sunflowers, with candidate solutions drawn to the best-known solution, while also repelling each other to maintain diversity. Although SFO is exceptional at exploration, it lacks the necessary exploitation for rapid convergence. To overcome these limitations, this study proposes a hybrid algorithm [1, 13] that combines the exploitation ability of BOA with the exploration of SFO to provide improved speed and accuracy in solving complex optimization problems [6].

2 Overview

2.1 Butterfly Optimization Algorithm (BOA)

The Butterfly Optimization Algorithm (BOA) is a technique based on the search patterns of butterflies as they forage for food. In this algorithm, each butterfly represents a potential solution, and its emitted fragrance indicates how well it fits the problem. BOA strikes a balance between exploring the entire solution space and refining nearby solutions, mimicking the behavior of butterflies. Due to its straightforward nature and minimal need for parameters, BOA can be easily applied to a wide variety of optimization challenges. Nevertheless, it may encounter premature convergence, particularly in intricate or multi-peaked landscapes, leading to potential entrapment in local solutions. Despite this drawback, BOA is known for its rapid convergence and ability to effectively exploit high-quality solutions.

2.2 Sunflower Optimization Algorithm (SFO)

The Sunflower Optimization Algorithm (SFO) draws inspiration from the heliotropic movement of sunflowers, each sunflower symbolizing a potential solution [10]. Sunflowers are drawn towards the optimal solution (the sun) while also maintaining distance from each other to prevent clustering and maintain exploration capability. This interplay of attraction and repulsion enables SFO to strike a balance between exploration and convergence, allowing it to thoroughly explore the search space while steering clear of local optimum points. Although SFO is proficient in exploration, it may lack the aggressive exploitation required for swift convergence, potentially slowing down the refinement of solutions. However, its robust exploration capabilities and ability to forestall premature convergence make it effective for tasks that require thorough exploration.

2.3 A Hybrid Approach

The proposed hybrid approach integrates the exploration strengths of SFO with the exploitation capabilities of BOA through a dynamic combination of their update mechanisms. Incorporating BOA's local and global search strategies into SFO's orientation update step improves the hybrid algorithm's convergence speed, solution accuracy, and robustness, making it highly effective for multimodal optimization problems [3, 4]. The subsequent section presents a detailed explanation of the hybrid algorithm.

3 Hybrid ButterFlower Algorithm (HBFA)

Algorithm 1 Hybrid ButterFlower Algorithm (HBFA)

```
Objective function: f(X), X = (x_1, x_2, ..., x_dim), dim = no. of dimensions
Initialize population matrix N_pop, x_i = (i = 1, 2, ..., n)
Define ϵ, c, a and p
Calculate fitness and fragrance (f) for sun (best solution)
while stopping criteria not satisfied do
    Calculate S_i for all butterflowers
    Remove m% butterflowers furthest from the sun
    Pollination probability (pp) = fitness_i / Σ_{i=1}^{n} fitness_i
    Step size (D_i) = λ · pp · ‖x_i^t + x_{i-1}^t‖
    Maximum step size (d_max) = ‖X_max − X_min‖ / (2×N_pop)
    Revised step size (D_i) = min((D_i), d_max)
    if r < p then
        x_i^{t+1} = (x_i^t + (r^2 · g* − x_i^t) · f_i) · ϵ + (1 − ϵ) · D_i S⃗_i
    else
        x_i^{t+1} = (x_i^t + (r^2 · x_j^t − x_k^t) · f_i) · ϵ + (1 − ϵ) · D_i S⃗_i
    end if
    Pollinate with best b% flowers to replace the removed plants
    Re-evaluate the population and update sun if new global best found
    Generate new random value of a and ϵ
end while
Return the best solution
```

HBFA begins with the exploration capability of the Sunflower Optimization Algorithm (SFO). SFO is inspired by the heliotropic behavior of sunflowers, which orient themselves towards the sun, being the best solution among the population. The exploration phase is one of the key components in SFO, allowing the algorithm to search broadly across the solution space while avoiding local optima. This phase emphasizes diversifying the search by repelling plants that are far from the best known solution, preventing premature convergence, and ensuring a thorough investigation of potential solutions.

As with any population-based optimization algorithm, the random population of N_{pop} butterflowers is initialized within the bounds of the search space using a uniform distribution. Each butterflower represents a potential solution for the given optimization problem. Let each butterflower in the population be represented by X_i, where $i = 1, 2, 3, \ldots, n$. Each flower has a position $X_i \in \mathbb{R}^d$ within the d-dimensional search space. The best solution of the population X_{best}, referred to as the "sun" , is identified as the flower with the best fitness value as given by Eq. (1):

$$X_{\text{best}} = \arg\min_{X_i} f(X_i), \quad i = 1, 2, \ldots, n \tag{1}$$

where f(X) is the objective function to be optimized.

All the flowers are then reoriented towards the direction of the best solution. This allows the algorithm to start from a broad search area and gradually converge towards the optimal regions in the space. This reorientation is calculated using Eq. (2):

$$\vec{S}_i = \frac{X_{\text{best}} - X_i}{\|X_{\text{best}} - X_i\|} \tag{2}$$

Once the orientation vectors are calculated for each butterflower, $m\%$ of the plants that are farthest from the sun are removed. This is done to focus computation on the promising regions of the search space and maintain diversity. These eliminated plants are replaced through pollination with the remaining b butterflowers as parents to give rise to new offspring within the vicinity of the parent plants, ensuring fresh exploration.

Let R_i be the distance between the i-th plant and the best solution X_{best} which is calculated using Eq. (3):

$$R_i = \|X_{\text{best}} - X_i\| \tag{3}$$

The $m\%$ plants with the highest distances R_i are removed from the population and replaced with the offspring X_{new}. This process promotes diversity by ensuring that individuals who are far away from the best solution, and likely to be sub-optimal, are replaced with new, unexplored points in the solution space.

After removing the worst plants, the step size D_i is calculated for the remaining plants. This step size determines how much the position of each sunflower is to be updated in the search space during each iteration. It is calculated using Eq. (4):

$$D_i = \lambda \cdot P_i \left(\|X_i + X_{i-1}\|\right) \cdot \|X_i + X_{i-1}\| \tag{4}$$

where λ is a constant value defining the "inertial" displacement of the plants, and $P_i \left(\|X_i + X_{i-1}\|\right)$ is the pollination probability, i.e., pollination will occur between flower i and its neighbor $i-1$, which generates an offspring in a random position within the vicinity of the parents. Hence, plants that are closest to the sun will have a smaller step size in search of a local refinement, whereas plants farther from the sun will move normally. There is also a need to restrict the maximum step of each individual to prevent overshooting and potentially skipping regions that may harbor the global minimum candidate. The maximum step size is given by Eq. (5):

$$d_{\max} = \frac{\|X_{\max} - X_{\min}\|}{2 \times N_{\text{pop}}} \tag{5}$$

where $X_{\max}$ and $X_{\min}$ are the maximum and minimum bounds of the search space, and each row in N_{pop} represents an entity, i.e. hybrid butterflower in the total population. The newly revised step size is the minimum of D_i and d_{max}.

The best performing $b\%$ of flowers are allowed to "pollinate" around the best solution X_{best}. These flowers undergo a finer search near the best solution by moving towards the best-known solution while ensuring local exploration. The fragrance-based movement mechanism from BOA is integrated to enhance the exploitation capabilities of SFO. The key idea is to use the fragrance intensity derived from BOA to influence the movement of the elite flowers during their update step. This combines SFO's local refinement with BOA's robust search strategies, improving the convergence towards the global optimum. To dynamically determine the type of pollination, a predefined switch probability p is used along with a randomly generated variable r in the interval [0, 1]. These parameters dictate the likelihood of selecting local pollination over global pollination.

CASE (I) : $r < p$: Global Pollination

$$X_{new} = X_i + \epsilon \left(r^2 X_{\text{best}} - X_i\right) f_i + (1-\epsilon) D_i \vec{S}_i \tag{6}$$

CASE (II) : $r >= p$: Local Pollination

$$X_{new} = X_i + \epsilon \left(r^2 X_{\text{j}} - X_k\right) f_i + (1-\epsilon) D_i \vec{S}_i \tag{7}$$

where $j, k \in [1, n]$ and ϵ is a random value in the interval [0,1]. In Eqs. (6) and (7), f_i is the fragrance parameter taken from the Butterfly Optimization Algorithm.

$$f_i = c \cdot \left(f\left(X_i\right)\right)^a \tag{8}$$

where c is the sensor modality kept at 0.01 and a is the power exponent in the interval [0,1].

Following the determination of the pollination strategy and updating the flower positions, the fitness of the population is evaluated. This evaluation is crucial for identifying potential improvements to the current global best solution, i.e. the sun. This process of fitness evaluation and updating the sun ensures that the algorithm maintains the most optimal solution found during its search. By continually comparing the fitness values of the updated butterflower positions with the current global best, the algorithm dynamically adapts to improvements in the solution space, enhancing its overall effectiveness in solving complex optimization problems.

4 Results on Benchmark Functions

In order to evaluate the proposed hybrid's performance, experimentation is done on 10 widely recognized benchmark functions, namely Goldstein & Price, Shcwefel, Schaffer, Rosenbrock, Sphere, Ackley, Restrigin, Martin & Gaddy, Easom and Griewank. These functions represent a range of unimodal and multimodal optimization problems (Fig. 1).

Type	**Function Name and Dimensions**	**Function**	**Range**	**Global Minimum (f_{min})**
Fixed-dimensional	Goldstein-Price, 2D	$f(x) = [1 + (x_1 + x_2 + 1)^2(19 - 14x_1 + 3x_1^2 - 14x_2 + 6x_1x_2 + 3x_2^2)] \times [30 + (2x_1 - 3x_2)^2 \times (18 - 32x_1 + 12x_1^2 + 48x_2 - 36x_1x_2 + 27x_2^2)]$	[-2,2]	3
Fixed-dimensional	Schaffer, 2D	$f(x) = 0.5 + \frac{\sin^2(x_1^2 - x_2^2) - 0.5}{[1 + 0.001(x_1^2 - x_2^2)]^2}$	[-100,100]	0
Fixed-dimensional	Easom, 2D	$f(x) = -\cos(x_1)\cos(x_2)\, e^{-(x_1-\pi)^2-(x_2-\pi)^2}$	[-100,100]	-1
Multimodal	Schwefel, 2D to 30D	$f(x) = 418.9829 \cdot n - \sum_{i=1}^{n} -x_i sin\sqrt{\lvert x_i \rvert}$	[-500,500]	0
Multimodal	Ackley, 30D	$f(x) = -20e^{-0.2\sqrt{\frac{1}{n}\sum_{i=1}^{n} x_i^2}} - e^{\frac{1}{n}\sum_{i=1}^{n}\cos(2\pi x_i)} + 20 + e$	[-100,100]	0
Multimodal	Rastrigin, 2D to 30D	$f(x) = 10n + \sum_{i=1}^{n}[x_i^2 - 10\cos(2\pi x_i)]$	[-5.12,5.12]	0
Multimodal	Griewank, 30D	$f(x) = \frac{1}{4000}\sum_{i=1}^{n} x_i^2 - \prod_{i=1}^{n}\cos\left(\frac{x_i}{\sqrt{i}}\right) + 1$	[-30,30]	0
Unimodal	Rosenbrock, 10D	$f(x) = \sum_{i=1}^{n-1}[100(x_{i+1} - x_i^2)^2 + (x_i - 1)^2]$	[-30,30]	0
Unimodal	Sphere, 2D	$f(x) = \sum_{i=1}^{n} x_i^2$	[-100,100]	0
Unimodal	Martin and Gaddy, 2D	$f(x) = (x_1 - x_2)^2 + \left(\frac{x_1 + x_2 - 10}{3}\right)^2$	[-20,20]	0

Fig. 1 Benchmark functions

In addition to testing the hybrid algorithm (HBFA), its performance is also compared with standard optimization algorithms such as Particle Swarm Optimization (PSO)[9], Artificial Bee Colony Optimization (ABC) [7], Flower Pollination Algorithm (FPA) [15] and also the individual components of the hybrid, i.e. Butterfly Optimization Algorithm and Sunflower Optimization.

4.1 Convergence Speed

The hybrid algorithm demonstrates a significantly faster convergence speed compared to PSO, FPA, and ABC. Observing the convergence plots in Fig. 2 for various benchmark functions, HBFA consistently displays a smooth and rapid decrease in the objective function value, indicating its effectiveness in reaching the global optimum. In contrast, the convergence curves [8] of the PSO and ABC algorithms often show oscillatory behaviors, signifying a slower refinement process and susceptibility to entrapment in local optima. The hybrid approach results in a more stable search trajectory, enabling faster convergence to optimal solutions without significant fluctuations.

4.2 Exploration-Exploitation Balance

A key advantage of HBFA is its effective balance between exploration and exploitation. The integration of the attraction-repulsion dynamics of SFO with BOA's fragrance-based search mechanism gives the hybrid the ability to navigate the search landscape efficiently, ensuring broad exploration of potential solutions and focused exploitation of promising regions (Table 1).

In contrast, algorithms such as PSO and ABC often exhibit premature convergence due to their limited search diversity during the later stages of optimization. HBFA maintains a dynamic search strategy, allowing it to explore diverse areas of the solution space while refining its focus on high-quality solutions. This balance is evident in Fig. 2 where the hybrid algorithm consistently outperforms PSO and ABC in terms of solution accuracy on the benchmark functions.

4.3 Robustness Across Different Problems

The robustness of HBFA is demonstrated through its performance across a diverse set of benchmark functions, including both unimodal and multimodal landscapes. The hybrid approach consistently achieves superior results, showcasing its ability to handle a wide range of optimization challenges. For instance, when tested against complex functions like the Griewank and Rosenbrock, HBFA not only converges to

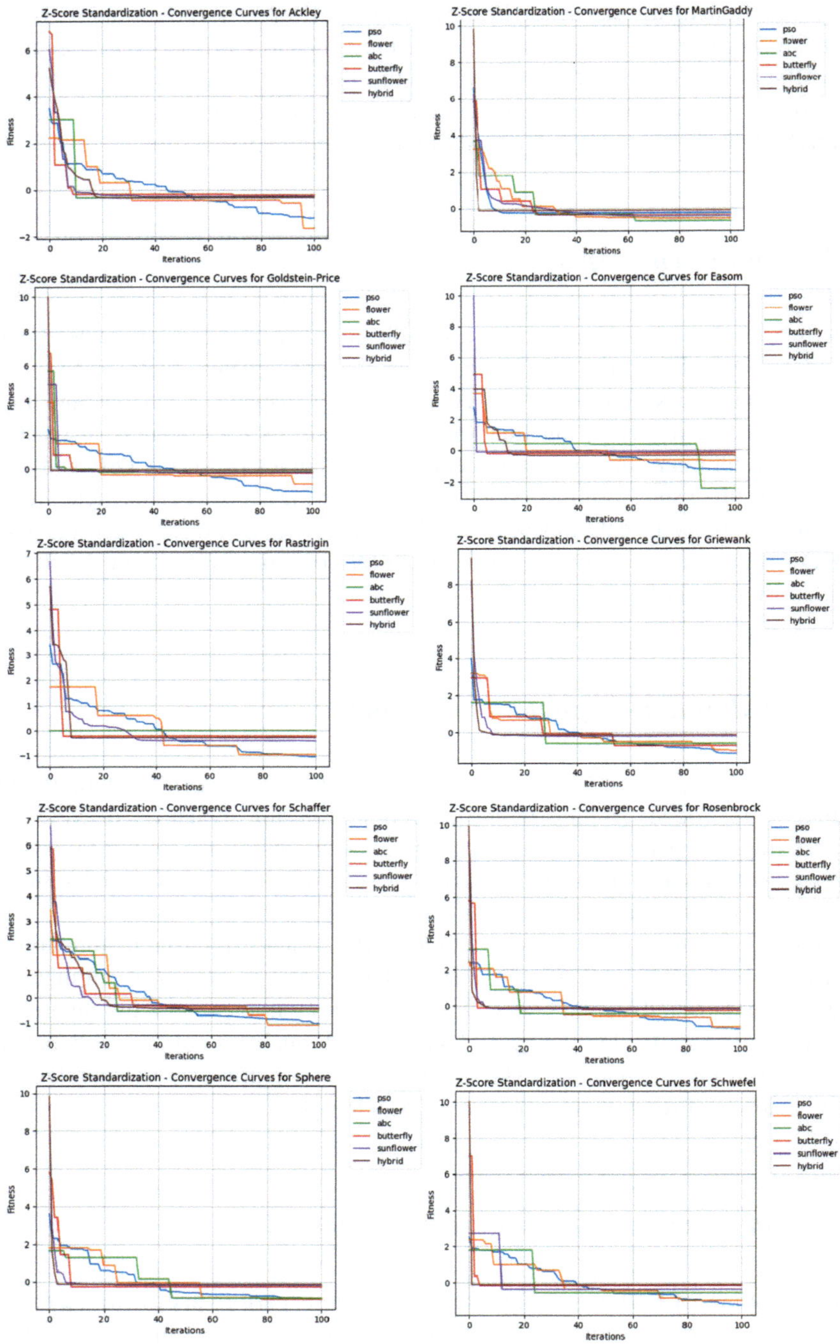

Fig. 2 Algorithm convergence curves for the given objective functions

Table 1 Comparison of HBFA based on mean and standard deviation

Objective function	Stats	BOA	SFO	PSO	FPA	ABC	HBFA
Sphere	Mean	0.0527	22.2167	60.8600	77.1420	0.0026	2.3652
	SD	0.1530	20.3475	51.0358	35.7444	0.0001	15.7824
Rastrigin	Mean	−0.9601	64.5969	72.3561	103.0176	0.0019	−157.1406
	SD	0.2714	27.9024	45.3414	28.4905	0.0001	72.6714
Rosenbrock	Mean	0.0359	99.1731	4.0832	5.3357	0.0002	46.5617
	SD	0.0757	104.6987	2.5411	1.7661	0.0000	101.4278
Schwefel	Mean	0.0449	10155.7913	5.9e+05	9.4e+05	0.0001	9817.7266
	SD	0.1699	246.5194	4.5e+05	5.0e+05	0.0000	15.9881
Schaffer	Mean	0.0617	155.6983	32604.48	28783.033	2.2e-03	18.4918
	SD	0.0815	23.6672	14396.66	13380.701	4.4e-07	45.6445
Ackley	Mean	4.5e-02	14.5548	2297.945	3186.744	0.0442	0.6377
	SD	9.2e-02	0.9174	1280.508	1528.171	0.0001	2.6114
Griewank	Mean	3.6e-02	64.3337	8.1e+05	1.4e+06	0.0008	7.9941
	SD	1.2e-01	47.0834	5.7e+05	6.4e+05	0.0000	62.6313
Goldstein-Price	Mean	0.0388	31.7340	12.1733	11.850	1.2e-06	16.0557
	SD	0.0240	0.6889	7.9876	5.8200	1.7e-07	13.6909
MartinGaddy	Mean	0.0402	2.6e-02	130.0518	65.5674	0.0106	0.1005
	SD	0.1143	1.3e-01	102.3529	149.7554	0.0005	0.0384
Easom	Mean	−0.0196	9.9e-06	18451.31	34157.051	0.9987	1.0e+02
	SD	0.0066	9.9e-05	15185.33	11549.135	0.0004	10.0e-05

optimal solutions but does so with a high degree of stability. The comparative plots indicate that while PSO and FPA may struggle in highly multimodal landscapes, leading to fluctuations in their convergence paths, the hybrid algorithm remains resilient, successfully avoiding local optima and delivering more accurate results.

Moreover, the Friedman test [11] ranks shown in Table 2, further highlights the effectiveness and reliability of HBFA. Compared to the other standard algorithms, our hybrid method achieves the lowest rank values across multiple test cases, reflecting its superior performance. The low rank signifies the algorithm's ability to outperform competing methods consistently. This ranking is particularly significant, as it highlights the capacity of the hybrid to generalize across various problem landscapes, handling both unimodal simplicity and multimodal complexity with equal efficacy.

4.4 Hypothesis Testing for Similarity Between Algorithms

To assess the similarity between the algorithms, i.e. BOA, SFO, PSO, FPA, ABC, HBFA, hypothesis testing is conducted on their performance distributions across various benchmark functions with the null hypothesis being that the distributions of these algorithms are identical. For each objective function, the chi-square [12]

Table 2 Comparison of HBFA based on minimum fitness and rank of friedman test

Objective function	Stats	BOA	SFO	PSO	FPA	ABC	HBFA
Sphere	Rank	2.85	4.18	4.93	5.89	1.84	1.31
	Min	0.0055	18.7957	10.1971	41.7704	0.0026	0.0001
Rastrigin	Rank	1.92	4.81	4.50	5.62	2.90	1.25
	Min	−1.0405	52.6947	17.7133	57.1303	0.0018	−197.5119
Rosenbrock	Rank	2.00	6.00	3.12	3.88	1.00	5.00
	Min	0.0086	76.9859	0.8614	2.6455	0.0002	28.9348
Schwefel	Rank	2.00	4.00	5.01	5.99	1.00	3.00
	Min	0.0059	9987.6892	1.3e+05	3.8e+05	0.0001	9816.1356
Schaffer	Rank	1.50	3.99	5.56	5.44	1.81	2.69
	Min	0.0257	149.0784	17063.12	13909.859	2.2e-03	0.0008
Ackley	Rank	1.84	4.00	5.00	6.00	2.67	1.49
	Min	2.2e-16	14.3114	750.675	1914.096	0.0441	0.0001
Griewank	Rank	1.91	4.00	5.20	5.80	2.15	1.94
	Min	2.2e-16	55.4818	2.0e+05	7.0e+05	0.0007	0.0001
Goldstein-Price	Rank	2.00	5.93	3.98	4.86	1.00	3.23
	Min	0.0154	31.4579	2.5947	5.0255	1.1e-06	13.9667
MartinGaddy	Rank	4.03	2.06	1.45	5.98	4.74	2.74
	Min	0.0031	1.8e-09	100.00	0.0002	0.0104	0.0952
Easom	Rank	1.93	2.95	5.17	5.83	4.00	1.12
	Min	−0.0256	−1.3e-119	4284.04	16630.459	0.9986	8.1e-08

values are converted to p-values that are significantly less than 0.05 to reject the null hypothesis, indicating statistically significant differences in the distributions of the algorithms which can also be observed in Fig. 3. This consistent rejection as shown in Table 3 in all functions tested suggests that the algorithms behave differently, with variations in performance depending on the specific function. These results emphasize that the hybrid algorithm, along with the others, has distinct optimization characteristics.

4.5 Global Optimum Proximity

The proximity to the global optimum throughout the optimization process is another critical metric in which HBFA excels. The convergence curves in Fig. 2 indicate that the hybrid algorithm maintains closeness to the best-known solution, exhibiting a trajectory that consistently approaches the global optimum without significant divergence. In benchmark tests, this was particularly evident in functions such as Sphere and Easom, where HBFA not only achieved optimal solutions but did so with minimal error margins compared to competing algorithms. The superior global

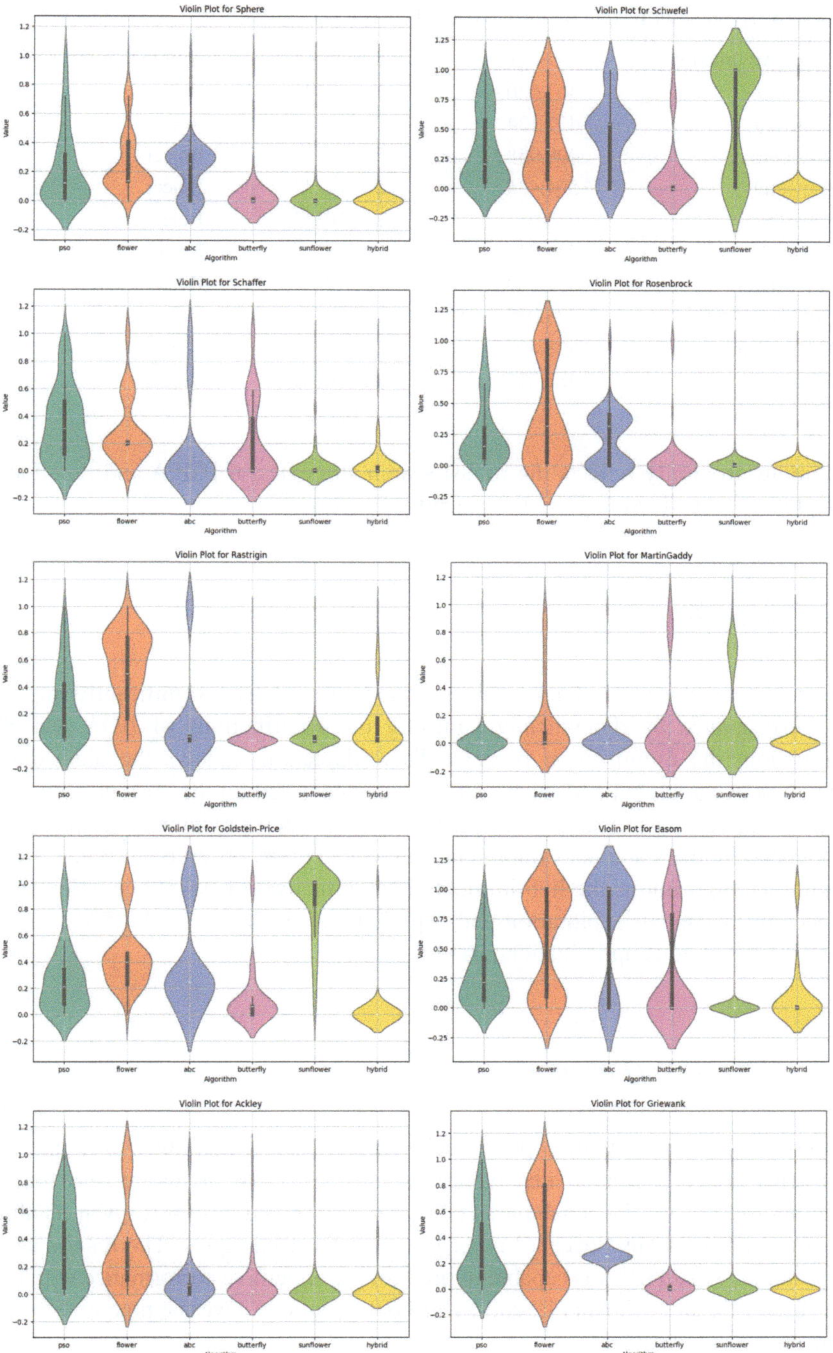

Fig. 3 Violin plots of the algorithms for the given objective functions

Table 3 Hypothesis Testing for the similarity between algorithms

Objective function	Chi-Square	P-value	Null hypothesis
Sphere	467.620	2×10^{-10}	Reject
Rastrigin	437.124	2×10^{-10}	Reject
Rosenbrock	498.958	2×10^{-10}	Reject
Schwefel	504.434	2×10^{-10}	Reject
Schaffer	453.911	2×10^{-10}	Reject
Ackley	468.740	2×10^{-10}	Reject
Griewank	439.088	2×10^{-10}	Reject
Goldstein-Price	478.058	2×10^{-10}	Reject
MartinGaddy	428.406	2×10^{-10}	Reject
Easom	487.823	2×10^{-10}	Reject

optimum proximity reflects the robust search strategies of HBFA and its effective adaptation to the landscape of the objective function.

4.6 Adaptiveness

The adaptive control mechanism incorporated within HBFA significantly improves its performance by dynamically adjusting the contributions of BOA and SFO throughout the optimization process. This adaptability allows the algorithm to switch between global and local search strategies based on the current state of the search. The results indicate that this adaptability prevents premature convergence, as evidenced by the stable and smooth convergence curves observed in Fig. 2. Unlike traditional algorithms, the hybrid method can effectively navigate complex search spaces by balancing exploration and exploitation, making it a robust and flexible solution for various optimization problems.

4.7 Overall Performance

The results of the simulation presented in Table 1, show that our hybrid approach performs better comparatively, being the most effective in finding the function minima for 5 of the 10 benchmarks considered as shown in Fig. 1, given its ability to converge the fastest over all other optimization techniques. The results shown here indicate that HBFA is capable of providing competitive performance with other metaheuristic algorithms [14], and hybrid optimizations have the potential to solve various real-life problems.

5 Conclusion

The proposed hybrid optimization algorithm, Hybrid ButterFlower Algorithm, integrates the exploration strengths of Sunflower Optimization Algorithm and the exploitation capabilities of Butterfly Optimization Algorithm to deliver improved convergence speed and robustness across a range of benchmark functions, being the most effective at 5 of the 10 benchmarks. The results of the Friedman Test and Hypothesis testing demonstrate that the hybrid algorithm outperforms traditional optimization techniques such as PSO, FPA, and ABC, particularly in handling multimodal landscapes and avoiding premature convergence.

References

1. Ahmadianfar, I., Bozorg-Haddad, O., Chu, X.: Gradient-based optimizer: A new metaheuristic optimization algorithm. Information Sciences **540**, 131–159 (2020)
2. Arora, S., Singh, S.: Butterfly optimization algorithm: a novel approach for global optimization. Soft computing **23**, 715–734 (2019)
3. Dutta, D., Rath, S.: Innovative hybrid metaheuristic algorithms: exponential mutation and dual-swarm strategy for hybrid feature selection problem. International Journal of Information Technology **16**(1), 77–89 (2024)
4. Fuladi, S.K., Kim, C.S.: Dynamic events in the flexible job-shop scheduling problem: Rescheduling with a hybrid metaheuristic algorithm. Algorithms **17**(4) (2024)
5. Gomes, G.F., da Cunha, S.S., Ancelotti, A.C.: A sunflower optimization (sfo) algorithm applied to damage identification on laminated composite plates. Engineering with Computers **35**, 619–626 (2019)
6. Halim, A.H., Ismail, I., Das, S.: Performance assessment of the metaheuristic optimization algorithms: an exhaustive review. Artificial Intelligence Review (2021)
7. Karaboga, D., Basturk, B.: On the performance of artificial bee colony (abc) algorithm. Applied Soft Computing **8**(1), 687–697 (2008)
8. Klerkx, J., Verbert, K., Duval, E.: Enhancing Learning with Visualization Techniques, pp. 791–807. Springer New York, New York, NY (2014)
9. Marini, F., Walczak, B.: Particle swarm optimization (pso). a tutorial. Chemometrics and Intelligent Laboratory Systems **149**, 153–165 (2015)
10. Shah, I.K., Rathaur, N.S., Dohare, Y.S., Maity, T.: Optimizing cluster head selection for energy efficiency in wireless sensor networks: a hybrid algorithm combining grey wolf and enhanced sunflower optimization. Internet Technology Letters (2024)
11. Sheldon, M.R., Fillyaw, M.J., Thompson, W.D.: The use and interpretation of the friedman test in the analysis of ordinal-scale data in repeated measures designs. Physiotherapy Research International **1**(4), 221–228 (1996)
12. Tallarida, R.J., Murray, R.B.: Chi-Square Test, pp. 140–142. Springer New York, New York, NY (1987)
13. Xia, X., Gui, L., He, G., Xie, C., Wei, B., Xing, Y., Wu, R., Tang, Y.: A hybrid optimizer based on firefly algorithm and particle swarm optimization algorithm. Journal of Computational Science **26**, 488–500 (2018)
14. Yang, X.S.: Metaheuristic optimization: Algorithm analysis and open problems. In: Pardalos, P.M., Rebennack, S. (eds.) Experimental Algorithms. pp. 21–32. Springer Berlin Heidelberg, Berlin, Heidelberg (2011)
15. Yang, X.S.: Flower pollination algorithm for global optimization. In: Durand-Lose, J., Jonoska, N. (eds.) Unconventional Computation and Natural Computation. pp. 240–249. Springer Berlin Heidelberg, Berlin, Heidelberg (2012)

HEAVY-SARS: Harnessing Explainable AI and Generative Models for Visual Synthetic Data Generation of SARS-CoV-2

Jyoti Maggu

Abstract Technological advancements have facilitated the creation of extensive real-world data, although deriving valuable insights from these substantial datasets continues to be labour-intensive and expensive. The production of synthetic data with Generative Adversarial Networks (GANs) provides a viable approach. This study integrates the network architecture of DCGAN with the parameter optimisation technique of WGAN to produce synthetic CT scan data. We assess the effectiveness of deep learning models trained on synthetic data by juxtaposing them with those trained on actual data, utilizing XAI LIME. The experimental findings illustrate the efficacy of the suggested model in producing high-quality synthetic data. The model demonstrates similar performance on both actual and synthetic datasets, achieving accuracy values of 0.9818 for real data and 0.9745 for synthetic data under identical conditions. These findings highlight the applicability of our technique in real-world contexts, especially in situations with restricted access to extensive datasets.

Keywords Synthetic data generation · SARS-CoV-2 · Generative Adversarial Networks (GANs) · Explainable AI (XAI) · CT scan data · Deep learning models

1 Introduction

One of the long-standing problems is the need for labelled data, the required data needs to be improved, or there are concerns about safely sharing it with the stakeholders. Especially in the healthcare industry, where data privacy is a vital requirement, and the generated data is prone to cyber-attacks from assailants with

J. Maggu (✉)
Thapar Institute of Engineering and Technology, Patiala, India
e-mail: jyotim@iiitd.ac.in

K. Deep et al. (eds.), *Artificial Intelligence, Machine Learning and Intelligent Systems*, Springer Proceedings in Mathematics & Statistics 520,
https://doi.org/10.1007/978-981-95-3841-6_4

lots of computational power, the options to cultivate data for training become limited [2, 6, 9]. To address this, we may employ synthetic data creation techniques [11] to create replacement data that mimics real data in data-scarce scenarios, hence reducing the necessity for obtaining precise data to facilitate informed decision-making. Specifically, one can employ generative adversarial networks (GANs) [8], a neural network that creates novel data from scratch based on the original dataset. These techniques generate data that is synthetic. Synthetic data are economical, safeguard privacy, and improve security and data augmentation. Nonetheless, artificial data-generating models possess inherent limits. Although synthetic data can replicate numerous characteristics of genuine data, it is acknowledged that these models can not perfectly reproduce the original data. All synthetic data analyses must be validated against the actual dataset. A conventional GAN comprises a generator and a discriminator, with both networks engaged in competition. The generator attempts to deceive the discriminator into believing that the data it examines originates from the authentic dataset, whilst the discriminator endeavours to distinguish the created data from the genuine. Upon completion of training and the attainment of satisfactory data quality from the generator, it can be employed as a substitute for real data in the training of deep Convolutional Neural Networks (CNNs) [19]. Simultaneously, fundamental GAN models often produce low-quality pictures, yielding subpar samples or failing to converge. Despite its theoretical foundation, the vanilla GAN has demonstrated significant training difficulties. General convergence is significantly influenced by hyperparameter optimisation to prevent vanishing or expanding gradients, and it is susceptible to mode collapse [5]. To mitigate these obstacles, the deep convolutional generative adversarial network (DCGAN) [13] demonstrates greater adaptability to intricate image distributions compared to alternative approaches. DCGAN explicitly employs convolutional and transposed convolutional layers in both the discriminator and generator. The discriminator consists of strided convolution layers, batch normalisation layers, and LeakyReLU activation functions. The generator comprises convolutional transposition layers, batch normalisation layers, and ReLU activations. Research indicates that the DCGAN model effectively addresses the issue of low-quality images produced by GANs, yielding favourable outcomes. Regrettably, there exists an additional array of issues associated with DCGAN. The model parameters fluctuate, destabilise, and fail to converge. The generator fails, resulting in a restricted range of samples and exhibiting significant sensitivity to hyperparameter choices. The discriminator has significant success, resulting in the generator's gradient vanishing and receiving no feedback. An imbalance between the generator and discriminator leads to overfitting. We provide a new model called HEAVY-SARS designed to utilise explainable AI and generative models for the creation of visual synthetic data from SARS-CoV-2 CT scans. In our initial model, we employ the network architecture and parameters of DCGAN (Deep Convolutional GAN) and adopt a novel method for updating the parameters of the discriminator and generator as introduced in WGAN (Wasserstein GAN) [12] for the second model, wherein we generate synthetic data from a dataset of CT scan images. This work makes two contributions:

- Produce synthetic data that resembles authentic data.
- Contrast models developed using synthetic data with those trained using authentic data.

The remainder of the paper is structured as follows. Section 2 examines the prior research pertinent to the proposed subject. Section 3 delineates the dataset utilised for experiments and the recommended approach. Section 4 elucidates the findings from the experimental results, while Sect. 5 ultimately finishes the study with a discussion of its future implications.

2 Literature Review

Recent systems have been created to diagnosis the novel COVID-19 virus [14] using X-ray and CT-scan pictures obtained from multiple afflicted patients [3]. This is accomplished through a crucial branch of deep learning known as DCGAN (Deep Convolutional Generative Adversarial Network), which is utilised for identifying various emerging diseases in the medical field where datasets are notably limited. DCGAN serves as an effective instrument for generating synthetic data due to the scarcity of available data for diagnosing coronavirus. Gastric cancer presents a significant global fatality rate. Early diagnosis with regular gastroscopy can facilitate prevention. A deep learning-based computer-aided diagnosis (CADx) system utilising data augmentation to assist physicians in classifying gastroscopy images as normal or abnormal [16]. However, a substantial quantity of training data is necessary to enhance deep learning efficacy. GAN is a nontraditional method for generative modelling challenges aimed at learning the distribution of real data points. Numerous studies have demonstrated that they are challenging to train, as GANs operate differently from many machine learning techniques. They function similarly to a two-player game with a discriminator and a generator. This can result in both inaccuracy in the training process and a general lack of comprehension of the convergence of GANs [1].

The recently introduced Wasserstein GAN (WGAN) [4] advances the stability of GAN training; nonetheless, it occasionally produces subpar samples or fails to converge. The issues frequently arise from weight clipping in WGAN to impose a Lipschitz constraint on the critic, potentially resulting in undesirable behaviour.

The capacity of GANs to generate images with exceptional realism offers optimism that the persistent shortage of labelled data in the medical domain may be addressed by these generative models. Initial investigations on GAN-based MR reconstruction concentrate on the DAGAN architecture. This method incorporates a perceptual loss alongside adversarial and pixel-wise losses to evaluate deep retrieved features in both real and produced data, hence improving the model's stability [10]. A sequence of studies enhances the DCGAN design progressively by adjusting loss functions to maintain frequency information. The benefit is that it can enhance the deficiencies of training instability and mode collapse via the Cucconi test.

3 Proposed Methodology

3.1 Dataset Description

We employ the publicly available dataset [15], which comprises a multiclass CT scan dataset for the detection of SARS-CoV-2 infections in PNG format. The dataset comprises 4173 CT scans from 210 unique patients, including 2168 scans from 80 patients confirmed to be infected with SARS-CoV-2 using RT-PCR, utilised for synthetic data synthesis. The pictures were acquired from the Public Hospital of Government Employees of Sao Paulo (HSPM) and the Metropolitan Hospital of Lapa, both situated in Sao Paulo, Brazil. The preprocessing procedures involved scaling photos to a uniform resolution of 256×256 pixels, normalising pixel values to a range of [0,1], and executing data augmentation techniques like as rotations, flips, and zooms to improve model resilience. The 2168 CT images, with an average of 27 scans per patient, constitute the foundation for the generation of synthetic data (Fig. 3).

3.2 HEAVY-SARS

We produce synthetic data utilising a fundamental aspect of deep learning called DCGAN (Deep Convolutional Generative Adversarial Network). The dataset for identifying various emergent diseases in the medical area is quite limited, and DCGAN serves as an effective instrument for producing synthetic data because to the lack of data available for diagnosing coronavirus. However, as anticipated, DCGAN encountered Vanishing Gradient and mode collapse problems. We present a novel model called HEAVY-SARS for utilising explainable AI and generative models to produce visual synthetic data of SARS-CoV-2 CT images. In our initial model, we employ the network architecture and parameters of DCGAN (Deep Convolutional GAN), while in the second model, we adopt a novel method for updating the parameters of the discriminator and generator as introduced in WGAN (Wasserstein GAN), generating synthetic data from a dataset of CT scan images. The Wasserstein loss function approximates the Earth Mover's Distance, which quantifies the work required to transform one distribution into another. DCGAN employs a convolutional layer in both the Generator and Discriminator, whereas WGAN modifies the loss function, optimiser, clipping mechanism, and final sigmoid function. WGAN employs a Gradient Penalty that is incorporated into the loss function as a regularisation technique for the discriminator's gradient. Implementing 1-L continuity is essential for the application of Wasserstein loss. The discriminator must be 1-Lipschitz continuous, meaning the gradient norm must not exceed one at any point. We assess the efficacy of the deep learning ResNet model, illustrated in Fig. 2, which was trained on synthetic data using XAI LIME [18]. LIME is a technique for elucidating models by providing individual predictions

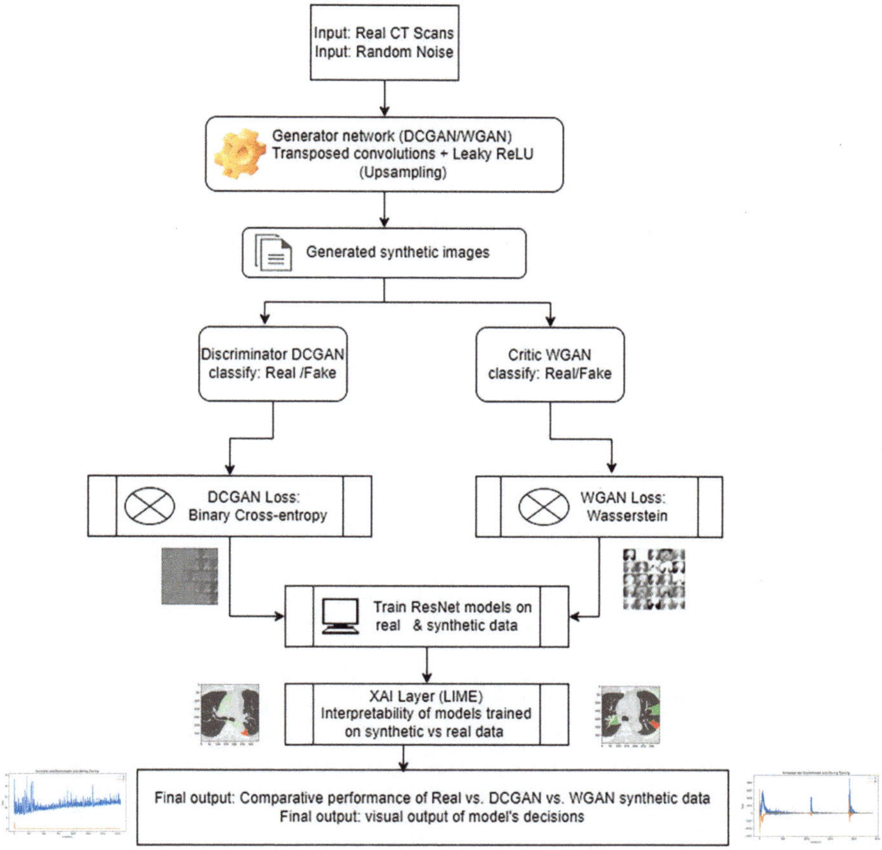

Fig. 1 Workflow of the proposed method HEAVY-SARS

alongside their interpretations in a comprehensible format. In image classification, numerous features substantially influence a prediction. To facilitate comprehension, we employ binary vector representation that signifies the "presence" or "absence" of an adjacent patch of analogous pixels (a super-pixel). It elucidates the targets in the photos by accentuating super-pixels with positive weights for a certain class in green, whereas super-pixels with negative weights are marked in red. This provides natural rationales for the model's assessment of the potential presence or absence of the class. The complete workflow is given in Fig. 1.

4 Results and Discussion

We have conducted both qualitative and quantitative assessments of SARS-CoV-2 data (Fig. 4). The visual comparison of the images produced by DCGAN and

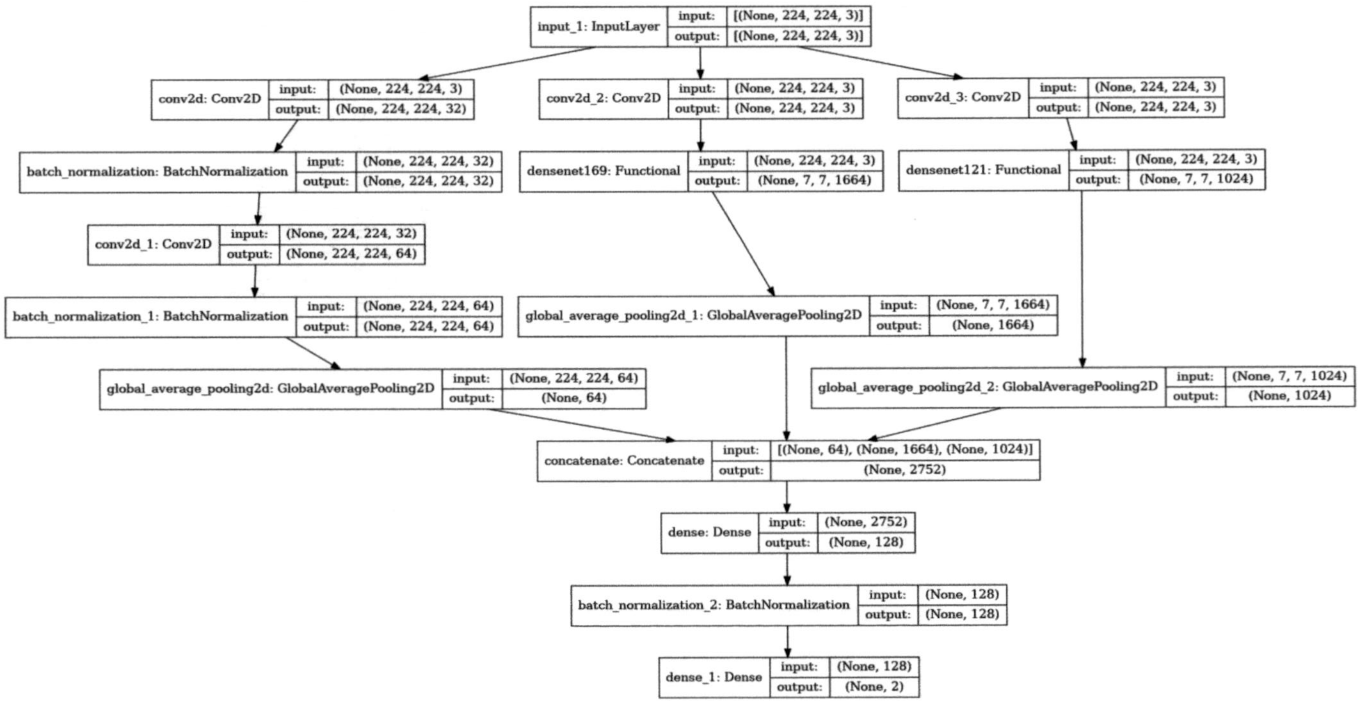

Fig. 2 The ResNet architecture for classification of SARS-CoV-2 infection

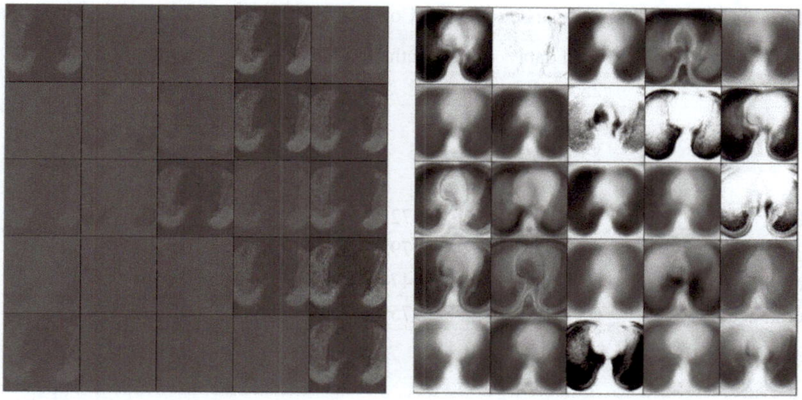

(a) Sample images generated with DCGAN

(b) Sample images generated with WGAN

Fig. 3 Visual comparison of the images generated from (**a**) DCGAN and (**b**) WGAN

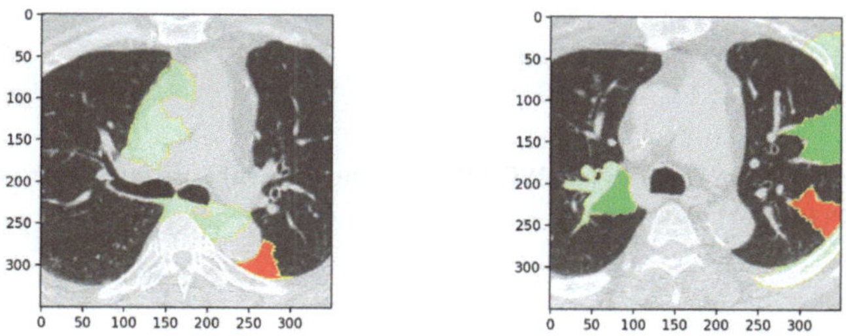

(a) XAI on image from DCGAN

(b) XAI on image from WGAN

Fig. 4 Visual comparison of a sample image generated from (**a**) DCGAN and (**b**) WGAN using XAI:LIME

WGAN is presented in Figs. 3 and 4. The generated images were evidently more realistic with WGAN compared to DCGAN. The comparative comparison of classification performance between real photos and synthetic images is presented in Table 1. The comparative measures include True Positives (TP), True Negatives (TN), False Positives (FP), False Negatives (FN), True Positive Rate (TPR), False Negative Rate (FNR), Overall Loss, and Overall Accuracy [17]. The optimal performance is evidently attained with real data. Due to the limited availability of medical data, generating synthetic data for analysis is advantageous, as the outcomes derived from synthetic data closely resemble those of actual data performance. The data from WGAN demonstrates superior performance compared to that from DCGAN. The plots for generator and discriminator loss during training of DCGAN and WGAN are given in Fig. 5. The final convergence plots for ResNet are given in terms of model loss and model accuracy at training time in Fig. 6. The results of ablation experiments are given in Table 2. We have performed experiments on different combinations of real and synthetic data generated from WGAN and

Table 1 Classification performance of SARS-Cov-2 infection from real data vs. synthetic data

	Real data	Synthetic DCGAN	Synthetic WGAN
TP	253	248	250
FN	2	7	5
TN	235	235	236
FP	7	5	6
TPR	0.9922	0.9725	0.9803
TNR	0.9710	0.9793	0.9752
Overall loss	0.0564	0.0474	0.0601
Overall accuracy	0.9819	0.9759	0.9778

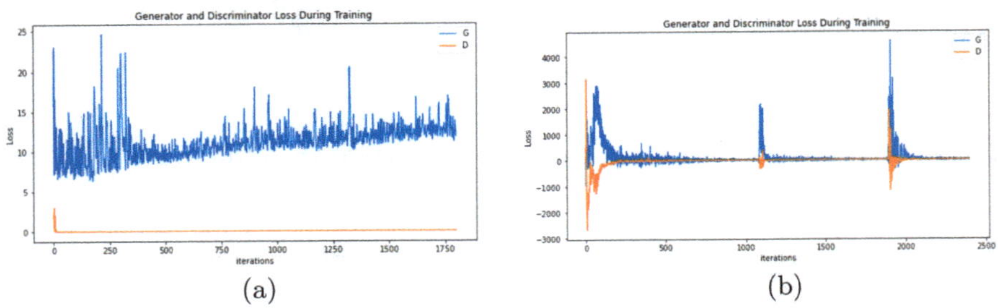

Fig. 5 Loss function from DCGAN and WGAN at training time. (**a**) Loss function from DCGAN. (**b**) Loss function from WGAN

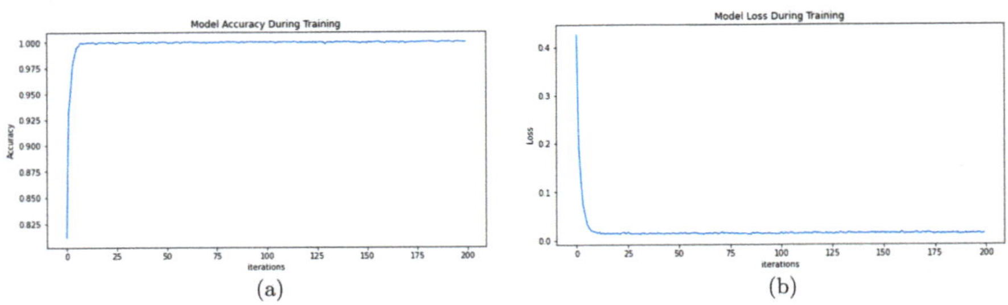

Fig. 6 ResNet model convergence at training time. (**a**) Model accuracy during training. (**b**) Model loss during training

computed model accuracy. It is observed that the best results are obtained when we have both training and test set as real images or both as synthetic images. For Real data as training, and synthetic data as test case, the model still performs close to the best results. For the last case, when the model is trained on synthetic data and tested upon the real data, the accuracy falls by nearly 1%. But, considering the scenario, when we have limited data for rare diseases, this approach can be used to generate more data and give preliminary analysis prior to clinical process.

Table 2 Ablation study: Classification performance of SARS-Cov-2 infection from real data vs. synthetic data

Training images	Test images	Accuracy
Real	Real	0.9818
Real	Synthetic	0.9781
Synthetic	Real	0.9655
Synthetic	Synthetic	0.9745

5 Conclusion and Future Work

Since the inception of Generative Adversarial Networks (GANs) in 2016 [7], their efficacy and applicability have significantly enhanced, creating new opportunities across multiple fields, including medicine. Notwithstanding these developments, dependence on vast training datasets continues to be a considerable constraint, especially in the medical domain where obtaining comprehensive datasets can prove difficult.The findings underscore the model's robustness, achieving accuracy scores of 0.9818 for real data and 0.9745 for synthetic data, so illustrating its efficacy in producing high-quality synthetic images. These results validate the model's capacity to connect actual and synthetic data while preserving performance, offering a promising solution for mitigating data scarcity in medical imaging. Future research should concentrate on enhancing GAN architectures to function efficiently with constrained input while preserving output quality. Two intriguing avenues for additional investigation involve modifying the loss function and gradient penalties of WGAN within the DCGAN framework and exploring alternative generative models, including Variational Autoencoders and Diffusion Models. By tackling these problems, we may enhance the capabilities of generative models, rendering them more varied and influential for practical applications.

References

1. Ahmad, Z., Jaffri, Z.u.A., Chen, M., Bao, S.: Understanding gans: fundamentals, variants, training challenges, applications, and open problems. Multimedia Tools and Applications pp. 1–77 (2024)
2. Bansal, M.A., Sharma, D.R., Kathuria, D.M.: A systematic review on data scarcity problem in deep learning: solution and applications. ACM Computing Surveys (Csur) 54(10s), 1–29 (2022)
3. Bhatele, K.R., Jha, A., Tiwari, D., Bhatele, M., Sharma, S., Mithora, M.R., Singhal, S.: Covid-19 detection: A systematic review of machine and deep learning-based approaches utilizing chest x-rays and ct scans. Cognitive Computation 16(4), 1889–1926 (2024)
4. Bobadilla, J., Gutiérrez, A.: Wasserstein gan-based architecture to generate collaborative filtering synthetic datasets. Applied Intelligence 54(3), 2472–2490 (2024)
5. Ding, Z., Jiang, S., Zhao, J.: Take a close look at mode collapse and vanishing gradient in gan. In: 2022 IEEE 2nd International Conference on Electronic Technology, Communication and Information (ICETCI). pp. 597–602. IEEE (2022)

6. Dlamini, W.M., Dlamini, S.N., Mabaso, S.D., Simelane, S.P.: Spatial risk assessment of an emerging pandemic under data scarcity: A case of covid-19 in eswatini. Applied Geography 125, 102358 (2020)
7. Goodfellow, I.: Nips 2016 tutorial: Generative adversarial networks. arXiv preprint arXiv:1701.00160 (2016)
8. Goodfellow, I., Pouget-Abadie, J., Mirza, M., Xu, B., Warde-Farley, D., Ozair, S., Courville, A., Bengio, Y.: Generative adversarial networks. Communications of the ACM 63(11), 139–144 (2020)
9. Hoyler, M., Finlayson, S.R., McClain, C.D., Meara, J.G., Hagander, L.: Shortage of doctors, shortage of data: a review of the global surgery, obstetrics, and anesthesia workforce literature. World journal of surgery 38, 269–280 (2014)
10. Kim, S., Park, H., Park, S.H.: A review of deep learning-based reconstruction methods for accelerated mri using spatiotemporal and multi-contrast redundancies. Biomedical Engineering Letters pp. 1–22 (2024)
11. Murtaza, H., Ahmed, M., Khan, N.F., Murtaza, G., Zafar, S., Bano, A.: Synthetic data generation: State of the art in health care domain. Computer Science Review 48, 100546 (2023)
12. Patil, M., Patil, M.M., Agrawal, S.: Wgan for data augmentation. In: GANs for Data Augmentation in Healthcare, pp. 223–241. Springer (2023)
13. Phukan, S., Singh, J., Gogoi, R., Dhar, S., Jana, N.D.: Covid-19 chest x-ray image generation using resnet-dcgan model. In: Advances in Intelligent Computing and Communication: Proceedings of ICAC 2021, pp. 227–234. Springer (2022)
14. Sadeghi, A., Sadeghi, M., Sharifpour, A., Fakhar, M., Zakariaei, Z., Sadeghi, M., Rokni, M., Zakariaei, A., Banimostafavi, E.S., Hajati, F.: Potential diagnostic application of a novel deep learning-based approach for covid-19. Scientific Reports 14(1), 280 (2024)
15. Soares, E., Angelov, P.: A large dataset of real patients CT scans for COVID-19 identification (2020), https://doi.org/10.7910/DVN/SZDUQX
16. Uema, R., Hayashi, Y., Kizu, T., Igura, T., Ogiyama, H., Yamada, T., Takeda, R., Nagai, K., Inoue, T., Yamamoto, M., et al.: A novel artificial intelligence-based endoscopic ultrasonography diagnostic system for diagnosing the invasion depth of early gastric cancer. Journal of Gastroenterology pp. 1–13 (2024)
17. Vanacore, A., Pellegrino, M.S., Ciardiello, A.: Fair evaluation of classifier predictive performance based on binary confusion matrix. Computational Statistics 39(1), 363–383 (2024)
18. Vimbi, V., Shaffi, N., Mahmud, M.: Interpreting artificial intelligence models: a systematic review on the application of lime and shap in alzheimer's disease detection. Brain Informatics 11(1), 10 (2024)
19. Yamashita, R., Nishio, M., Do, R.K.G., Togashi, K.: Convolutional neural networks: an overview and application in radiology. Insights into imaging 9, 611–629 (2018)

Infected Hyena Optimizer: A Novel Bio-inspired Metaheuristic Algorithm

Shreyansh Tiwari, Ratnesh Kherudkar, Yash Kumar, Enduri Jahnavi, Aditya Patil, and Utkarsh Mahadeo Khaire

Abstract This paper introduces the Infected Hyena Optimization (IHO), an innovative hybrid algorithm that effectively combines the exploratory strength of the Virus Colony Search (VCS) with the exploitation prowess of the Spotted Hyena Optimization (SHO). This synergistic integration allows the IHO to enhance search efficiency and accuracy across complex optimization landscapes. A comprehensive comparative analysis with its progenitor algorithms demonstrates the IHO's superior capabilities in achieving global optima, highlighting notable improvements in both convergence rates and solution precision.

Keywords Hybrid optimization techniques · Metaheuristics · Evolutionary computation · Swarm intelligence · Exploration-Exploitation trade-off

1 Introduction

In recent decades, the increasing complexity of real-world problems has created a higher demand for more advanced metaheuristic methods to address engineering design challenges [1]. These methods are becoming more popular due to their efficiency and complexity compared to traditional approaches. Metaheuristic algorithms are typically categorized based on their search strategy, structure, complexity, solution representation, and application domain. They are generally

S. Tiwari (✉) · R. Kherudkar · A. Patil
Computer Science and Engineering, Indian Institute of Information, Dharwad, India
e-mail: 21bcs114@iiitdwd.ac.in

Y. Kumar
Electronics and Communications Engineering, Indian Institute of Information, Dharwad, India

E. Jahnavi · U. M. Khaire
Data Science and Artificial Intelligence, Indian Institute of Information, Dharwad, India

K. Deep et al. (eds.), *Artificial Intelligence, Machine Learning and Intelligent Systems*, Springer Proceedings in Mathematics & Statistics 520,
https://doi.org/10.1007/978-981-95-3841-6_5

divided into three main types: Evolutionary Algorithms (EA), Swarm Intelligence (SI) algorithms, and Physics-Based (PB) algorithms [2].

Evolutionary-based metaheuristic algorithms, including Genetic Algorithm (GA), Evolutionary Strategies (ES), and Genetic Programming (GP), are optimization methods inspired by natural evolutionary processes such as natural selection. Li et al. [3] They employ techniques like selection, crossover, and mutation to evolve solutions across generations. Similarly, Swarm Intelligence Algorithms, exemplified by Particle Swarm Optimization (PSO), Ant Colony Optimization (ACO), and Cuckoo Search Algorithm, draw from the coordinated behaviors of self-organized systems found in nature, leveraging collective intelligence or swarm intelligence to solve complex optimization challenges. Additionally, Physics-Based algorithms like Gravitational Search Algorithm (GSA) [4], Simulated Annealing (SA), and Electromagnetic Field Optimization (EFO) [5] utilize principles from physical sciences such as thermodynamics and quantum mechanics to explore and optimize solution spaces effectively. These diverse classes of algorithms adapt various natural phenomena to address intricate problems by finding optimal or near-optimal solutions. [6] Meta-heuristic optimization algorithms consist of two essential components: exploration and exploitation. Exploration allows for a global search of the solution space, helping to avoid local optima, while exploitation focuses on improving promising solutions through local search. Effective optimization depends on striking a balance between these two processes. Although all population-based algorithms incorporate both, they use different operators and mechanisms to achieve this balance.

For meta-heuristic algorithms to be effective, there must be a balance between exploration and exploitation. Excessive focus on exploration can result in wasting time on unpromising areas, while too much exploitation can lead to early convergence and getting trapped in local optima. To address this, many algorithms use adaptive strategies that adjust exploration and exploitation rates based on performance feedback. This adaptability enhances the algorithm's robustness and improves its efficiency across different optimization scenarios.

This paper introduces a novel metaheuristic algorithm termed as Infected Hyena Optimization (IHO), for optimizing constraint and unconstrained design problems. It is inspired by two existing metaheuristic algorithms, namely Virus Colony Search Optimizer (VCS) and Spotted Hyena Optimizer (SHO). In IHO, the exploration capabilities are enhanced by VCS's diffusion strategies, which utilize Gaussian walks to effectively traverse the solution space and discover diverse potential solutions while accounting for immunity and reinitialisation of the population. This allows the algorithm to escape local optima and explore new areas. On the other hand, IHO leverages the cooperative hunting behavior of spotted hyenas, as seen in SHO, to focus on refining and improving promising solutions through localized searches.

2 Parent Algorithms

The Infected Hyena Optimization (IHO) algorithm integrates the social hierarchy and cooperative hunting behavior of spotted hyenas, as modeled by the Spotted Hyena Optimizer (SHO), with the diffusion strategies and infection dynamics from Virus Colony Search. In Spotted Hyena Optimizer, tightly coordinated hyena groups enhance exploitation by collaboratively tracking and capturing prey, mirroring efficient local search in the solution space. IHO adopts this behavior, where agents simulate hyena packs to optimize exploitation around high-potential solutions. The algorithm's core operations are inspired by the hunting dynamics of spotted hyenas, balancing exploration and exploitation to enhance convergence towards optimal solutions.

Building on the Virus Colony Search Optimizer, IHO employs two key strategies—Gaussian walks for virus diffusion and evolutionary mechanisms for immune response—to boost performance. The Gaussian walks ensure thorough exploration of the search space, while evolutionary strategies replace less optimal individuals with better agents, enhancing efficiency. In addition, Random perturbations, such as mutation and infection, governed by mutation and infection rates, help the algorithm escape local optima, further improving optimization.

2.1 Virus Colony Search Optimizer

The Virus Colony Search (VCS) algorithm is inspired by the biological behaviors of viruses, particularly their mechanisms for diffusion, infection, and evasion of host immune responses. This optimization technique leverages the strategies used by viruses to efficiently solve complex optimization problems by exploring and exploiting the solution space [3]. The VCS algorithm draws inspiration from the way biological viruses invade host cells, replicate themselves, and manage to survive attacks by the host's immune system. This biological analogy is used to design an optimization process that mimics these efficient survival strategies.

2.2 Spotted Hyena Optimizer

The Spotted Hyena Optimizer (SHO) is a bio-inspired metaheuristic algorithm based on the social dynamics and hunting strategies of spotted hyenas. Known for tactics like encircling prey and collaborative pack hunting, these behaviors are abstracted into iterative mathematical steps to solve optimization problems. Simulated agents, representing hyenas, explore and exploit the search space, balancing the discovery of new possibilities with refinement of known solutions. This approach enhances the algorithm's efficiency and effectiveness in navigating complex optimization challenges [2].

3 Infected-Hyena Optimizer: The Hybrid Algorithm

The Infected-Hyena Optimization (IHO) algorithm merges the adaptive search strategies of the Spotted Hyena Optimizer (SHO) with the mutation and infection mechanisms of the Virus Colony Search (VCS). This hybrid approach enhances the balance between exploration and exploitation, improving convergence in complex optimization landscapes. By combining the global search capabilities of VCS with the local search strengths of SHO, IHO addresses the limitations of each algorithm individually, especially in tackling multi-modal functions. It employs a population of agents (hyenas) that evolve through processes of encircling and attacking prey, undergoing infections and mutations, and adapting via an immune response. Pseudocode 1 outlines the detailed implementation of the IHO process.

3.1 Algorithm Overview

IHO combines the dynamic, cooperative behavior of hyenas in SHO with the random infection and robust mutation strategies of VCS to create a versatile optimization framework. This hybrid structure enables the algorithm to maintain a diverse population of solutions while efficiently focusing computational efforts on promising areas of the search space. The following equations involved in making IHO are inspired from both VCS and SHO.

3.1.1 Prey Encircling

$$D_{\text{prey}} = \left|C \cdot \text{Prey} - \text{Agent}\right| \tag{1}$$

$$\text{New Position} = \text{Prey} - A \cdot D_{\text{prey}} \tag{2}$$

This mechanism allows agents to dynamically adjust their positions relative to the best current solution or 'prey.' In Eq. (1), D_{prey} defines the distance between the agent and prey, with C adding randomness. Equation (2) updates the agent's position, using A to control movement intensity. These equations balance exploration and exploitation, mimicking the hunting behavior of hyenas in the SHO algorithm.

3.1.2 Gaussian Diffusion

$$\text{New Position} = Gaussian(Prey, \tau) + (r1 \cdot Prey - r2 \cdot Agent) \tag{3}$$

In Eq. (3), the agent's position is updated via Gaussian diffusion, where Gaussian(Prey, τ) adds stochastic perturbations to enhance local exploration. The difference term adjusts the position relative to both the prey and agent with random factors between (0,1). This helps the algorithm escape local optima, mimicking the random search behavior of viral mutation seen in VCS.

3.1.3 Virus Infection and Mutation

$$\text{New Position} = \text{Agent} + \epsilon \cdot \text{Random Infection Term} \tag{4}$$

$$\text{New Position} = \text{Agent} + \delta \cdot \text{Random Mutation Term} \tag{5}$$

In Eq. (4), the agent's position is slightly adjusted by a random infection term, introducing minor variations. In Eq. (5), mutation causes larger changes, increasing population diversity. Together, these processes ensure genetic variation, helping the algorithm avoid premature convergence and enhancing its overall search effectiveness.

3.1.4 Attack on Prey

$$G_{\text{prey}} = \frac{1}{n} \sum_{i=1}^{n} New\_Position_i \tag{6}$$

$$\text{New Position} = \frac{G_{\text{prey}}}{n} \tag{7}$$

This phase models N hyenas pursuing prey, with Eq. (1), the distance between the i-th agent and prey is calculated, with C adding randomness for variable movement intensity. Equation (2) updates each agent's position, with A controlling the movement's aggression. Equation (6) computes the group's centroid, representing the collective movement of n hyenas, and in Eq. (7), agents adjust their positions toward the centroid, promoting cohesive group movement. This simulates cooperative hunting, balancing convergence and exploration.

Immune Response In the final phase, IHO simulates the immune response by ranking agents by fitness, selecting the top 50%, and reinitializing the rest. This evolutionary pressure ensures only the strongest solutions survive, mimicking natural selection, and improving solution quality in future generations.

Pseudocode 1: Pseudocode for Infected-Hyena Optimization (IHO)

Initialization:

Initialize population within $[lower\_bound, upper\_bound]$;
Determine fitness of each agent;
Determine best agent as prey;

while *iteration* $\leq$ max_iter **do**

  Adjust a linearly from $a_{\max}$ to 0 $a \leftarrow a_{\max} - \left(\frac{\text{iteration}}{max\_iter}\right) \cdot a_{\max}$

  foreach *agent* **do**

    if *rand(0,1)* < $P_{Encircle}$ **then**
      Encircle prey using Eqs. (1) & (2)
      if *rand(0,1)* < P_{Diff} **then**
        Apply **Gaussian diffusion** using Eq. (3)

    else
      Infect agent using Eq. (4)
      if *rand(0,1)* < *mut_rate* **then**
        Mutate agent using Eq. (5)

    if *rand(0,1)* < P_{Attack} **then**
      Collective **attack on prey** using Eqs. (1), (2), (6) & (7)
    Clip position within bounds
    Re-evaluate fitness
    if *fitness improved* **then**
      Update position

  Immune response: Retain top 50% agents
  Introduce new agents with random positions as needed
  if *new global best found* **then**
    Update global best (prey)

Store fitness history;
return the best found solution and its fitness;

4 Experiment and Visualisations

4.1 Benchmark Functions

This study utilizes a set of well-known benchmark functions to assess the performance of the Infected Hyena Optimization (IHO) algorithm in comparison to other optimization algorithms. These benchmark functions are standard tools for evaluating optimization algorithms, offering a variety of challenges that test the algorithm's ability to manage different landscape types, including multi-modal, uni-modal, separable, and non-separable features. These functions present a wide

Table 1 Functions and their details

Function	Bounds	Global optimum	Dim	Equation
Goldstein-Price	[−2,2]	3	2	$f(x, y) = (1 + \frac{(x+y+2)^2}{4}) \cdot (1 + \frac{(x+y2)^2}{4})$
Schwefel	[−500, 500]	−837.658	2	$f(x, y) = 418.9829 \cdot n - \sum_{i=1}^{n} x_i \cdot \sin(\sqrt{\lvert x_i \rvert})$
Schaffer	[−100, 100]	0	2	$f(x, y) = 0.5 + \frac{\sin^2(\sqrt{x^2+y^2}-0.5)}{(1+0.001(x^2+y^2))^2}$
Sphere	[−5.12, 5.12]	0	10	$f(x) = \sum_{i=1}^{n} x_i^2$
Ackley	[−32.768, 32.768]	0	10	$f(x) = -20 \cdot \exp\left(-0.2 \cdot \sqrt{\frac{1}{n} \sum_{i-1}^{n} x_i^2}\right) -$
				$\exp\left(\frac{1}{n} \sum_{i=1}^{n} \cos(2\pi x_i)\right) + 20 + c$
Rastrigin	[−5.12, 5.12]	0	10	$f(x) = An + \sum_{i=1}^{n} \left(x_i^2 - A \cdot \cos(2\pi x_i)\right)$ where $A = 10$
Martin-Gaddy	[0,10]	0	2	$f(x, y) = \sin^2(x, y) - \frac{1}{2}(x^2 + y^2)$
Easom	[−100, 100]	−1	2	$f(x, y) = -\cos(x) \cdot \cos(y) \cdot \exp\left(-\left((x-\pi)^2 + (y-\pi)^2\right)\right)$
Griewank	[−600, 600]	0	10	$f(x) = 1 + \frac{1}{4000} \sum_{i=1}^{n} x_i^2 - \prod_{i=1}^{n} \cos\left(\frac{x_i}{\sqrt{i}}\right)$

range of challenges for optimization algorithms like IHO, including deceptive local minima, flat surfaces, narrow valleys, and complex multi-modal landscapes. IHO's performance is measured by its ability to navigate these landscapes and converge to the global optimum. Table 1 shows all the benchmark functions that were used, detailing their specific properties, bounds, global optima, dimensionality, and mathematical formulations, which collectively illustrate the diverse challenges posed to the Infected Hyena Optimization algorithm during the evaluation process.

4.2 Convergence Analysis of IHO

Figure 1 assesses the convergence of the IHO algorithm using metrics such as Search History, Trajectory in the 1st Dimension, and Optimization History [6]. While some functions operate in higher dimensions, the two-dimensional view offers valuable insights into the domain's topology. The search history reveals the exploration patterns used by the particles. Circular patterns in the graphs indicate particle aggregation around an optimal point, highlighting the efficiency of IHO's exploitation abilities. Additionally, the distribution of particles throughout the search space demonstrates the algorithm's exploration capabilities, clearly visible in the search history graphs.

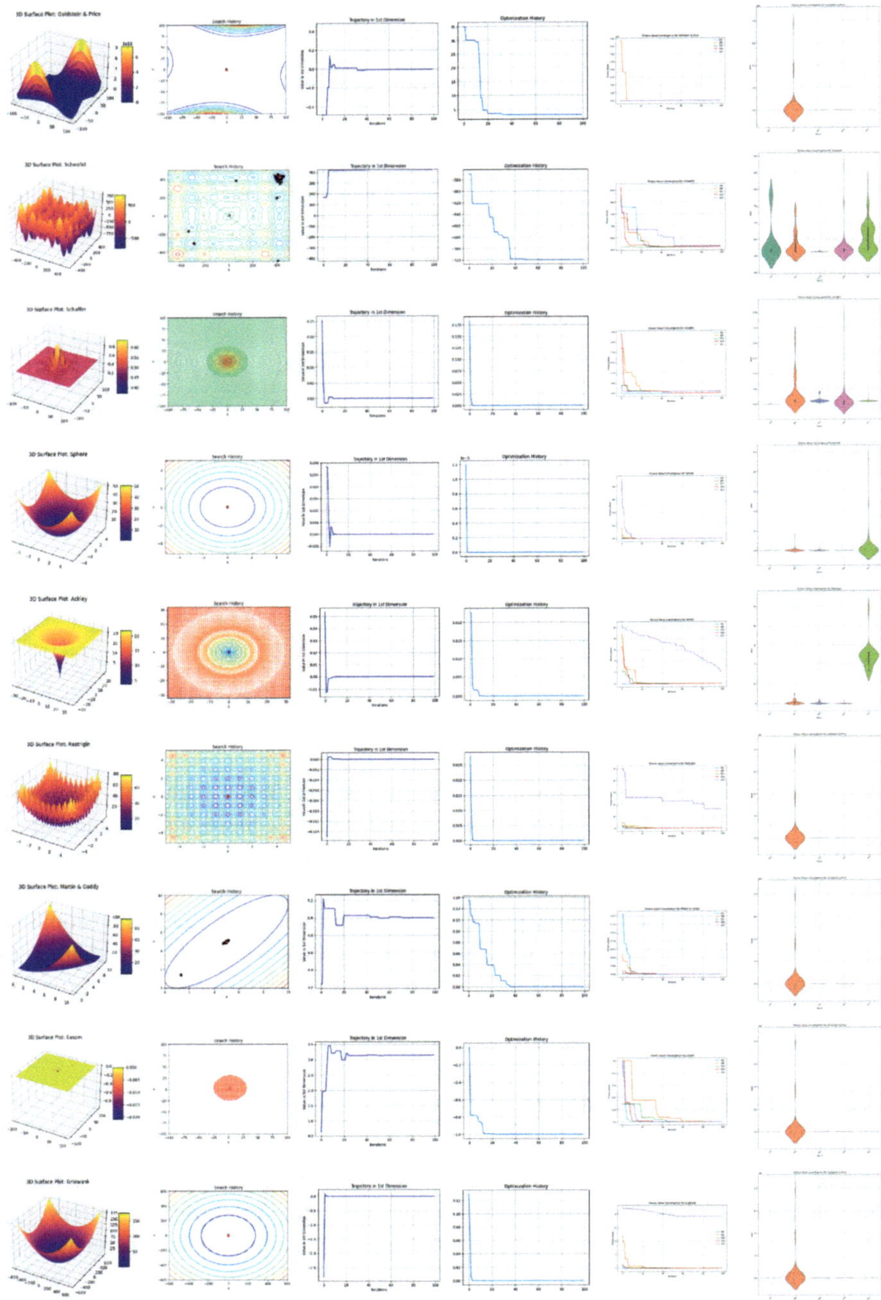

Fig. 1 Qualitative metrics: search history, trajectory in 1st dimension and optimization history

The particle trajectory provides a key metric, showing the positional changes of the first particle in the first dimension. Early iterations often display sudden fluctuations due to global exploration, which transitions into local search for convergence. In more complex search domains, such as Schwefel, these fluctuations are more pronounced, as reflected in the detailed search history graphs. In contrast, simpler search spaces demonstrate fewer fluctuations, indicating smoother and more direct trajectories. The optimization history reflects the fitness of the best particle across all iterations, with the graphs showcasing IHO's convergence for various benchmark functions. IHO achieves effective convergence on functions like Goldstein and Price, Schwefel, Martin and Gaddy, and Easom, while faster convergence is observed in other cases due to a smaller initial search space. The violin plots show that IHO generally has a tighter distribution of fitness values, indicating more consistent performance and lower variance than other algorithms. It excels on functions like Goldstein and Price, Schwefel, Martin and Gaddy, and Easom. However, for Sphere, Ackley, and Rastrigin, the wider distribution suggests higher sensitivity to search space complexity, with faster convergence due to a smaller initial population spread.

The convergence line plots highlight IHO's efficient early exploration followed by smooth convergence to optimal solutions. For functions like Goldstein and Price, Schwefel, Martin and Gaddy, and Easom, IHO shows a steady decline in fitness, indicating strong optimization. In contrast, Sphere, Ackley, and Rastrigin exhibit faster initial convergence due to a smaller search space, needing fewer iterations.

4.3 Statistical Analysis

The aim of the statistical analysis is to assess IHO's performance compared to other algorithms across 9 objective functions, focusing on the Friedman test rankings and pairwise comparisons. Figure 2 shows the rank tables, pairwise comparisons, and Friedman test results. IHO ranks 1st in 6 of the 9 objective functions and places 2nd, 3rd, and 4th for the Goldstein and Price, Martin Gaddy, and Schwefel functions, respectively. This consistent performance highlights IHO as a top-performing algorithm in most cases.

Null Hypothesis The null hypothesis for each objective function assumes no significant performance differences between the algorithms (Infected Hyena Optimization, Artificial Bee Colony [7], Particle Swarm Optimization [8], Spotted Hyena Optimizer, Virus Colony Search).

Friedman Test Results The null hypothesis was rejected for all objective functions, indicating statistically significant performance differences among the algorithms. This suggests that the observed performance variations are unlikely to be due to random chance.

Pairwise comparisons show that IHO shares some similarities with its parent algorithm, SHO, suggesting it retains beneficial traits while enhancing performance.

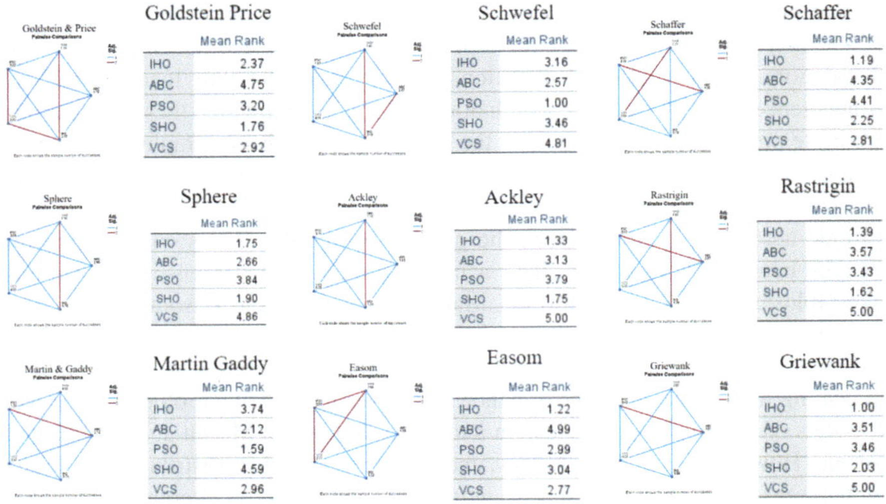

Goldstein Price	Mean Rank
IHO	2.37
ABC	4.75
PSO	3.20
SHO	1.76
VCS	2.92

Schwefel	Mean Rank
IHO	3.16
ABC	2.57
PSO	1.00
SHO	3.46
VCS	4.81

Schaffer	Mean Rank
IHO	1.19
ABC	4.35
PSO	4.41
SHO	2.25
VCS	2.81

Sphere	Mean Rank
IHO	1.75
ABC	2.66
PSO	3.84
SHO	1.90
VCS	4.86

Ackley	Mean Rank
IHO	1.33
ABC	3.13
PSO	3.79
SHO	1.75
VCS	5.00

Rastrigin	Mean Rank
IHO	1.39
ABC	3.57
PSO	3.43
SHO	1.62
VCS	5.00

Martin Gaddy	Mean Rank
IHO	3.74
ABC	2.12
PSO	1.59
SHO	4.59
VCS	2.96

Easom	Mean Rank
IHO	1.22
ABC	4.99
PSO	2.99
SHO	3.04
VCS	2.77

Griewank	Mean Rank
IHO	1.00
ABC	3.51
PSO	3.46
SHO	2.03
VCS	5.00

Fig. 2 Rank tables for friedman test and pairwise comparisons

Notably, IHO is statistically distinct from other algorithms, indicating its superior performance in comparison.

5 Summary and Conclusions

The Infected Hyena Optimization (IHO) algorithm effectively combines the adaptive search strategies of the Virus Colony Search (VCS) and the Spotted Hyena Optimizer (SHO), creating a robust hybrid framework. By integrating the infection and mutation dynamics of VCS with the cooperative hunting behavior of SHO, IHO excels in balancing exploration and exploitation. This balance enables the algorithm to efficiently navigate through complex optimization landscapes, avoiding local optima and converging towards global solutions. Evaluations using a variety of benchmark functions showcased IHO's superior performance compared to its parent algorithms. Convergence analysis revealed that IHO maintains a strong exploration phase early on, followed by a smooth transition to exploitation, resulting in stable and consistent convergence. The algorithm demonstrated strong adaptability across different landscapes, particularly excelling in complex multi-modal and deceptive environments. Benchmark tests like Goldstein and Price and Schwefel functions revealed IHO's efficient convergence, while simpler functions like Sphere displayed more variability, indicating potential sensitivity to search space complexity. The Friedman test further confirmed IHO's competitive edge, with statistically significant differences in performance compared to other algorithms such as ABC, PSO, SHO, and VCS. The test results rejected the null hypothesis, indicating that IHO consistently outperformed these algorithms across all tested objective functions. This statistical validation underscores the algorithm's capability to handle a wide range of optimization challenges.

In conclusion, IHO not only offers a substantial performance boost over its parent algorithms, but also competes effectively with other well-established optimization techniques. Its robust mechanisms for adaptive search, infection-based diversification, and cooperative exploitation make it a highly versatile and powerful tool for solving complex optimization problems. The algorithm's proven efficiency and consistency in handling both simple and challenging landscapes highlight its potential as a reliable and effective solution for a broad spectrum of optimization tasks.

References

1. W. Wong *et al.*, "A Review on Metaheuristic Algorithms: Recent Trends, Benchmarking and Applications" in *7th International Conference on Smart Computing & Communications (ICSCC), Sarawak, Malaysia*, (2019). https://doi.org/10.1109/ICSCC.2019.8843624..
2. G. Dhiman and A. Kaur, "Spotted hyena optimizer: A novel bio-inspired based metaheuristic technique for engineering applications," in *Advances in Engineering Software, Volume 114*, (2017). https://doi.org/10.1016/j.advengsoft.2017.05.014
3. Mu Dong Li *et al.*, "A novel nature-inspired algorithm for optimization: Virus colony search," in *Advances in Engineering Software, Volume 92*,, (2016). https://doi.org/10.1016/j.advengsoft.2015.11.004.
4. Esmat Rashedi *et al.*, "GSA: A Gravitational Search Algorithm, " in *Information Sciences, Volume 179, Issue 13*,, (2009). https://doi.org/10.1016/j.ins.2009.03.004.
5. Hosein Abedinpourshotorban *et al.*, "Electromagnetic field optimization: A physics-inspired metaheuristic optimization algorithm, " in *Swarm and Evolutionary Computation, Volume 26*, , (2016). https://doi.org/10.1016/j.swevo.2015.07.002.
6. Afshin Faramarzi *et al.*, "Equilibrium optimizer: A novel optimization algorithm," in *Knowledge-Based Systems, Volume 191*,, (2020). https://doi.org/10.1016/j.knosys.2019.105190.
7. Karaboga *et al.*, "Artificial Bee Colony (ABC) Optimization Algorithm for Solving Constrained Optimization Problems," in *Melin, P., Castillo, O., Aguilar, L.T., Kacprzyk, J., Pedrycz, W. (eds) Foundations of Fuzzy Logic and Soft Computing*,, (2007). https://doi.org/10.1007/978-3-540-72950-1_77
8. J. Kennedy *et al.*, "Particle swarm optimization," in *Proceedings of ICNN'95 - International Conference on Neural Networks*, (1995). https://doi.org/10.1109/ICNN.1995.488968

Socio-Nomadic Learning Optimization

Smruti Patil, Pratik Pakhale, Ashish Gidijala, Chinmay Shewale, Shashwat Chandel, and Utkarsh Mahadeo Khaire

Abstract This paper introduces a new hybrid metaheuristic algorithm called Socio-Nomadic Learning Optimization (SNLO), which combines the strengths of two existing algorithms: the Socio Evolution and Learning Optimization Algorithm (SELO) and the Nomadic People Optimizer (NPO). SELO takes inspiration from how families learn socially, helping it explore and make decisions effectively. On the other hand, NPO is based on how nomadic tribes adapt and lead, using a group approach to search large areas for solutions. When tested on standard problems, this algorithm showed faster and better results than other leading methods. The results indicate that SNLO effectively manages to avoid local optima adapting well to various problem landscapes, making it a promising tool for complex optimization tasks.

Keywords Metaheuristics · Optimization · SELO · NPO

1 Introduction

Optimization has always as always been an essential aspect of human life and has become a cornerstone in solving complex, high-dimensional problems across numerous fields such as engineering design, operations research, and machine learning. Traditional deterministic optimization techniques often struggle with non-linear, multi-modal problems due to their tendency to get trapped in local optima and their high computational cost [1]. Consequently, metaheuristic algorithms, which provide approximate solutions through iterative improvement and stochastic

S. Patil (✉) · P. Pakhale · C. Shewale · S. Chandel
Computer Science and Engineering, Indian Institute of Information Technology, Dharwad, India

A. Gidijala · U. M. Khaire
Data Science and Artificial Intelligence, Indian Institute of Information Technology, Dharwad, India

K. Deep et al. (eds.), *Artificial Intelligence, Machine Learning and Intelligent Systems*, Springer Proceedings in Mathematics & Statistics 520,
https://doi.org/10.1007/978-981-95-3841-6_6

processes, have emerged as powerful alternatives. In this context, many meta-heuristic algorithms have proven to successfully achieve the goal, namely Particle Swarm Optimization [2], Genetic Algorithm [3], etc. This paper presents a novel hybrid metaheuristic that synergizes two recent and effective algorithms: the Socio Evolution and Learning Optimization Algorithm (SELO) and the Nomadic People Optimizer (NPO). By merging SELO and NPO, the hybrid approach aims to leverage the strengths of both algorithms, potentially leading to more robust and efficient solutions.

2 Parent Algorithms

2.1 Nomadic People Optimizer (NPO)

The Nomadic People Optimizer (NPO) [4] is a novel swarm-based metaheuristic algorithm designed to address complex optimization problems, particularly those involving high-dimensional search spaces. Drawing inspiration from the behaviors of nomadic tribes, this migration-based behavior is simulated in the algorithm as a form of search strategy, where multiple clans (swarms) explore the solution space, guided by their clan leaders. The algorithm incorporates hierarchical leadership and migration patterns in a parameter-free multi-swarm approach to model the way nomadic tribes adapt and move based on environmental changes. NPO addresses balance of exploration and exploitation by modeling each clan's leader as the best solution, with families searching locally and the clans globally.

2.2 Socio Evolution and Learning Algorithm

The Socio-Evolution Learning Optimization (SELO) [5] algorithm is a socio-inspired optimization technique modeled after the social learning processes seen in human families. SELO captures the essence of how individuals improve by learning from the experiences and knowledge of others, specifically within a structured family unit. In SELO, the members of the family learn from the leader, i.e. the best performing member, adjusting their positions in the search space to move closer to the optimal solution. The learning process is social, meaning that individuals do not act independently but tend to rely on knowledge sharing within the group. The leader's position influences the movements of other family members, but randomness is also introduced to promote the exploration of the solution space and prevent being trapped in local optima.

3 Overview and Implementation of SNLO

The proposed hybrid algorithm, Socio-Nomadic Learning Optimization (SNLO), integrates the decentralized social learning dynamics of SELO with the adaptive migratory behavior and leadership dynamics of NPO. This integration is achieved through a coordinated framework where SELO's family units operate within NPO's clan-based multi-swarm environment, allowing families to both compete and cooperate in their search for optimal solutions. The hybrid algorithm adapts NPO's exploratory abilities and SELO's exploitative abilities.

3.1 *Initialization*

The algorithm begins by initializing multiple clans. Each clan is characterized by a leader, or a "sheikh", responsible for guiding the clan's search for optimal solutions with every leader assigned randomly. Each family consists of two parents and is also assigned a number of children for local search in the space of the family.

Parameters

- $MaxIter$: Maximum number of iterations.
- C: Number of clans.
- F: Number of families per clan.
- c: Number of children per family.
- D: Dimensions of the problem.
- LB, UB: Lower and upper bounds of variables.

Position Initialization For each clan c_i (where $n = 1$ to C):

– Initialize the leader's position L_n randomly within bounds:

$$L_n = \text{rand}(LB, UB)$$

– Evaluate the fitness $f(L_n)$:

$$f(L_n) = \text{objective function}(L_n)$$

3.2 *Family Dynamics Adapted from SELO*

Each parent p in clan c_i can decide to learn either from another parent within the same clan or from a leader of another clan who are chosen using the roulette wheel method. The update equation is given by:

$$x_{p,t+1} = x_{p,t} + \beta_1(x_{p',t} - x_{p,t}) + \beta_2(x_{l',t} - x_{p,t}) + \gamma \cdot N(0, 1) \tag{1}$$

where $x_{p',t}$ denotes the position of the selected parent within the same clan, $x_{l',t}$ represents the position of the leader from another clan, β_1 and β_2 are learning coefficients, γ is the random step size, and $N(0, 1)$ indicates a normally distributed random vector.

Each child c in clan C decides to learn either from his parents, siblings, or peers, guided by the probabilities $P_{\text{parent}}, P_{\text{siblings}}, P_{\text{peers}}$. The update equation is given by:

$$x_{c,t+1} = x_{c,t} + \alpha \left(x_{s,t} - x_{c,t}\right) + \delta \cdot N(0, 1) \tag{2}$$

where $x_{s,t}$ is the position of the selected individual, α is the learning coefficient, and δ is the random step size. In case of worse fitness, the individual can correct their behavior by reverting to their previous best or the children can seek guidance from their parents. This enables the algorithm to maintain overall optimization in the search space. This form of local search helps the families to improve their individual fitness.

3.3 Movement Strategies Adapted from NPO

Semicircular Distribution Families are distributed around their leader in a semicircular pattern. The equation governing this distribution is given as follows:

$$X_i = L_n + r \cdot \cos(\theta) \tag{3}$$

where X_i is the position of family member i, L_n indicates the position of the leader, r is a random radius, and θ is the random angle chosen from the range $[0, \pi]$.

By focusing on local search around their leader, families can refine their solutions and improve their fitness values while exploring their surroundings to avoid local optimum.

Families Searching To enhance the clan's best solution, the leader selects certain families from within the clan and encourages them to engage in a more extensive exploration of the search space using Lévy flights. This strategy prompts these families to search beyond their usual areas, encouraging them to escape the local optimum.

$$X_{\text{new}} = X_{\text{old}} + \lambda \cdot L \cdot (X_{\text{old}} - L_c) \tag{4}$$

where X_{new} is the new position of the family, X_{old} is the current position, L is a Lévy distributed random step, and λ serves as a scaling factor to adjust the step size.

3.4 Behavior Correction

In case of worse fitness, the individuals can correct their behavior by reverting to their previous best or seeking guidance from their parents. This enables the algorithm to maintain the overall optimization in the search space. The parents can also decide to self-contemplate in such cases. This form of local search helps families escape potentially sub-optimal spaces. As they explore the space to escape local optima, children learn how to explore from their parents, siblings, or peers. Over time, the influence of parents declines, and children gradually learn more from their peers and siblings.

3.5 Leadership Transition

The objective of this step is to maintain the best possible leader within each clan. After the families have moved (either by exploitation or exploration), families that have better solutions than the current leader are searched for:

$$f(X_i) < f(L_c) \implies L_c = X_i \tag{5}$$

where X_i is the position of family member i. When a suitable family is found, the leader is updated accordingly (Fig. 1).

3.6 Periodical Meetings (Cooperation Among Clans)

At regular intervals, clan leaders convene to share information regarding their respective best solutions. The leader with the best solution influences other leaders to adjust their search directions. This cooperation promotes a coordinated effort in exploring promising regions of the solution space, thereby improving the overall convergence of the algorithm. The best leader is identified from all clans as follows:

$$L_{\text{best}} = \arg\max_{c} f(L_c) \tag{6}$$

Update the positions of the other leaders based on the best leader's fitness. We ensure that all updated positions are within the defined bounds. Thus, the leader with the best solution influences other leaders to adjust their search directions, furthering the coordinated effort.

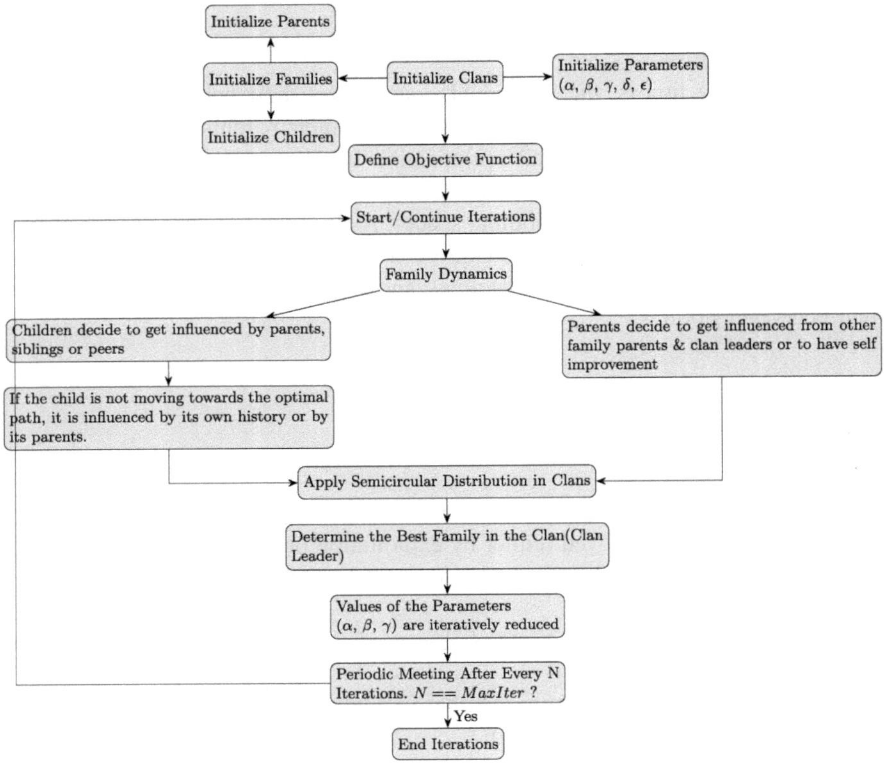

Fig. 1 SNLO algorithm flowchart

3.7 Termination

The algorithm concludes when the maximum number of iterations, MaxIter, is reached, or when another stopping criterion is met.

4 Results

To validate our algorithm's effectiveness, we conducted experiments using well-known benchmark functions, including unimodal, multimodal, separable, and non-separable problems. We compared our algorithm's performance against state-of-the-art socio-inspired metaheuristics like PSO, ABC [6], SELO, and NPO. Metrics such as convergence speed, solution quality, and computational efficiency were used for evaluation. We chose the functions mentioned in Table 1 to test the effectiveness of our algorithm. Figure 2 shows a 3-dimensional view of the chosen benchmark functions for understanding the distribution.

Pseudocode for Socio-Nomadic Learning Optimization (SNLO)

Input : Number of clans C, Parents per clan F_p, Children per clan F_c, Learning coefficients α, β_1, β_2, γ, δ, ϵ, Selection probabilities P_{parent}, $P_{siblings}$, P_{peers}, Lévy flight parameters λ, μ, σ

Output: Best solution found

Initialization:

```
Randomly initialize clan leaders r_c for each clan c;
Distribute families around leaders using a semicircular distribution;
while not converged do
    foreach clan c in C do
        foreach parent p in clan c do
            Decide whether to learn from another parent or a leader from another clan;
            Select influencing individual using roulette wheel selection;
            Update parent's position using Parent Influence Function;
            Apply self-contemplation if necessary;
            Correct behavior if the new position is worse;
        foreach child i in clan c do
            Decide whether to learn from parent, sibling, or peer using
            P_parent, P_siblings, P_peers;
            Select influencing individual;
            Update child's position using Child Influence Function;
            Apply behavior correction if necessary;
        Apply semicircular distribution to redistribute families around leader;
        Perform Families Searching using Lévy flights for global exploration;
        Update clan leader if a better solution is found within the clan;
    if periodical meeting condition is met then
        Leaders gather for a periodic meeting;
        Update the influence or direction of clan leaders using the Inter-Clan Learning
        Equation;
    Evaluate fitness of all individuals;
    Check convergence criteria;
return the best solution found;
```

Our initial hypothesis posited that the distributions of the algorithms (ABC, PSO, SELO, NPO, and SNLO) were uniform. However, Friedman's Two-Way Analysis of Variance, illustrated in Fig. 4, led us to reject this hypothesis, indicating distinct distribution patterns among the algorithms. As depicted in Fig. 3, the SNLO algorithm generally demonstrated strong convergence, with the exception of the Ackley function (Fig. 3a), where SELO exhibited superior performance. This observation underscores SNLO's robust capability to reach global optima across various scenarios. Pairwise comparisons further revealed that SNLO shares similarities with its parent algorithms on certain functions. SNLO's adaptability has potential for improved performance in diverse optimization tasks (Fig. 4).

Table 1 Functions and their details [7]

Function	Bounds	Global optimum	Dim	Equation
Goldstein-Price	[−2, 2]	3	2	$f(x, y) = (1 + \frac{(x+y+2)^2}{4}) \cdot (1 + \frac{(x+y-2)^2}{4})$
Schwefel	[−500, 500]	−837.658	2	$f(x, y) = 418.9829 \cdot n - \sum_{i=1}^{n} x_i \cdot \sin(\sqrt{\lvert x_i \rvert})$
Schaffer	[−100, 100]	0	2	$f(x, y) = 0.5 + \frac{\sin^2(\sqrt{x^2+y^2})-0.5}{(1+0.001(x^2+y^2))^2}$
Rosenbrock	[−1.2, 1.2]	0	10	$f(x, y) = (a - x)^2 + b \cdot (y - x^2)^2$
Sphere	[−5.12, 5.12]	0	10	$f(x) = \sum_{i=1}^{n} x_i^2$
Ackley	[−32.768, 32.768]	0	10	$f(x) = -20 \cdot \exp\left(-0.2 \cdot \sqrt{\frac{1}{n}\sum_{i=1}^{n} x_i^2}\right) - \exp\left(\frac{1}{n}\sum_{i=1}^{n} \cos(2\pi x_i)\right) + 20 + e$
Rastrigin	[−5.12, 5.12]	0	10	$f(x) = An + \sum_{i=1}^{n} \left(x_i^2 - A \cdot \cos(2\pi x_i)\right)$ where $A = 10$
Martin-Gaddy	[0, 10]	0	2	$f(x, y) = \sin^2(x + y) - \frac{1}{2}(x^2 + y^2)$
Easom	[−100, 100]	−1	2	$f(x, y) = -\cos(x) \cdot \cos(y) \cdot \exp\left(-\left((x - \pi)^2 + (y - \pi)^2\right)\right)$
Griewank	[−600, 600]	0	10	$f(x) = 1 + \frac{1}{4000}\sum_{i=1}^{n} x_i^2 - \prod_{i=1}^{n} \cos\left(\frac{x_i}{\sqrt{i}}\right)$

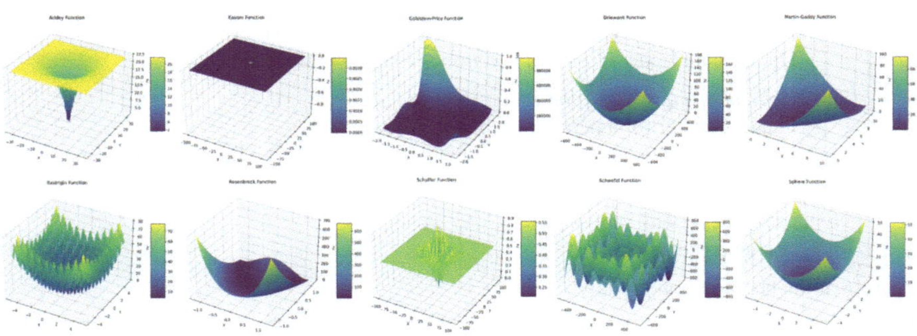

Fig. 2 A 3-dimensional perspective for the mathematical benchmark functions in Table 1

5 Conclusion

In conclusion, by harmoniously combining the social learning dynamics of SELO with the adaptive migratory behavior of NPO, SNLO effectively balances exploration and exploitation in the search space. The algorithm's design benefits from decentralized learning within families, dynamic leadership models, and cooperative interactions among clans, which collectively enhance its capacity to navigate complex, high-dimensional optimization landscapes. The integration of SELO's robust social learning mechanisms with NPO's adaptive exploration strategies

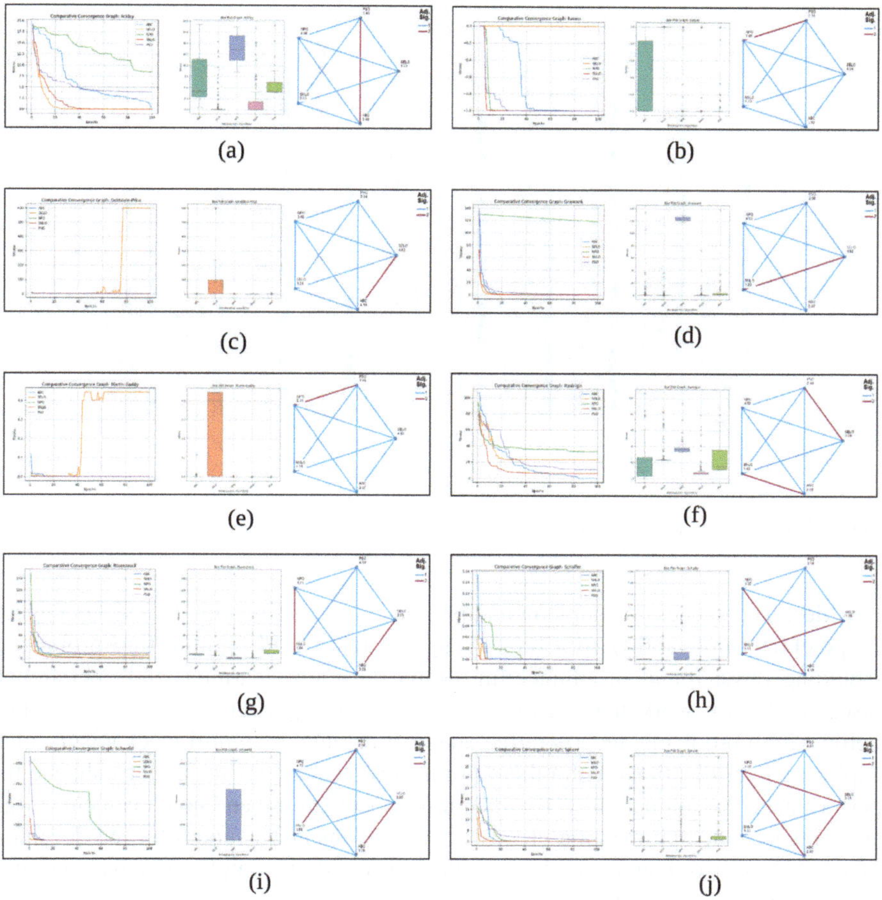

Fig. 3 Comparative and qualitative metrics of different algorithms on standard benchmark functions. (**a**) Ackley. (**b**) Easom. (**c**) Goldstein-Price. (**d**) Griewank. (**e**) Martin-Gaddy. (**f**) Rastrigin. (**g**) Rosenbrock. (**h**) Schaffer. (**i**) Schwefel. (**j**) Sphere

resulted in an algorithm capable of addressing complex optimization problems with enhanced performance.

The empirical evaluations on a diverse set of benchmark functions revealed that SNLO not only surpasses its parent algorithms but also outperforms other well-established metaheuristic algorithms like PSO and ABC in terms of convergence speed as well as solution quality. This performance is heavily attributed to its ability to maintain diversity in the population through global exploration mechanisms such as Lévy flights, while also refining solutions via local exploitation inspired by social learning processes.

SNLO effectively balanced exploration and exploitation, avoided premature convergence, and exhibited robustness across various problem dimensions and com-

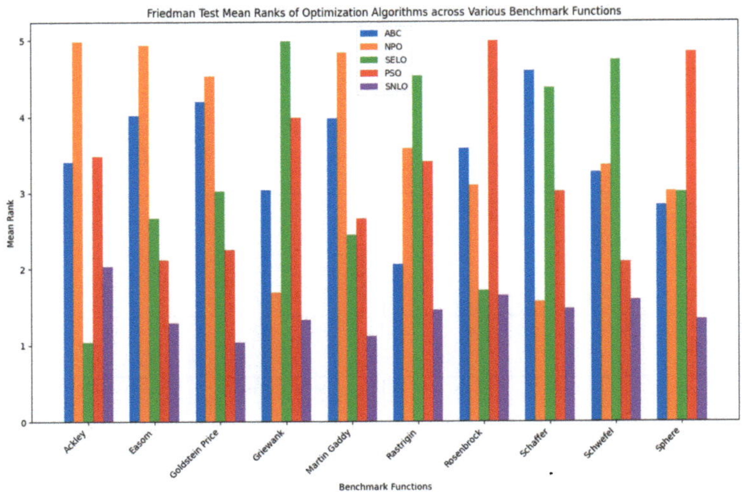

Fig. 4 Friedman test mean ranks

plexities. The successful implementation and promising results of SNLO suggest its potential applicability to a broad range of real-world optimization problems, including those found in engineering design, machine learning, and operational research. Future work may involve extending the algorithm to handle dynamic optimization problems, incorporating adaptive parameter control mechanisms, and exploring hybridizations with other optimization techniques to further enhance its performance and applicability. By integrating sociocultural and nomadic behavioral concepts into the optimization process, SNLO opens new avenues for the development of innovative metaheuristic algorithms inspired by complex adaptive systems in nature and human societies.

References

1. Gendreau, M., Potvin, J.-Y. 2010, Handbook of Metaheuristics, 2nd edn. Springer. Available at: https://EconPapers.repec.org/RePEc:spr:isorms:978-1-4419-1665-5
2. Kennedy, J., Eberhart, R. 1995, Particle Swarm Optimization, Proceedings of ICNN'95 - International Conference on Neural Networks, 4, 1942–1948. https://doi.org/10.1109/ICNN.1995.488968
3. Lambora, A., Gupta, K., Chopra, K. 2019, Genetic Algorithm: A Literature Review, 2019 International Conference on Machine Learning, Big Data, Cloud and Parallel Computing (COMITCon), 380–384. IEEE. https://doi.org/10.1109/COMITCon.2019.8862255
4. Salih, S. Q., Alsewari, A. A. 2020, A New Algorithm for Normal and Large-Scale Optimization Problems: Nomadic People Optimizer, Neural Computing and Applications, 32(14), 10359–10386. https://doi.org/10.1007/s00521-019-04575-1
5. Kumar, M., Kulkarni, A. J., Satapathy, S. C. 2018, Socio Evolution & Learning Optimization Algorithm: A Socio-Inspired Optimization Methodology, Future Generation Computer

Systems, 81, 252–272. Available at: https://www.sciencedirect.com/science/article/abs/pii/S0167739X17317259
6. Karaboga, D., Basturk, B. 2007, A Powerful and Efficient Algorithm for Numerical Function Optimization: Artificial Bee Colony (ABC) Algorithm, Journal of Global Optimization, 39, 459–471. Available at: https://link.springer.com/article/10.1007/S10898-007-9149-X
7. Jamil, M., Yang, X.-S. 2013, A Literature Survey of Benchmark Functions for Global Optimization Problems, International Journal of Mathematical Modelling and Numerical Optimisation, 4(2), 150–194. https://doi.org/10.1504/IJMMNO.2013.055204

Binary Tree-Based Key Management Scheme for a Ground Station Managing IoT Devices in a Satellite Multicast Communication

Priyanka Ahlawat, Mayank Dave, and Supriya Mishra

Abstract Satellite-terrestrial networks are crucial in supporting global applications like environmental monitoring, disaster response, and military communication. These networks are essential for global communication as well as connecting underserved and isolated locations where traditional terrestrial infrastructure is too costly, difficult to establish, or prone to collapse in an emergency. The implementation of a binary tree key based scheme in satellite- terrestrial network is examined in this research. With a focus on satellite links for connectivity and secure communication between IoT device and ground stations. In comparison to other schemes such as Distributed logical key hierarchy (DLKH), Logical key Hierarchy (LKH) and Hyper graph schemes, binary tree key based scheme is evaluated based on its efficient key generation and distribution, improved security features, and minimized communication overhead. For large-scale IoT networks, binary trees provide a more durable and balanced structure. The binary tree architecture is used in this paper to offer an efficient algorithm for network routing and key distribution. Case studies and simulation results are presented and provide how binary tree can be effectively deployed to manage secure communication, achieving scalable, efficient for satellite terrestrial communication systems.

Keywords Satellite communication · Rekeying · Group key · Key management · Latency · Scalability · Processing cost · IoT

P. Ahlawat (✉) · M. Dave · S. Mishra
Department of Computer Engineering, National Institute of Technology, Kurukshetra, Haryana, India
e-mail: priyankaahlawat@nitkkr.ac.in

K. Deep et al. (eds.), *Artificial Intelligence, Machine Learning and Intelligent Systems*, Springer Proceedings in Mathematics & Statistics 520,
https://doi.org/10.1007/978-981-95-3841-6_7

1 Introduction

Satellite networks provide communication in difficult-to-reach and distant locations and offer worldwide communication services. Satellites are divided into three categories based on their orbital altitude: Geostationary Earth Orbit (GEO) satellites with an altitude above 35,786 km, commonly used for communication and meteorological observation; Medium Earth Orbit (MEO) satellites with an altitude between 2000 and 35,786 km, commonly used for GPS and satellite navigation; Low Earth Orbit (LEO) satellites with an altitude between 500 and 2000 km, commonly used for Earth observation and communication. These networks bridge the digital divide but face challenges like security, latency, and bandwidth limitations. Terrestrial networks, using technologies like Ethernet and cellular, cover extensive areas. Integrating satellite and terrestrial networks enhances secure global communication. This architecture focuses on efficient communication between IoT devices and ground stations, using XOR-based operations and a hierarchical binary tree for scalable and secure group key management with minimal overhead.

1.1 Importance of Binary Tree Based Key Management

In satellite communication systems, binary trees play a vital role in managing cryptographic keys efficiently. By allowing key recovery using XOR operations, it minimizes communication and computation costs and reduces the impact on IoT devices with limited resources.ECC significantly improves security by reducing complexity and key sizes. The binary tree structure ensures high scalability and supports dynamic networks, allowing seamless addition or removal of devices with minimal overhead. It also reduces latency by enabling efficient group key computation, requiring updates only for unmodified xor values. Using a public bulletin board eliminates the need for direct communication during updates, improving efficiency and reducing transmission and processing time.

The rest of the paper is organized as follows. Section 2 provides a briefs overview of the literature on binary tree key-based schemes in dynamic joining and leaving of IoT Devices. Section 3 details the proposed Binary key-based scheme, including group setup, join, and leave operation of user members. Section 4 presents an analysis of the proposed scheme with examples for group setup, join, and leave operations, followed by a performance comparison of proposed scheme with existing schemes. Finally, the paper concludes with conclusion.

2 Related Work

In dynamic contexts like the internet of things (IOT), secure communication depends on efficient group key management (GKM). Many methods have been put out to deal with this difficulty. An elliptic curve-based GKM technique without an encryption key was presented by Seo et al. [1]. In contexts with limited resources, this approach enhances security and computing efficiency. Vijay Kumar et al. and Kumar et al. [2] presented how to effectively generate and distributed group keys using the Chinese remainder theorem (CRT), building on mathematical fundamentals. For dynamic group communication, the CRT-based method is appropriate since it streamlines the key distribution procedure and improves scalability. The logical key hierarchy (LKH) model was improved by Xu et al. [3] by substituting a ternary key tree for the conventional binary tree. In dynamic multicast groups, this change is especially helpful because it lowers the height because it lowers the height of the key tree, which in turn minimizes the number of the key update operations needed during user join or exit events. In a comparable manner, Nafi et al. [4] introduced a matrix-based key management approach that prioritized organized and effective key changes to accommodate dynamic situations. Rodeh et al. [5] proposed a distributed logical key hierarchy (D-LKH) does not require a central group controller (GC), unlike the centralized LKH framework. In order to prevent a one entity from ever obtaining all the keys, the key hierarchy is instead divided among users. Security and scalability are improved by this decentralized method, but controlling the distributed key changes becomes more difficult. Ding et al. [6] proposed the hyper graph model, another novel structure. Because hyper edges can connect more than two vertices, a hyper graph extends conventional graph architectures. The overhead required to update many connections for every device added or removed makes managing dynamic IoT devices with hyper graphs difficult, even if hyper graph based key management may support complicated relationships and linkages (Table 1).

3 Proposed Scheme

Numerous tree-based group key management systems exist for wireless contexts, but their rekeying operations are often inefficient. This work introduces a lightweight, scalable, low latency, and low-storage cost binary tree-based key management scheme that employs XOR operations and elliptic curve cryptography. By utilizing a binary tree design, the system maintains constant computations costs while significantly reducing rekeying expenses related to computation, storage and transmission.

The proposed strategy aims to establish secure group communication within a terrestrial network where IoT devices serve as leaf nodes and ground station act as the key management system (KMS). The KMS is critical for creating, distributing, updating and revoking session keys, ensuring secure communication amidst dynamic membership changes. This approach is perfect for internet of things situations since it allows dynamic membership changes, or devices joining and leaving, with no overhead. The group KMS including initial setup, dynamic key recovery, adding a new device and removing a device, utilizing a binary tree key based management structures as a representation.

Table 1 Examining many group key management techniques, including Binary Tree topologies, distributed logical key hierarchy (DLKH), hyper graph, and logical key hierarchy

Criteria	Hyper graph [6]	Distributed logical key hierarchy(DLKH)	Logical key hierarchy (lkh) [1, 7]	Binary key tree [8, 9]
Overview	Generalized graph with hyper edges	Multiple hierarchical key structures managed by distributed controllers. Reduces load on central authority	Hierarchical tree, group key at root	Complete binary tree structure
Overhead	High	Moderate overhead.	Moderate	Minimizes overhead
Latency	High(depends on hyper graph size)	Moderate latency. Distributed key updates reduce the load on a single central controller, improving response times	Moderate (rekeying intermediate keys)	Less latency (logarithmic depth reduces rekeying operations)
Scalability	High scalability for large and complex networks	The load on each controller is limited to a subset of the hierarchy	Good, but tree height increases with users.	High scalability with logarithmic growth, making it highly scalable.
Dynamic IOT Add/Remove	Complex to manage IOT Changes	Distributed controllers handle rekeying locally, which allows for quicker addition/removal of users	Moderate overhead when adding/removing users	Allow seamless addition and removal of IoT device as users

STIN Architecture is discussed as follows:

Satellite-terrestrial networks—which combine satellite communication with internet of things devices and ground stations—are the focus of this research. Figure 1 shows the network architecture of STIN.

It guarantees seamless data transfer between an IoT device and a ground station. Its architecture involves safe communication and effective data flow between ground station and IoT devices via a satellite link. The ground station in this network acts as an intermediary between satellite and IoT devices. Using scalable key management systems that can effectively generate, distribute and update cryptographic keys for secure communication links is essential to managing this complexity. This infrastructure also ensures low latency and high reliability in real time, maximizing the amount of data transmitted between IoT devices and ground stations.

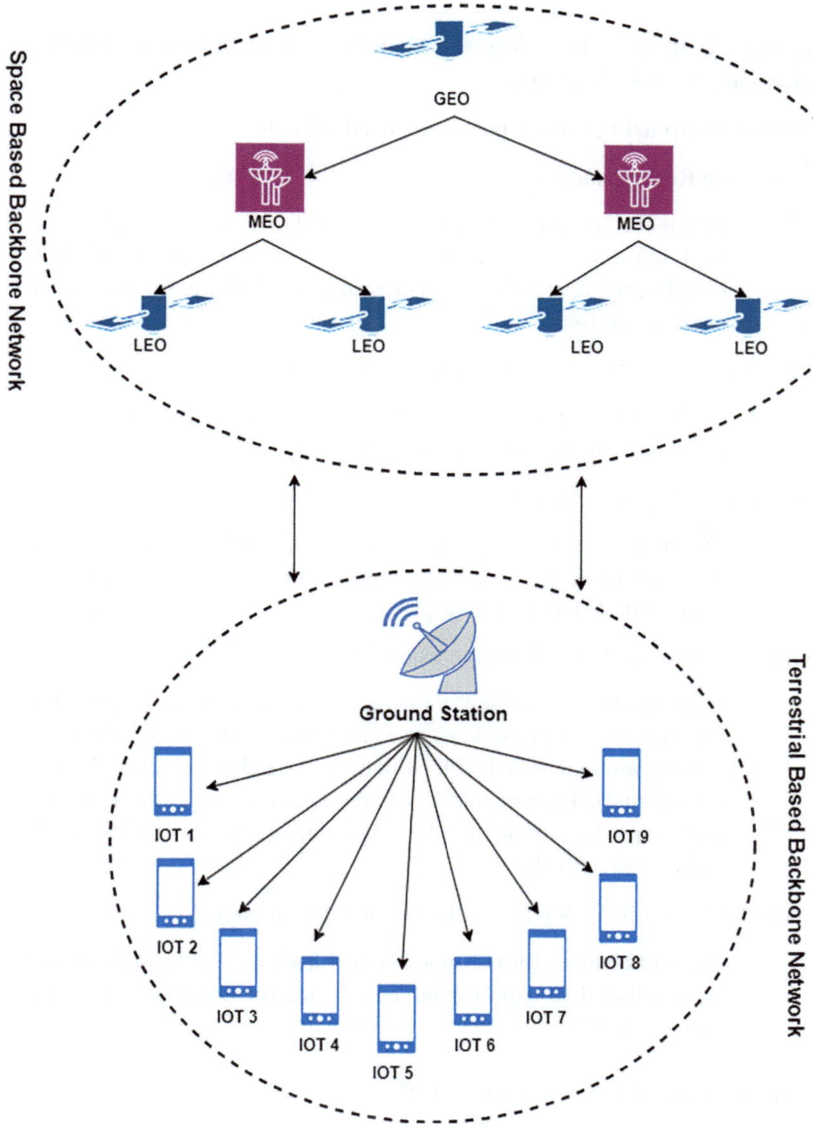

Fig. 1 Network architecture of STIN

Algorithm 1 Binary Tree Key-Based scheme and Dynamic Joining and Leaving for IOT Networks

(a) **Initial Setup using binary tree key-based scheme**

(i) Node Representation:

- Define the set of IoT device as N = $\{IOT_1, IOT_2 \ \ldots \ IOT_n\}$.
- Each IoT device is represented as a leaf node in the binary tree key based scheme, which establishes structured hierarchy for secure communication.

(ii) Random Number generation for each device:

- For each IoT device i, generate a unique random number ri using a one-way function F, such that: ri = F (i)

(iii) Group Key Generation:

- The root node of binary tree is the ground station, which contain the master key K_{master}. GS securely stores K_{master}. And uses for internal node and leaf nodes.

(iv) Internal node key calculation using XOR:

- Internal nodes in the binary tree key based scheme are responsible for representing group keys and managing groups of IoT devices.
- For any internal node j with child nodes j_1 and j_2, calculate the key for the internal node using XOR operations:$K_j = r_{j1} \oplus r_{j2}$Here, r_{j1} and r_{j2} are the random numbers associated with the child nodes j_1 and j_2, respectively.

(v) Publishing XOR Values on the Public bulletin board:

- The XOR Values for the edges in the binary tree key based scheme are published to a public bulletin board for transparency: Edge Value = k $\oplus r_j$.

(b) **Dynamic Key Recovery Using XOR**

(i) Group key recovery by IOT Device i:

- Each IoT device retrieves the group key k by following the path from the root to its leaf node, XOR-ing the values along the path; $k = (K_j \oplus r_{j1}) \oplus (k_j \oplus r_{j3}) \oplus \ \ldots \ \oplus (k_n \oplus r_n)$.

(c) **Adding a New IOT Device**

(i) Generating a random number for the new device:

(continued)

- For a new IoT device Iot_{n+1}, generate a new random number using the one-way function: r_{n+1}= F (n + 1).
- Update the group key to new value k' using the function G that depends on the existing key and the new random number: K′ = G (k,r_{n+1})

(ii) **Recalculating XOR Values for Affected Edges**: Recalculate the XOR values for the affected internal nodes due to the addition of the new device: $K'_j = r_{j1} \oplus r_{j2} \oplus r_{n+1}$

(d) **Removing an IOT Device**

(i) When an IoT device IoT_m leaves the group, the group key is updated to a new value k" using the function H: $k'' = \text{H}\,(\text{k},r_m)$

- The function H ensures that the new group key does not depend on the random value r_i of the departed device.

(ii) Rekeying the internal nodes:

- Updates the XOR values for the affected internal nodes after removing the device.$k''_j = r_{j1} \oplus r_{j2} \oplus r_m$
- Publish the updated XOR values on the public bulletin board to inform the remaining IoT devices of the changes to maintain secure communication.

4 Result and Discussion

A binary tree key-based scheme offers an efficient framework for secure communication in large, dynamic networks, optimizing group key management. Each leaf node represents an IoT device, and the root node (ground station or KMS) manages group key generation and distribution. Rekeying is handled efficient by updating only the affected portion of the tree rather than the entire network. Leaf nodes have unique keys, while internal nodes derive their keys from xor operation on child nodes, ensuring secure communication between devices. When nodes join or leave, only the relevant internal nodes are recalculated, maintaining security without disrupting the network. The scheme includes protocols for group setup, join and leave to handle key generation, updates and rekeying, ensuring both forward and backward secrecy.

4.1 Example Group Setup with Ground Station (GS) as Key Management System (KMS)

In Initial Setup using a Binary Tree: We have taken eight IOT devices:IOT_1, IOT_2, … IOT_8.The Ground Station (GS) sets up these IoT devices as leaf nodes of a binary tree. The GS is responsible for generating and distributing the keys. The GS uses a One-Way-Function F (r_i) to create random values r_1, r_2, r_3 … r_8 for each IOT device. In Group Key Generation, the GS generates a group key k, which serves as the shared secret key for all the IoT devices in the network. Then assigns XOR values for the connections between nodes in the tree. XOR Values for Key Recovery are stored. These XOR-ed values are stored on the public bulletin board. This allows each IoT device to retrieve relevant encrypted values for secure communication. During dynamic Key Recovery Using XOR and ECC, when an IOT device needs to recover the group key k, it traces its path from the root of the binary tree to its corresponding leaf node. By accessing the XOR-ed values stored on the public board and performing. Figure 2 is given below.

In dynamic Join Operation, Suppose we need to add four new IOT Devices (IOT9, IOT10, IOT11 and IOT12) to the existing binary tree. For creating New Random Numbers, the GS generates new random numbersr_9, r_{10}, r_{11}, r_{12} for the new devices. During the updating of the Group Key, the GS updates the group key to a new value k' to ensure backward secrecy (new devices cannot access previous communication data).The new IOT Devices are attached to the binary tree at their appropriate locations. The GS calculates the new XOR Values for affected nodes. And post these values on the public bulletin board. This allows the new devices to recover the updated group key k' using the same XOR operation technique, without disrupting other devices (Fig. 3).

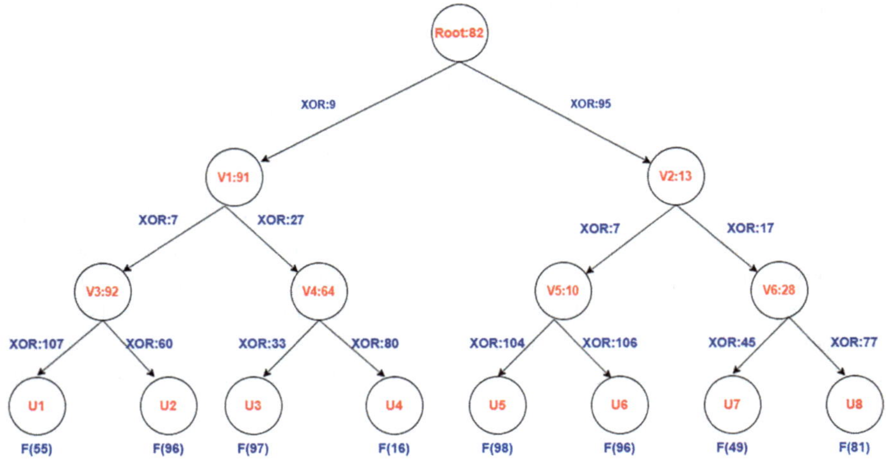

Fig. 2 Initial set up phase

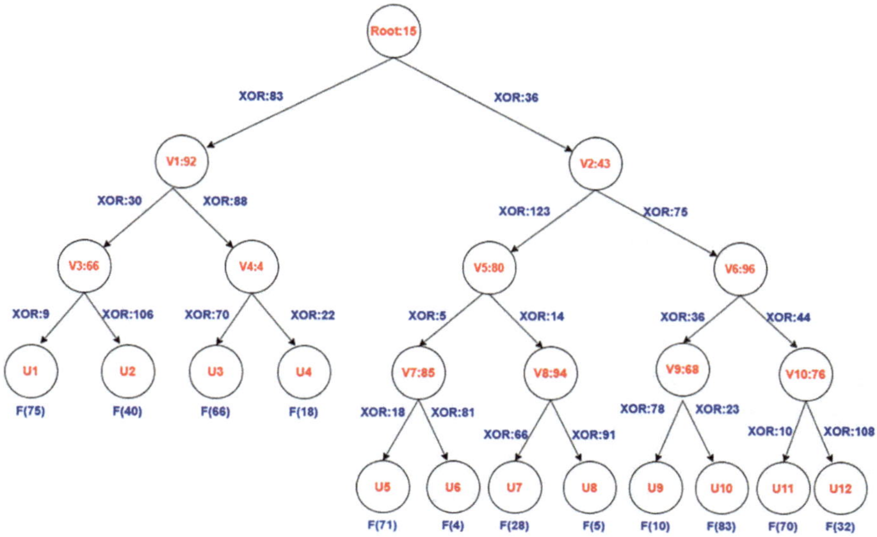

Fig. 3 Dynamic Join Operation

In Dynamic Leave Operation, Let's consider the scenario where IOT Devices IOT10, IOT11 and IOT12 leave the network. The nodes corresponding to IOT10, IOT11 and IOT12 are removed from the tree, and the binary tree structure is updated accordingly. The GS generates a new group key k" to ensure forward secrecy (departed devices cannot access future communication). The updated XOR Values of the affected edges are published on the public bulletin board, allowing the remaining IOT Devices to recover the new group key. The remaining IOT devices can instantly update their keys using the XOR values form the public bulletin board, significantly reducing processing time by using pre-computed values (Fig. 4).

4.2 *Time Comparison with Other Algorithms*

In this section, the comparative analysis of existing binary tree key based scheme is discussed with proposed scheme, which is designed to be light weight. The Comparison focuses on the rekeying cost dynamic join and leave operations. We also compare use of between XOR operation and one way functions. This study includes scheme like DLKH, LKH and hyper graph, highlighting the advantages of XOR operations over the one way functions. The proposed scheme shows improved performance in computation cost.

Figure 5 illustrates the join and leave phased of the scheme, indicating that as the number of users increase, the time required to join or remove nodes also rises. The x-axis represents the number of users (0–1000), while the y-axis shows the time in milliseconds.

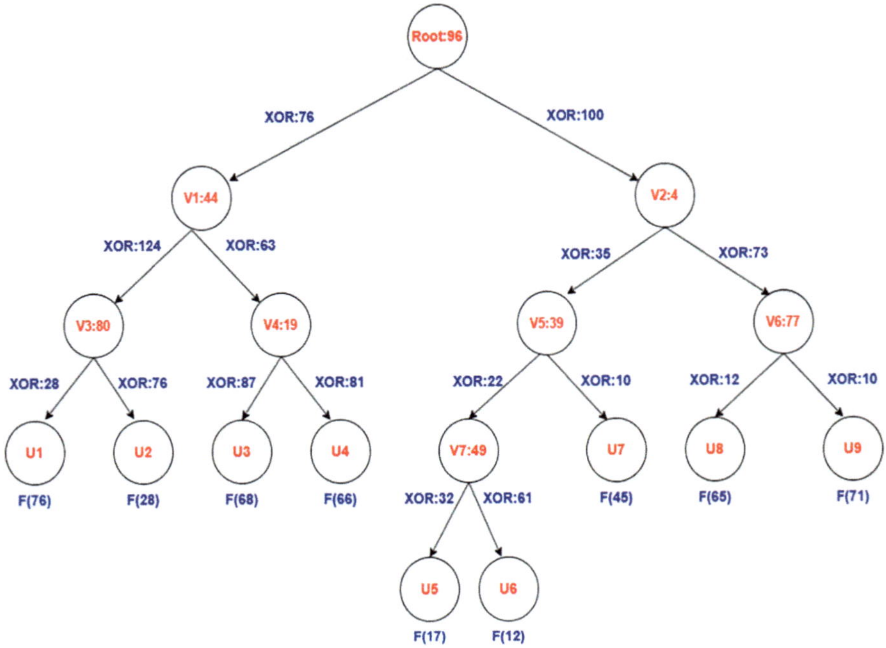

Fig. 4 Dynamic Leave Operation

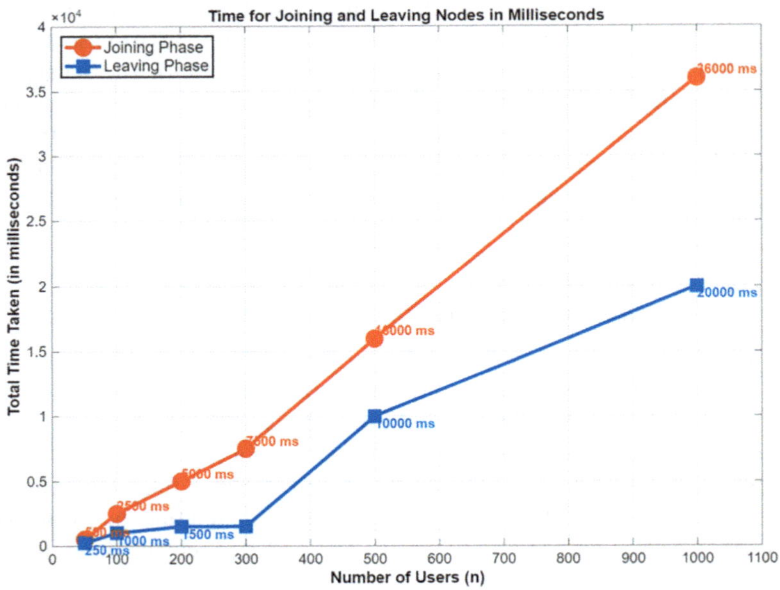

Fig. 5 Time of Dynamic Joining and Leaving Phase

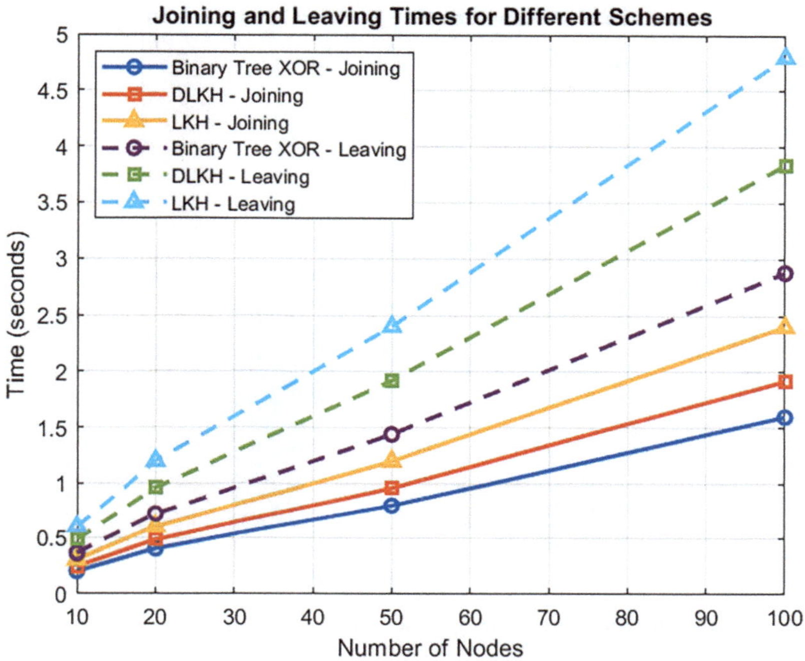

Fig. 6 Time of Joining or Leaving for Various Key Management Plans

Figure 6 show a comparative analysis of the joining and leaving times for various numbers of nodes (10, 20, 50, 100) across the proposed scheme, the existing scheme, DLKH and LKH. The XOR operations-based scheme consistently shows the quickest joining and departing times, with a slow increase as the number of nodes increases. This shows how well it manages dynamic node memberships with minimal overhead. The DLKH scheme, in contrast, shows modest joining and departing rimes, which similarly rise as the number of nodes increases. The longest joining and departing times, however, are displayed by the LKH scheme, which in indicative of its more intricate key management procedure and greater processing requirements. The outcomes clearly highlight the suggested binary tree key-based scheme's scalability and effective methods by utilizing XOR operations.

Figure 7 shows a comparative analysis of the computational efficiency of XOR operations versus one-way functions when applied to 1 node and 10 nodes. The computation times for XOR operation times are represented by the blue bars, and those for one-way function times are shown by the orange bars. The horizontal axis indicates the number of nodes involved, and the vertical axis indicates the calculation time in seconds.

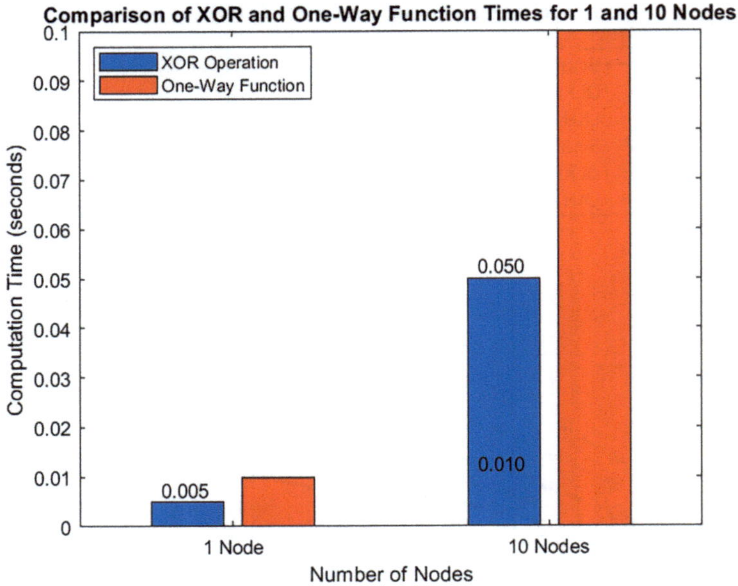

Fig. 7 Time taken for one-way function or XOR operation computations

5 Conclusion and Future Scope

The proposed approach uses a binary tree key based scheme to build a dynamic, effective and scalable key management system for a terrestrial Internet of things network. The system achieves minimum computational and communication overhead, guaranteeing great scalability and minimal processing or transmission delays, with the Ground Station (KDC) handling safe key distribution through XOR operations and ECC for cryptographic security. For real-time applications in dynamic IoT contexts like smart cities and sensor networks, this makes it extremely suitable. For secure communication between internet of things devices and ground stations, the suggested satellite-terrestrial integration architecture provides an effective and scalable solution. The system guarantees secure and transparent key generation, distribution, updates and revocation by utilizing a hierarchical binary tree key based scheme with xor operations for group key management. By facilitating safe key exchange and reducing risks related to dynamic IoT device membership, the usage of a public bulletin board further improves security.

In the future scope of the proposed scheme, additional research can concentrate on more improving rekeying procedures to lower overhead during dynamic membership changes, particularly in extensive IoT networks. Performance may be further improved by incorporating machine learning techniques to anticipate changes in group membership and proactively manage important updates. Furthermore, a crucial area for development involves ensuring compatibility among different IoT devices with different computational and energy constraints. Testing and real-world implementation in extensive IoT settings, like smart cities or industrial networks, will confirm the viability of the suggested solution and point out areas that require more improvement.

References

1. Wallner, D., Harder, E., & Agee, R. (1999). Key management for multicast: Issues and architectures. *IETF RFC 2627.*
2. Kumar, V., Kumar, R., & Pandey, S. K. (2020). A secure and robust group key distribution and authentication protocol with efficient rekey mechanism for dynamic access control in secure group communications. *International Journal of Communication Systems*, 33(14), 1–22.
3. Li, J., Yao, S., Liu, J., & Wu, Y. (2021). A hierarchical multicast key distribution protocol. *Electronics*, 10(9), 995. Retrieved from https://www.mdpi.com/2079-9292/10/9/995.
4. Wong, C. K., Gauda, M., & Lam, S. S. (2000). Secure group communications using key graphs. *IEEE/ACM Transactions on Networking*, 8(1), 16–30.
5. Rodeh, O., Birman, K., & Doley, D. (1999). Optimized group rekey for group communication systems. *Technical Report 2*, Department of Computer Science, Hebrew University.
6. Ding, Y., Zhou, X., Cheng, Z., & Wu, Q. (2013). Key management in secure satellite multicast using key hyper graphs. *Wireless Personal Communications*, 70, 1859–1883.
7. Pande, A. S., & Thool, R. C. (n.d.). Survey on Logical Key Hierarchy for secure group communication. *IEEE Xplore*. Retrieved from https://ieeexplore.ieee.org/document/7877763.
8. Pareek, G., & Purushothama, B. R. (2018). Provably secure group key management scheme based on proxy re-encryption with constant public bulletin size and key derivation time. *Sadhana - Academy Proceedings in Engineering Sciences*, 43(9), 1–13. https://doi.org/10.1007/s12046-018-0917-8.
9. Sharma, P., & Purushothama, B. R. (2024). A lightweight group key management scheme with constant rekeying cost and public bulletin size. *Information Security*. https://doi.org/10.1080/19393555.2023.2198737.

AI-Driven Deterrence System Utilizing YOLO Algorithm for Real-Time Wildlife Detection and Mitigation

G. Spoorthy, L. Nikhil, A. Sravya, R. Venu Gopal, and V. Rohit

Abstract Animals cause more destruction on crops because there are animals proven to disturb crops, such as buffaloes, sheep, cows, and tapirs, forcing big losses to the farmer. Crop watching requires all-day care due to the vast size of acreage. This can be relieved through physical barriers, like walls or motorized fences, colorful enclosures, or cameras monitoring fields to scare off animals. Other people have attempted using decoys or scarecrows, though their effectiveness to ward off animals has remained questionable. Therefore, an AI-based system is proposed here, through which images would be fed to cameras with the object detection algorithms like YOLOv3 and OpenCV, cv2, in order to prevent the intrusions of animals on crops. It will detect the animals based on the movement and behavioral characteristics. If it detects the clapping or whistling sounds being produced by humans, it will raise an alarm to alert the people around as well as make quite a ruckus to scare away those animals. As soon as any stray animal is spotted entering the field, instant messaging will be sent to the farmer. This is a totally full-proof system of crop protection against the most destructive animals because it captures through live video feeds.

Keywords YOLOv3 · OpenCV · Cv2 · OBJECT detection · AI-based system · Crop protection · Intrusion detection · Behavioural characteristics · Sound detection (clapping whistling) · Scarecrows

G. Spoorthy (✉)
Assistant Professor, Department of CSE—(CyS, DS) and AI&DS, VNRVJIET, Hyderabad, India
e-mail: spoorthy_g@vnrvjiet.in

L. Nikhil · A. Sravya · R. Venu Gopal · V. Rohit
Cyber Security, Department of CSE—Cyber Security, VNRVJIET, Hyderabad, India

K. Deep et al. (eds.), *Artificial Intelligence, Machine Learning and Intelligent Systems*, Springer Proceedings in Mathematics & Statistics 520,
https://doi.org/10.1007/978-981-95-3841-6_8

1 Introduction

Animal Inroads to the farm now become the main threat that crop suffers. In southern India, the deer and lizards and monkeys constituted a heavy threat to crop production. With this issue pressurizing at the heart of the farmer, it was my duty to come up with a solution to this problem. Tens of thousands of undetected wild invasions into crops in India are actually responsible for considerable crop damage due to their failure to notice these incursions through old detection systems [12]. In their rural backyard, poor defense leaves farmers with no other choice but to cope with the situation that arises. Also, the urban greenery is losing themselves to alien species and open ways are given to wild animals in the wheat and rice fields.

Such protection is important because animals such as deer, wild boars, rabbits, and monkeys and many others continually destroy crops [1, 7]. The project idea is to design the application of the tool for real-time monitoring using YOLOv3 that, for the detection of objects, will send audio alarms and SMSs via Twilio in case animals are present in the field.

This would be developed as computer vision applications where wild animals will be protected from crops using object detection in live video feeds using YOLOv3 accompanied with audio alarms and alerts in real-time to the involved farmers.

This solution is intended to be built for use in the rural environment with energy efficiency [8, 14]. The developments of an integrated computing solution require infrastructures to be tested both in software and hardware components [13].

This work will be executed on two low-power computing devices, the Raspberry Pi 3 with Intel Movidius Neural Compute Stick (NCS), and NVIDIA Jetson Nano [8]. The detection of animals is done with DNN-based object detection models: YOLOv3 and Tiny-YOLOv3. Due to good balance of accuracy and efficiency, these models can be executed even on low-power processing platforms.

LoRa and LoRaWAN are also suggested for communication technologies as they are particularly designed for the rural areas with weak or problematic telecom infrastructures [8]. As the power consumption amount is relatively low, it has an ability to cover large area; hence such an approach provides safe and cost-effective surveillance connectivity for farmers to guard their produce effectively.

2 Literature Survey

Langalizadeh and Moghbeli (2016) evaluated the effectiveness of Naive Bayesian Networks (NBN) in analyzing mortality trends over time, particularly concerning vaccination complaints [1]. Their research aimed to improve NBN algorithms and compare them with other established methodologies. After reviewing 99 papers from PubMed (2005–2015), they removed duplicates and conducted a thorough examination of 38 studies. Ultimately, 23 papers met the inclusion criteria and were utilized in their analysis.

The study builds on the foundational work of pioneers like Yann LeCun and Andrew Ng in the realm of artificial intelligence. LeCun's contributions to Convolutional Neural Networks (CNNs) revolutionized computer vision by enabling the automatic extraction of features from images [5, 9, 13], which is essential for object detection in video frames [9, 10]. His research has played a crucial role in applications such as wildlife monitoring, which is vital for AI-driven scarecrow systems. Similarly, Andrew Ng's research into AI algorithms, [2–4] Bayesian networks, and data modelling has enriched the understanding and needed. With that, some limitations are accompanying size along with the needed memory and processing application of AI in agriculture. His findings demonstrate how AI can aid in managing and improving complex systems designed to protect crops from animal threats.

This study addresses the problem of crop damage caused by ungulates by integrating IoT and AI technologies [8, 13]. By employing the YOLOv3 and TinyYOLOv3 object detection models, along with hardware platforms like the Raspberry Pi 3B, Intel Movidius Neural Compute Stick (NCS), and NVIDIA Jetson Nano, the system can detect animals in real time. The AI-driven scarecrow system also explores the use of ultrasound technology, which has proven effective in deterring animals such as wild boars and deer without causing harm [11, 14]. Ultrasound has shown potential in discouraging certain species while minimizing environmental impact, making it a valuable component of crop protection strategies.

3 Methodology

3.1 Data Set Description

Thousands of images of wild boars, deer, elephants in very different conditions were used in the dataset with hopes of upgrading the performance of the model. Images were captured at different times, ranging from at a different season to a different hour of the day, so the model can learn to distinguish animals in changing light conditions like daylight, dusk, and nighttime.

To include diversity in the environment, images are taken from diverse habitats, such as forests, grasslands, and farmlands. Animals in the dataset appear to be standing, walking, or running in pictures captured from all sides, so that the model generalizes well to real situations. Variations in resolution, focus, and camera type of images within the dataset provide more robustness. The method of data mining selects 1000 images from the total number of 10,000 images, which contain the same number of images of both classes-wild boars and deer. The original diversity of conditions is retained.

3.2 Data Set Preparation

It mainly dealt with data preparation, ensuring richness and quality within the dataset. Image cleaning concerned the removal of blur or unnecessary images, size uniformity of all images by resizing, and normalization of pixel values for enhancing training. It ensured proper and accurate labeling for wild boars and deer-annotated markings on even partially overlapping instances of animals.

Technically, here, rotation, flipping, brightness adjustments, and cropping were used to increase diversity. This resulted in increased variability without requiring more data collection. In this case, the dataset was split into training, validation, and testing subsets with good class balance and diversity in all splits. All the selected images underwent quality control, and only the best samples were considered for determination of a good and representative dataset for training the Beast Discovery model, with no below-par or wrongly labeled images.

3.3 Image Processing and Image Recovery

Object detection does involve trying to match input images to pre-trained models in some specific resolution known as network resolutions. Models such as the You Only Look Once, short for YOLO, attempt to align the dimension of input images to those of the network, which are 416 × 416 pixels but which preserves the original aspect ratio when which subdivides the set batch. Subdivisions help when optimizing memory usage while training power, but it still allows you to enlarge the resolution up to 2048 × 2048 pixels.

Augmentation of Data before Training the Model—The data will be augmented so it can simulate and account for any variation in image quality and shape [5, 10]. This includes corrections for camera shakes, possible rotations in images, color noise, either Gaussian or achromatic, and multi-angle adjustments. All this enhances the robustness and performance when processing real-world data in less-than-ideal conditions.

3.4 YOLOv3 Algorithm and Annotation Tools

The YOLOv3 is a framework presented by Redmon et al. [6]. This one detects various objects including animals and even humans with the help of a neural network. In a nutshell, it is a grid-based system in which every cell of the grid represents an interest region. It enables speedy and accurate detection in object detection, which seems to be the reality concerning most applications coming from the real-time category.

The YOLO-Annotation-Tool is a tool used to annotate and label pictures in a direct manner while preparing datasets to train a model [9, 10]. This tool allows one to capture and annotate objects in just one step, thereby simplifying the process. Sources: One can download this source of annotation tool at GitHub (https://github.com/ManivannanMurugavel/YOLO-Annotation-Tool).

3.5 Memory Management

The parameters in the model's training involved the following:

Batch size: This is the number of images that the network will pass over in one cycle. Here, it has been set at 64. This means that once the network processes each batch, it updates the parameters. Subdivision parameter: This parameter helps prevent GPU memory overflow. In this example, it was set at 2, Then, the YOLOv3 model is trained on the annotated dataset for specific animal-detection functionality [9, 13].

3.6 Platform and Computational Requirements

Training deep neural networks is computationally intensive, especially in training. Inference is less computationally intensive but incurs considerable computation. As such, for the reasons above, such requirements cannot be fulfilled on any embedded platforms. Instead, powerful systems with a high-performance GPU or TPU would be suggested.

We used Google Cloud Platform virtual machines to conduct this research so we wouldn't rely on physical hardware. The virtual machine was configured with 16 vCPUs and 64 GB of memory and was running Ubuntu 18.04 LTS with the T4 NVIDIA Tesla GPU. The T4 GPU comes with 16 GB of GDDR6 memory coupled with Tensor Core as well as RTX accelerators for higher processing speeds in training as well as inferences. In addition, the specifically designed GPUs are optimized for all features such as active thermal management and high-speed data transfer, which is going to be essential when working with very good performance during large-scale model training.

3.7 Addressing Ethical and Social Concerns

One of the most significant social concerns in using the restraint sounds to keep animals away is the observation of noise and zoning laws. The intense sounds may, incidentally, disturb the human societies around the fields or affect the non-target species, resulting in unforeseen results. We ensure this by fine-tuning the speakers to make sure the system act in accordance with local noise control laws and regulations. Sounds are employed only on requirement and configured to stay within allowed decibel limits so that no interference occurs in the neighboring zones.

The second related issue is the stress that may be imparted to the animals due to the repeated utilization of loud deterrence sounds. Repeated noise can be irritating and disorienting for animals, which may eventually cause severe behavioral disorders. To limit this, we set limits on the length and occurrence of the noises to prevent them from getting overly stressed. We have selected predator noises that would best scare the intruder while not harming or panicking the animals to ensure their welfare although securing the crops.

4 System Architecture

YOLOv3 Model. Utilizing the YOLOv3 object detection model to scan a place for animals and birds will effectively prevent pests from farmland or yard without engaging the age-old scarecrow [11] Fig. 1 illustrates a convolutional neural network architecture for object detection. The process begins with an inputimage resized to 416 × 416 pixels. It progresses through several layers, each represented by 3D blocks with dimensions and convolutional operations specified. The layers include operations like convolution (Conv) and residual connections (Res) with varying filter sizes and strides. The network branches into two scales for detecting medium and small objects, leading to convolutional layers for detection and results.

The steps for building the AI-powered scarecrow are discussed below.

4.1 Environment Configuration

First of all, we need to install all requirements to work on deep learning and computer vision tasks. Requirements include OpenCV for processing images, CUDA, and cuDNN for GPU acceleration.

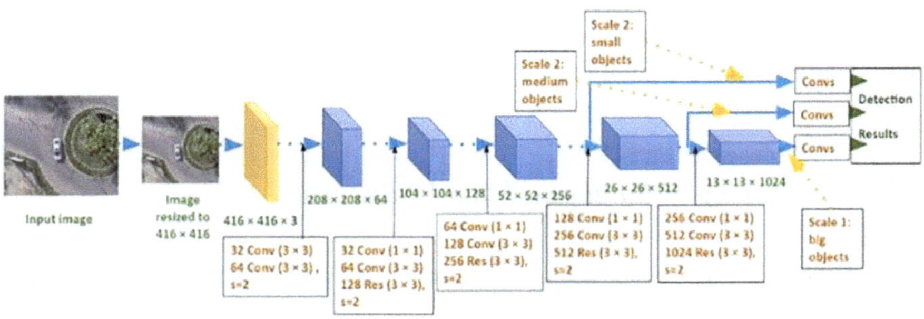

Fig. 1 System architecture

4.2 Train the YOLOv3 Model

To do that, we collect a dataset of images that contain the animals or birds we wish to detect. We then annotate images using annotation tools, helping us mark up the target creatures using bounding boxes.

This thus enables the model to function well when deployed in the real field so that the pests would be accurately identified.

4.3 Implementation of Real-Time Detection

Integrate the trained YOLOv3 model in a system that could process real-time feeds from videos or images. Capture footage from the area you want to be monitored using cameras or sensors. It will analyze the video feed and understand if there are animals around.

4.4 Response System

It should be coupled with an avoidance response mechanism so that the animals will stay away once detected. For instance, it may be in the form of loud noise, flashing lights, or moving mechanical parts with human-like activities. It should instill fear on the animal's side without harming it in any manner.

4.5 Feedback Loop and Optimization

Provide a feedback cycle so that inputs could be taken and performance of the system could be traced out over time. These inputs should be utilized to feed new data to the model for retraining purposes [5]. The response mechanism is fine-tuned, and afterwards, the entire system needs to be fine-tuned to make it more effective. This learning process is sure to train the AI-powered scarecrow in adapting according to time-changing conditions.

4.6 Deployment and Testing

After this, install the AI-powered scarecrow in the designated area and monitor its working performance for a long run. There must be more improvements following practical performance feedback and evaluation reviews. The system needs to be serviced regularly as well as updated so that the system is not demoted down and is very useful for controlling pests.

The effectiveness of the AI-powered scarecrow would then depend on the precise accuracy that the YOLOv3 model boasts for now and also on the reliability of the response mechanism. For a system like this to remain continually effective in keeping pest away from crops, constant maintenance and fine-tuning are required.

5 Algorithm

Creating AI-Based Scarecrow System Using the YOLOv3 Algorithm: An AI-powered scarecrow is essentially just a grouping of an object detection model with the capability to detect any form of animal that might destroy a crop, be it birds or whatever [7, 8]. Here's a rundown of the procedures to develop and implement such a system.

5.1 Data Collection

Then, the kind of animals that need to be detected in agricultural fields, such as birds, and other creatures, will be collected in the form of an image dataset. It needs to be diverse; therefore, it must contain images taken with backgrounds, illumination, and camera angle in different conditions for generalization.

5.2 Data Annotation

Collect the dataset and annotate it by putting bounding boxes on the animals that you want the system to detect. There are several tools which can be employed in annotating images, such as LabelImg, VGG Image Annotator (VIA), or RectLabel. Proper annotation of the dataset will help the model determine animals in any context.

5.3 Model Training

Train the YOLOv3 algorithm using this annotated dataset. One can run the model on the Darknet framework, which is the main implementation platform for YOLO. Alternatively, one could implement directly from deep learning libraries such as TensorFlow and PyTorch, which provide a direct access to YOLOv3 implementations. After training, the algorithm has learned to detect and classify the various animals from the annotated data.

5.4 Deployment

Now, after training the model, this should be deployed on a system that captures video footage in real time from the fields. It can be a camera connected to some tiny computer-say, a Raspberry Pi-or a full-fledged machine, based on how much computation is required.

5.4.1 In Larger Fields

In large fields, for it to work effectively, there is a need to mount more cameras strategically all over the field, hence it can provide wider coverage. The area should have an installation of cameras that can reduce blind spots to a minimal area and at times overlap one's field of view with that of the next one so as to enhance detection with accuracy. For broader coverage, wide-angle cameras are used; for far places, long-range cameras.

In addition to cameras, the whole field should have an installation of a distributed speaker system that will deter animals. The speakers must be spread throughout and cover larger areas that have a greater probability of animal incursion. The sounds released by the speakers should not be against the laws to avoid injuring animals and human beings. Such sounds can be automated to activate only in specific areas where the movement of animals is detected to avoid unnecessary disturbance.

This also ensures that the system is scalable and can work reliably over massive regions of agricultural fields, providing an ideal solution in safeguarding crops against animal intrusions.

5.5 Scarecrow Functionality

Every time the model detects unwanted animals; it tends to trigger scarecrow-like responses [11, 13]. In such a case, the responses would be in terms of noise alarms or motion-based deterrents, and otherwise in automatic response to scare away the birds or creatures. All these are meant for protection against invasion by animals of the crops.

5.6 Real-Time Monitoring and Maintenance

It will also implement an alert system automatically notifying the farmers if there is the detection of animals in the field. There is retuning of the model with new data on regular intervals so that it is aided in increasing the accuracy with better performance over time. In order to keep it operational over a long period, environmental factors like power requirements, weather proofing, and maintenance needs are all considered.

6 Results

6.1 Detection Examples

The detection results provided in this study demonstrate the effectiveness of the system in working in different real-world conditions. These include rural fields, areas with high vegetation, and areas with fluctuating lighting, such as overcast or bright sunlight. The adaptability of the YOLOv3-based detection model in cluttered and dynamic backgrounds has been emphasized in terms of accuracy in intrusion detection. For instance, the system demonstrates its efficacy in challenging settings where animals may enter farmlands amid dense foliage or crop layouts. These scenarios showcase the model's reliability in detecting threats and adapting to diverse environmental factors, thereby offering practical applicability for farmers in preventing crop damage Fig. 2 shows Four photographs of animals with labeled boxes: 1. Top left: A bear walking on grass, labeled "BEAR 99%." 2. Top right: An elephant near water, labeled "ELEPHANT 99%." 3. Bottom left: A goat standing on

Fig. 2 Animal detection examples across varied environments demonstrate model's robustness

grass, labeled "SHEEP 90%." 4. Bottom right: A deer walking near trees, labeled "SHEEP 24%." Each animal is outlined with a purple box and text indicating the label and confidence percentage.

6.2 *Detection of Failure Cases*

6.2.1 In Heavy Rainfall and Foggy Weather

The system's accurateness is compromised during heavy rainfall, as the alteration detections and reduced performance Fig. 3 depicts an image divided into two panels.

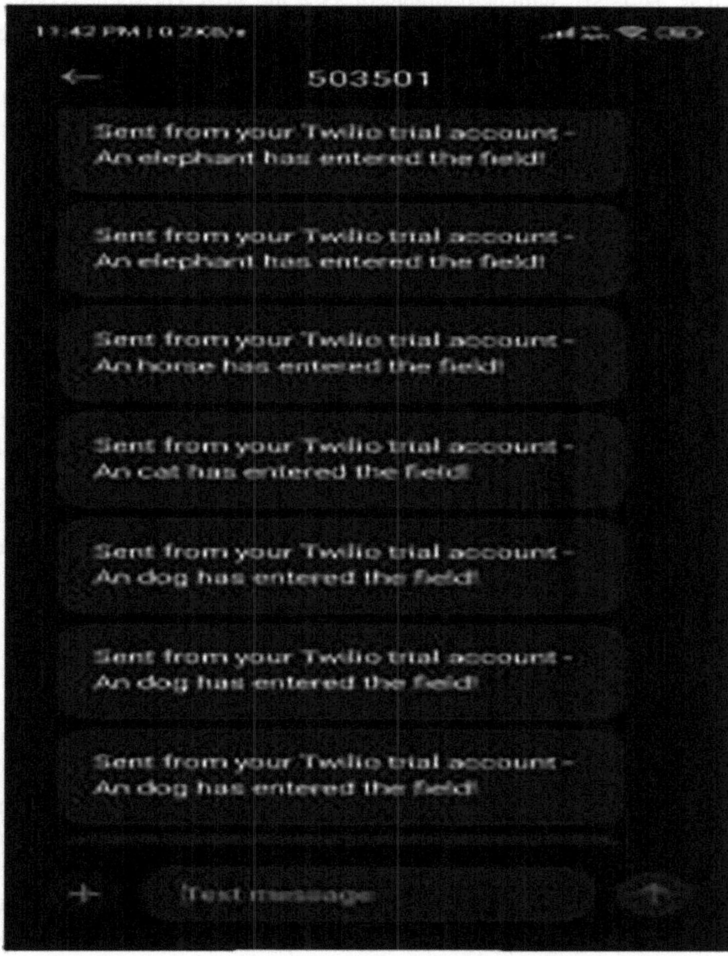

Fig. 3 Notification sent to user

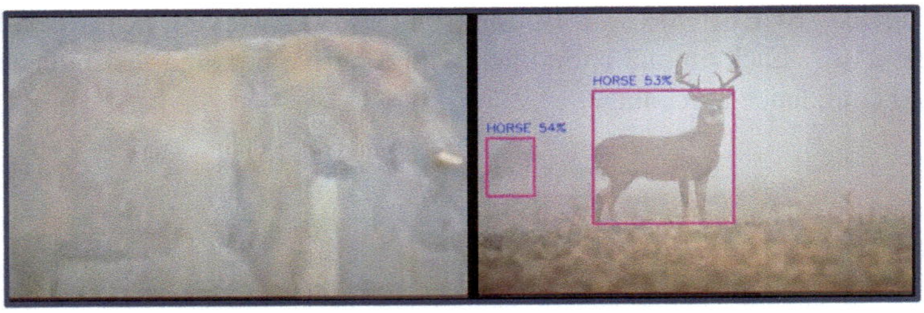

Fig. 4 Failure to detect in heavy rainfall conditions

The left panel shows an elephant in a foggy environment, with a faint overlay of text. The right panel features a deer standing in a misty field, with purple boxes and text labeling it as "HORSE" with percentages of 53% and 54%. The scene is obscured by fog, creating a mysterious atmosphere. Figure 4 shows A mobile phone screenshot showing a series of text messages. Each message states, "Sent from your Twilio trial account - An [animal] has entered the field!" with different animals mentioned, including elephant, horse, cat, and dog. The time displayed is 11:42 PM, and the signal strength and battery icons are visible at the top.

6.3 Managing Inaccurate Outcomes

To effectively lower the number of false positives, it's advisable to always be working on the diversity and variety of the training datasets. The dataset of such a kind is the perfect one for this model, because it is well-edited and represents different situations, weather conditions, and light making it the model to be well-aware of and decide which are the real threats and which are the false detections. Besides, this can also be done by tweaking YOLOv3 model through additional training on other datasets, especially those containing the false positives cases. The model will be more accurate in the identification of the objects and as a result, the number of false alarms will be reduced.

7 Conclusion

Intelligent Harvesters using YOLOv3 algorithm: The innovative advancement of YOLOv3 in AI and computer vision subtly drifted into agricultural technology, catapulting it to a rather classy platform. It identifies impending threats on farms due to pests and stray animals and, thus evades further damage to crops. It reduces the usage of harmful chemicals and promotes sustainable means of farming with quality products. Using such AI-based solutions would revolutionize the farming sector. Totally, the future of agriculture lies in this. Some even make use of scarecrows or symbolic items but these are much away from deterring any aggressive animals or birds. Now, based on this concept, we had established a smart video crop protection system that involves prevention through AI from animal interference. It continuously monitors fields by way of real-time feeds through cameras but then finds the suspicious animals or stray cats. After a potential threat has been detected by the system, it can pop out audible alarms or alerts to scare the intruder away.

8 Future Scope

Advanced Detection Capabilities: The system can be developed further to detect a greater variety of fauna, fowl, and even pests. Advanced versions of YOLO could help improve speeds and accuracies in detection, especially in challenging environments such as low-light conditions and dense vegetation, or with fast-moving animals.

This can be further enhanced in the future for reliable operation under adverse weather conditions like fog and rain using advanced preprocessing techniques that include image dehazing and removal of rain streaks. It may also further improve detection using Multi-Sensor Integration like thermal or infrared imaging is used. In this case, the dataset would be expanded to include photographs under all weather conditions, together with adaptive machine learning models that can self-adapt dynamically to changes in the environment, whereupon the system may perform robustly. The improvements thus made will ensure that the system repeatedly demonstrates consistent performance while at the same time broadening its applicability across different real-world settings

References

1. Langarizadeh, M., & Moghbeli, F. (2016). Applying Naive Bayesian Networks to Disease Prediction: A Systematic Review. Acta Informatica Medica, 24(5), 364.
2. Habibi, N., Hashim, S. Z. M., Norouzi, A., & Samian, M.R. (2014). A review of machine learning methods to predict the solubility of overexpressed recombinant proteins.
3. Bonasso, R.P. and Miller, D.P. (eds.), 1993. Fall Symposium on Instantiating Intelligent Agents. AAAI Press
4. Olaniyi, E. O., & Adnan, K. (2014). Onset diabetes diagnosis using artificial neural network. International Journal of Scientific and Engineering Research, 5(10).
5. Yu, and L. Deng, 2011, "Deep learning and its applications to signal and information processing," IEEE Signal Process. Mag., vol. 28, no. 1, pp. 145-154.
6. Jayalakshmi, T., & Santha kumaran, A (2010, February). A novel classification method for diagnosis of diabetes mellitus using artificial neural networks. OSDE, 159-163. (2010)
7. Bhusal, S., Bhattarai, U., Karkee, M.: Improving pest bird detection in a vineyard environment using super- resolution and deep learning. IFAC-Papers online 52(30), 18–23 (2019)
8. Bochkovskiy, A., Wang, C.-Y., Mark Liao, H.-Y.: YOLOv4: optimal speed and accuracy of object detection. CoRR, abs/2004.10934, 2020

9. Deng, J., Dong, W., Socher, R., Li, L.-J., Li, K., Fei, L.F.: ImageNet: a large-scale hierarchical image database. In: Proceedings of IEEE Conference on Computer Vision and Pattern Recognition, pp. 248–255 (2009)
10. He, K., Zhang, X., Ren, S., Sun, J.: Spatial pyramid pooling in deep convolutional networks for visual recognition. IEEE Trans. Pattern Anal. Mach. Intell. 37(9), 1904–1916 (2015)
11. Marcon. P., et al.: A system using artificial intelligence to detect and scare bird flocks in the protection of ripening fruit. Sensors 21(12), 4244 (2021). Suleiman, I., Babawuya, A., Oyewole, A., Salihu, B., Adeoti O., Alhaji, S.Y.: A review of bird pest repellent systems in farms (2021)
12. Abbas Zaidi, S.S., Ansari, M.S., Aslam, A., Kanwal, N., Asghar, M., Lee, B.: A survey of modern deep learning-based object detection models. Digit. Sig. Process. 126, 103514 (2022)
13. Zhao, Z.-Q., Zheng, P., Shou-Tao, X., Xindong, W.: Object detection with deep learning: a review. IEEE Trans. Neural Netw. Learn. Syst. 30(11), 3212–3232 (2019)
14. Zou, Z., Shi, Z., Guo, Y., Ye, J.: Object detection in 20 years: a survey. CoRR. (2019). https://arxiv.org/abs/1905.05055arXiv:1905.05055

Adaptive Social Robot Navigation Using Vision-Language Models

Ketan Anand, Daymond Chang, and Nathaniel John

Abstract Achieving socially aware navigation in robotics remains a challenge, particularly for integrating embodied agents into dynamic human environments. Building upon NaviSTAR and the MuSoHu dataset, we introduce a novel approach to social robot navigation using language-guided methods. Our methodology employs a multi-modal model leveraging Vision-Language Models (VLMs) to decode high-level text commands and social cues from human interactions. This approach embraces the complexity of human-environment interactions through a diverse array of sensors and datasets. A significant innovation is the dynamic adaptability to user preferences via instruction tuning mechanisms, enabling the system to encompass a wide range of human navigation styles and bridge the gap between robotic functionality and human expectations. Our system is rigorously tested using real-world datasets and simulated environments, employing qualitative assessments and quantitative metrics, including a 5-point Likert scale. The incorporation of VLMs enhances the robot's understanding of social nuances and human intentions, grounding comprehension in a rich Vision-Language context. This fosters improved situational awareness and nuanced decision-making capabilities, leading to more intuitive and responsive human-robot interactions. Our contributions include the Instruct-MuSoHu dataset, combining language guidance with multi-modal sensory data in dynamic environments, advancing socially aware navigation in complex social settings.

K. Anand (✉)
M. S. Ramaiah Institute of Technology, Bangalore, India

D. Chang
Georgia Institute of Technology, Atlanta, GA, USA
e-mail: dchang42@gatech.edu

N. John
R. V. College of Engineering, Bangalore, India
e-mail: nathanjn.is20@rvce.edu.in

K. Deep et al. (eds.), *Artificial Intelligence, Machine Learning and Intelligent Systems*, Springer Proceedings in Mathematics & Statistics 520,
https://doi.org/10.1007/978-981-95-3841-6_9

Keywords Vision-language models · Social robot navigation · Embodied systems · Human-robot interaction

1 Introduction

Achieving socially aware navigation is a key challenge in robotics, crucial for integrating robots into dynamic human environments. Current methods often fail to address the full spectrum of human social conventions in dense environments. Building on NaviSTAR [1] and the MuSoHu dataset [2], we introduce a novel approach using language-guided methods.

Our method employs a multi-modal model with Vision-Language Models (VLMs) to decode text commands and social cues. Unlike traditional systems, our approach leverages diverse sensors and datasets. A key innovation is dynamic user preference adaptation through instruction tuning [3], capturing a broad range of human navigation styles.

Incorporating VLMs enhances the robot's understanding of social nuances, leading to better situational awareness and decision-making. This significantly improves navigation and interaction in complex social environments, aiming for genuine human-robot collaboration. Our methodology, centralizing VLMs and instruction tuning, sets a foundation for future socially aware navigation systems. The code and models developed for this study can be accessed at:
https://github.com/anandketan/VLM_Social_Navigation.

The paper is structured as follows: Sect. 2 covers related works about instruction tuning, detailing the training of large-language models with (instruction, output) pairs to create high-level navigation instructions. Section 3 outlines the method, including the problem statement and the generation of language instructions using the MuSoHu dataset and GPT-4. It also discusses the process of aligning generated instructions with multimodal data samples and explains the model architecture for integrating vision, language, and point cloud data. Section 4 describes the simulation setup, sensor integration, and control paradigms for evaluating the system. Finally, Sect. 5 lists the basic and advanced tasks designed to test the model in real-world scenarios.

2 Related Work

2.1 Social Robot Navigation

Social navigation poses a unique set of challenges compared to static-environment navigation. An existing effort utilized a hybrid spatio-temporal graph transformer to understand interactions in human-rich environments [1]. However, this approach requires significant computation due to the large spatial and temporal attention blocks. To address this limitation, we propose semantically grounding the social navigation via language instruction.

The *Large-Scale Multi-Modal, Social Human Navigation Dataset* (MuSoHu) [2] provides an avenue for robots to learn human-like, socially compliant navigation behaviors. The dataset contains approximately 20 hours, 300 trajectories, and 100 kilometers of socially compliant navigation demonstrations collected by 13 human demonstrators in both indoor and outdoor environments. We propose Instruct-MuSoHu, which attaches generated instructions to semantically distinct subsets of the existing dataset. Section 4.1 provides further details.

2.2 Vision-Language Models

Vision-language models (VLMs) are adept at establishing accurate and generalizable alignments between visual concepts and textual semantics efficiently [4]. State-of-the-art advancements like CLIP [7] and Flamingo [8] can be applied to a variety of visual classification tasks. Social robot navigation is an enticing application for VLM approaches because it allows the robots to gain a semantic understanding of their surroundings. It is critical for language-guided navigation systems to have semantic understanding.

2.3 Instruction Tuning

Instruction tuning is a method of training large-language models on a dataset consisting of (instruction, output) pairs [6]. Instruction refers to the human instructions for the model to follow, and output is the human's desired response or execution of the model for that specific instruction. The goal is to produce an instruction-following dataset in which these pairs can be utilized for downstream tasks. *Visual Instruction Tuning* [3] extends this work by generating multimodal language-image instruction-following data. We leverage similar multimodal instruction tuning techniques on the MuSoHu dataset to create high-level, plausible navigation instructions.

3 Method

3.1 Problem Statement

Our goal is to create a socially aware navigation model that integrates human language instructions. This problem can be formulated as a standard ML problem. Given an observation horizon T_o and prediction horizon T_p, sensory inputs over T_o, such as camera images $I_o \in \mathbb{R}^{T_o \times C \times H \times W}$ and LiDAR point clouds $P_o \in \mathbb{R}^{T_o \times 3 \times N}$ along with a natural language navigation instruction L: find a navigation model $\mathcal{F}_\theta$ with parameters θ such that,

$$(v_p, \omega_p) = \mathcal{F}_\theta(I_o, P_o, L)$$

Where v_p and ω_p are the linear and angular velocity of the agent over the prediction horizon. We thus train the model $\mathcal{F}_\theta$ and obtain the optimal θ^* such that

$$\mathcal{L} = \sum_p (v_p - v_p^*)^2 + (\omega_p - \omega_p^*)^2$$

is minimized. Where, $v_p^* \in \mathbb{R}^{T_p}$ and $\omega_p^* \in \mathbb{R}^{T_p}$ are the linear and angular velocity of a socially compliant human over the prediction horizon, navigating in the same environment and given the same navigation instruction. This particular formulation allows the model to learn aspects of human navigation in social settings and due to the inclusion of multi-sensory data, it will enable the agent to process environmental cues and relate them to the human instruction.

So to summarize the problem, the inputs include multi-sensory data, like camera images and LiDAR point clouds, and the goal is to predict the optimal linear and angular velocity of the agent, **conditioned** on the language instruction given to the agent, such that the agent demonstrates human-like socially compliant navigation behavior.

In the following section, we shall discuss how the language instructions for the navigation data from the `MuSoHu` dataset are generated.

3.2 *Language Instruction Generation*

The `MuSoHu` dataset provides multi-modal data on human navigation in diverse social and crowded scenarios, encompassing LiDAR, IMU-based odometry, RGB images, audio, and depth modalities. Notably, language instructions are absent in the `MuSoHu` dataset.

While datasets like `VLN-CE` [5] include language instructions for navigation, they are limited to static scenarios. Datasets like `Talk the Walk` [9] do not offer sensor modality data. Other current datasets present a trade-off: those capturing dynamic, unstructured environments lack language instructions, whereas those with language instructions are recorded in static settings or may not offer sensor data. To address this gap, we propose generating instruction data using `GPT-4` [11], a foundational generative text model. Our approach aims to augment the `MuSoHu` dataset with language instructions, enhancing its suitability for developing our socially aware navigation model. This approach aligns with methodologies previously employed for visual instruction tuning of VLMs for complex Q/A tasks [3]. Shown below

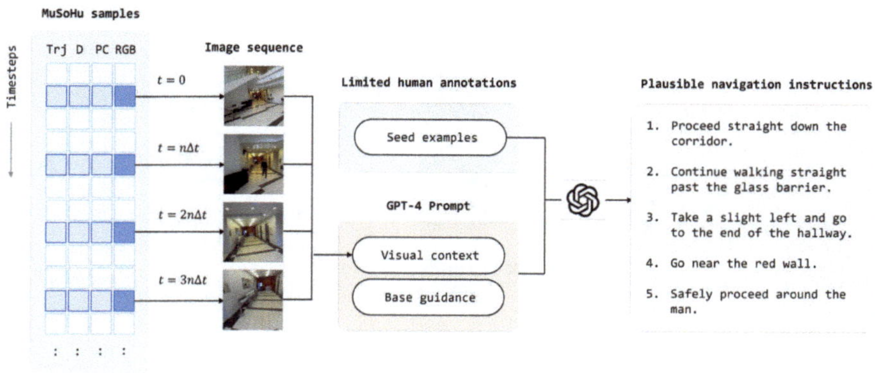

Fig. 1 Instruction generation pipeline for building the `Instruct-MuSoHu` dataset

is our pipeline to generate synthetic instruction data for the `MuSoHu` dataset. We present this augmented dataset as `Instruct-Musohu` (Fig. 1).

We leverage trajectory data (`Trj`), depth images (`D`), point clouds (`PC`), and RGB images (`RGB`) from the `MuSoHu` dataset, chosen for their rich navigation cues and information content. To generate instruction data, we systematically sub-sample sequential and semantically consistent images from the `MuSoHu` video corpus. These images serve as visual context for our foundational VLM of choice, `GPT-4`. Augmenting this process, we manually curate a set of seed examples for instruction generation, enhancing the diversity of generated instructions and minimizing the risk of task failure. The prompt is carefully designed to include basic task guidance, with an emphasis on human-like plausibility.

Utilizing the `GPT-4` API, we generate a set of five instructions per sampled image sequence within a given navigation scenario. This gets us multimodal sensor data that aligns semantically with the generated instructions, presenting an ideal training set for a language-guided social navigation model.

3.3 Multimodal Sample Modification

As previously discussed, the `MuSoHu` dataset did not originally include a language input modality, thus inspiring the creation of `Instruct-MuSoHu`. Consequently, the sampled data sequences must also reflect these changes. After the language instructions were created, they became decoupled from the original images that produced them. To account for this, we utilized the same timestep from the instruction generation process to systematically correlate the instructions to each sample (Fig. 2).

There are two approaches to incorporating the language instructions back into the dataset: (1) including the most recent instruction given, and (2) including a subset of past instructions given. For our approach, we chose option 2 due to the nuances

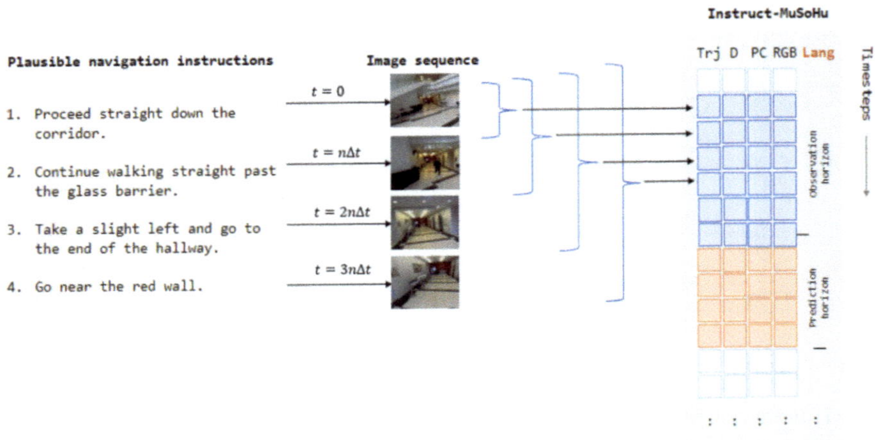

Fig. 2 MuSoHu sample modification pipeline

of model training. If we only include the most recent instruction, the model will be attuned to only learning short horizon instructions. Due to the long-horizon nature of our generated instructions, we randomly sampled a past instruction so that the model could learn a mix of short and long-horizon tasks.

3.4 Proposed Architecture

In this section, we discuss our model architecture for generating synthetic language instructions aligned with multi-modal sensor data. Our model (see Fig. 2) operates within an observation and prediction horizon paradigm. At the end of the observation horizon, language instructions become available, which, along with multi-timestep point cloud and RGB data [10], are passed through modality-specific encoders. The vision and language encoders belong to the same base VLM and have aligned representations.

The concatenated vision and point cloud embeddings are then passed through an LSTM, aimed capturing temporal and spatial evolution of entities within the environment. The language instruction is crucial for guiding the navigation behaviour and aligning it with human preferences. So we employ a cross attention mechanism that fuses the language embedding with the spatiotemporal information available as an output of the LSTM. Finally, we have two MLPs responsible for predicting the linear and angular velocities v and ω where we employ an MSE loss w.r.t the target trajectory. The formal description of our architecture can be summarized as follows:

Instruction embedding: $$\mathbf{e_L} = \mathbf{W_L} \cdot \phi_L(L) + \mathbf{b_L}$$

Table 1 Model parameters

Submodule	Trainable params	Non-trainable params	Total params
pc_encoder	2,803,529	0	2,803,529
vision_encoder	0	309,823,488	309,823,488
lang_encoder	0	1,674,121,216	1,674,121,216
lstm	142,671,872	0	142,671,872
env_fc	524,544	0	524,544
lang_fc	524,544	0	524,544
x_attention	263,168	0	263,168
lin_mlp	8224	0	8224
ang_mlp	8224	0	8224
Total	146,804,105	1,983,944,704	2,130,748,809

$$
\begin{aligned}
&\text{Vision embedding:} && \mathbf{e_I} = \phi_I(I_o)\\
&\text{Point cloud embedding:} && \mathbf{e_P} = \phi_P(P_o)\\
&\text{Temporal processing:} && \mathbf{e_{env}} = \mathbf{W_{env}} \cdot \text{LSTM}(\mathbf{e_I} \,\|\, \mathbf{e_P}, (h, c)) + \mathbf{b_{env}}\\
&\text{Cross-modal processing:} && \mathbf{e_{att}} = \sigma(\mathbf{W_{att}} \cdot \text{XAttention}(\mathbf{e_L}, \mathbf{e_{env}}, \mathbf{e_{env}}) + \mathbf{b_{att}})\\
&\text{Linear velocity prediction:} && v_p = \mathbf{W_v} \cdot \mathbf{e_{att}} + \mathbf{b_v}\\
&\text{Angular velocity prediction:} && \omega_p = \mathbf{W_\omega} \cdot \mathbf{e_{att}} + \mathbf{b_\omega}
\end{aligned}
$$

Where, $\{\phi_I, \phi_L\}$ is the `MobileVLM-1.7B` vision language model [12] consisting of an image encoder ϕ_I (`CLIPVisionTower`) and language tokenizer and encoder ϕ_L (`MobileLLaMA-1.4B`) [13]. ϕ_P is the `PointNet` encoder. Also σ is the softmax function. We employ dimensionality reduction for the language embedding and the environment embedding consisting of the point cloud and the image enbeddings since we observed that the `CrossAttention` block is an expensive operation in terms of the number of parameters. Additionally, we did not find any suitable pretrained weights for the PointNet encoder, with a small domain gap in the pretraining dataset, we plan to unfreeze it and train it on point clouds obtained from the `MuSoHu` data. Check the Fig. 2 for the model architecture. Check Table 1 for model parameter summary.

4 Simulation Setup

4.1 Hypotheses

– H1: Integrating language instructions with multi-modal sensory data enables robots to navigate more effectively in dynamic social environments compared to traditional navigation systems that do not utilize language-guided instructions.

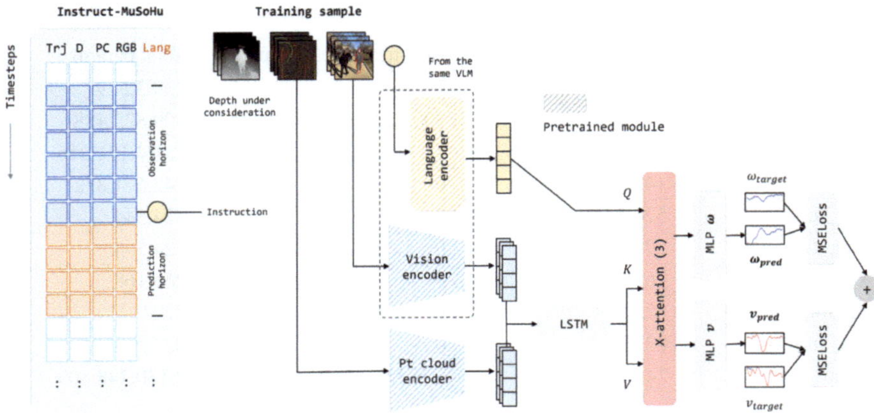

Fig. 3 Our proposed architecture, leveraging our dataset `Instruct-MuSoHu`

The use of Vision-Language Models (VLMs) for processing and understanding social cues and high-level textual commands improves the robot's ability to interpret complex social interactions and navigate accordingly.

– H2: Instruction tuning mechanisms that adapt to user preferences and navigation styles lead to higher satisfaction in human-robot interaction within social contexts.

4.2 Language Modality Data

The integration of language modality data with the VLM model facilitates training in language guided socially aware navigation. We use detailed language instruction as shown Fig. 3. which allows the model to interpret and act upon complex human like navigation commands within dynamic environments. The given instructions emulate a real-world path through a specific environment, denoted by markers such as "the red chair", "tiles colour" which provides contextual relevance to the training data.

This input data allows the model to align language with visual cues and environmental context. This helps the model learn to generate socially compliant navigation behaviours, improving its real world applicability where social norms and human interactions are unpredictable. Ultimately, the goal is to enhance the system's environmental awareness, decision-making capabilities, and alignment with human preferences, thus fostering intuitive human-robot interactions within complex social settings (Fig. 4).

- Enter the corridor with the black and white tiles.
- Walk straight down the hallway past the red chairs.
- Proceed towards the end where the corridor turns right.

Fig. 4 Parsed RGB data with its associated language instruction generated

4.3 Simulation Environment

There are several open-source simulation environments that can be used to model social robot navigation. We did a survey of robotic simulators and evaluated them in terms of sensor support, support for a large number of agents, realistic rendering and iterability. We considered Isaac Sim, Unity-ML Agents Toolkit, Gazebo and Coppelia sim. Finally, after careful consideration, we chose Nvidia Isaac Sim.

Since we did not use ROS integration with Isaac sim, we had to develop custom extensions and nodes that allowed for capturing raw data from the sensors available within the Isaac sim sensor suite and functionality like preparing batches of sensor data (i.e., camera and point clouds) in real-time, using custom queues. We also had to implement a custom Omnigraph extension for capturing and saving the pose history of any prim in the stage. For simulating humans and human crowds within the scene we leverage the `omni.anim.people` extension, which makes it convenient to spawn and control human characters on the stage using animation graphs with associated state machines.

For the robot agent, we chose the Nvidia CarterV1 robot, with a built-in camera sensor and a PhysX solid state Lidar sensor. We tried to match the Lidar specifications to the Velodyne lidar used for capturing the point clouds within the `MuSoHu` data. We opted for a physics based Lidar instead of an RTX based lidar due to bugs in Isaac sim leading to failures in saving the RTX Lidar visualization pipeline, which required us to repeated setup the lidar for the robot for every instance of the simulation stage. Shown in Fig. 5 is the robot hardware we chose too employ.

For controlling the robot we chose to design a action graph that implemented sensor integration, batching, model inference, and the control aspects of the robot including articulation. Shown below is a high level overview of our control paradigm within Isaac sim (Fig. 6).

Fig. 5 Chosen robot hardware for simulation

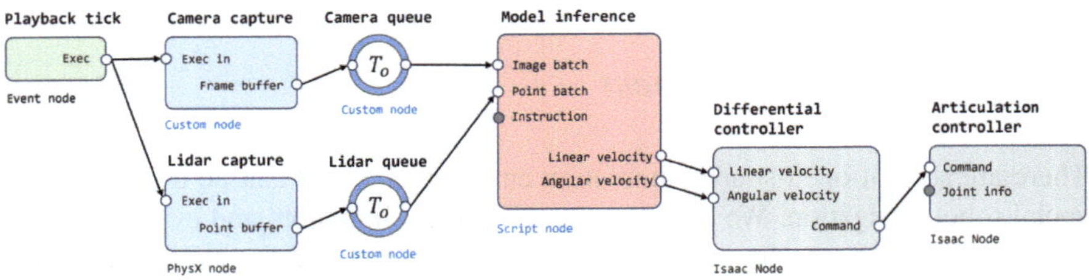

Fig. 6 High level action graph for our agent

5 Evaluation Tasks

Once we train our model, we aim to develop and assess our proposed VLM system's adeptness in dynamically populated environments. We intend to test our model on the following tasks, categorized from basic to advanced complexity, which are designed to test the system's performance in real-world scenarios with human and environmental variability.

Basic Tasks

- Single Human Avoidance: Agent approaches and avoids a single human agent.
- Single Human Avoidance + Instruction Following: In addition to single human avoidance, a language instruction is given to guide the agent towards a goal location.

Advanced Tasks

- Multi-Human Avoidance: Agent approaches and avoids multiple human agents along its trajectory.
- Multi-Human Avoidance + Instruction Following: In addition to multi-human avoidance, a language instruction is given to guide the agent towards a goal location.
- Multi-Human Avoidance + Complex Instruction Following: In addition to multi-human avoidance, a language instruction requiring the processing of multiple visual cues (colors, signs, etc) will guide the agent towards a goal.

These tasks are challenging due to the unpredictable nature of human behavior and the need for the robot to make real-time decisions that respect social norms and personal space. We did perform a basic level evaluation of our agent in these environments, but we found the results unreliable so we do not include it here. Performance is evaluated using a combination of quantitative metrics such as:

- Success rate: The percentage of tasks completed successfully without causing discomfort or collision.
- Time to goal: The time taken to reach the destination, indicating efficiency.
- Human comfort index: A Likert-scale based assessment of human viewer comfort levels during the robot's navigation.

6 Experiments and Results

We have accomplished the following dataset generation and evaluation:

- Demonstrate language-guided socially aware navigation on the test set as seen in the Likert analysis shown in Fig. 7.
- Release our dataset called `Instruct-MuSoHu` for the wider robotics community.
- Qualitatively evaluate the effect of particular phrases in the instruction like *proceed carefully, walk slowly, avoid collisions, turn right now, stop* etc. on the navigation behavior.

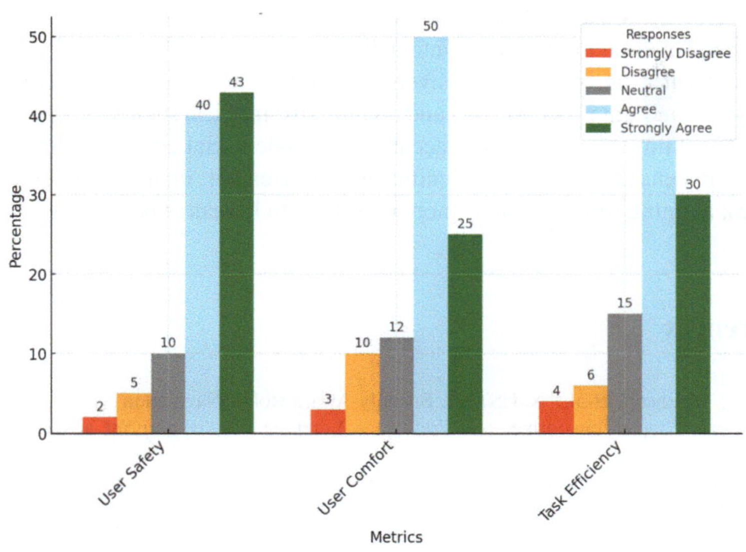

Fig. 7 5-point Likert parameter assessments on various robot tasks

We did observe that running a 1.9B parameter model leads to low performance and slow response times due to the complexity of the system and the substantial memory requirements. It will take careful system design and optimization to balance computational economy and performance in order to mitigate these dangers. We also found instances where the language instructions generated by `GPT-4` did not adhere to the specifications we set, which resulted us having to exclude the bag files corresponding to the particular trajectory and environment. This led to a loss in diversity of environments. We also found that certain bag files had very few time steps of data, which resulting in failure to generate instructions for the scenario due to the absence of enough visual and temporal context for prompting `GPT-4`.

7 Conclusion

This study enhances human-robot navigation by integrating Vision-Language Models (VLMs) to better navigate socially dense environments. Utilizing the MuSoHu dataset, our approach enables robots to interpret complex textual instructions and subtle social cues, facilitating smoother interactions within dynamic human settings. A key innovation of our system is the implementation of instruction tuning mechanisms, which dynamically adapt to user preferences and navigation styles, improving the responsiveness of robots to their social surroundings.

Our methodology builds upon existing language-guided navigation techniques, introducing a higher level of adaptability by embedding operational strategies within a comprehensive Vision-Language framework. This enhancement increases the robots' situational awareness and decision-making capabilities, leading to more intuitive and natural human-robot interactions.

We evaluated our system extensively in simulated and real-world environments using both qualitative assessments and quantitative metrics, such as a 5-point Likert scale. The ability of our robots to accurately decode social cues and mimic human navigation behaviors marks a significant advancement in the field of assistive robotics, aligning robotic actions more closely with human expectations.

References

1. Wang, Weizheng, et al. "NaviSTAR: Socially Aware Robot Navigation with Hybrid Spatio-Temporal Graph Transformer and Preference Learning." *arXiv preprint* arXiv:2304.05979 (2023).
2. Nguyen, Duc M., et al. "Toward Human-Like Social Robot Navigation: A Large-Scale, Multi-Modal, Social Human Navigation Dataset." *arXiv preprint* arXiv:2303.14880 (2023).
3. Liu, Haotian, et al. "Visual instruction tuning." *arXiv preprint* arXiv:2304.08485 (2023).
4. Wu, Hao, et al. "Unified visual-semantic embeddings: Bridging vision and language with structured meaning representations." *Proceedings of the IEEE/CVF Conference on Computer Vision and Pattern Recognition* (2019).

5. Krantz, Jacob, et al. "Beyond the Nav-Graph: Vision-and-Language Navigation in Continuous Environments." *European Conference on Computer Vision* (2020).
6. Zhang, Shengyu, et al. "Instruction tuning for large language models: A survey." *arXiv preprint* arXiv:2308.10792 (2023).
7. Radford, Alec, et al. "Learning transferable visual models from natural language supervision." International conference on machine learning. *PMLR, 2021* (2021).
8. Alayrac, Jean-Baptiste, et al. "Flamingo: a visual language model for few-shot learning." *Advances in Neural Information Processing Systems* 35 (2022): 23716–23736.
9. de Vries, Harm, et. al. "Talk the Walk: Navigating New York City through Grounded Dialogue." *arXiv preprint* arXiv:1807.03367(2018).
10. Anand, Ketan, et al. "Precision Testing of a Single Depth Camera System for Skeletal Joint Detection and Tracking." *Lecture Notes in Networks and Systems, vol 400. Springer* (2022): https://doi.org/10.1007/978-981-19-0095-2_37.
11. Achiam, Josh, et al. "Gpt-4 technical report." *arXiv preprint* arXiv:2303.08774 (2023).
12. Chu, X., Qiao, L. et. al. "MobileVLM: A Fast, Strong and Open Vision Language Assistant for Mobile Devices". *arXiv preprint*, arXiv:2312.16886 [cs.CV] (2023).
13. Touvron, H. et. al. "LLaMA: Open and Efficient Foundation Language Models". *arXiv preprint*, arXiv:2302.13971 [cs.CL] (2023).

Egocentric Perception for Open Vocabulary Object Rearrangement

Monica Anand, Seema Aggarwal, Ketan Anand, Abrar Omar, and Sri Selventhiran

Abstract With the increasing need for service robots in households, it is essential to develop systems that can function in open environments and perform tasks such as organizing spaces. The diverse and ever-changing set of objects, home arrangements, and personal preferences make tidying household spaces a challenging problem. While several formulations of object rearrangement have been proposed as embodied AI tasks, they are limited to recognizable objects and universal arrangement preferences. We propose a method that performs tidying tasks in a personalized manner with open-vocabulary generalization to unseen objects. By leveraging geometric features learned through viewpoint invariance, inspired by CSR, and visual features obtained using a frozen CLIP model, we can learn tidying preferences. Our method consists of a perception module to recreate a map of the environment and a ranking module to predict rearrangement sequences to tidy the environment. Both modules generalize to an open set of objects and are learned from environment observations. Our perception module achieves a 27.1% accuracy in predicting correct object locations, and our ranking module correctly identifies desired receptacles for 49.0% of objects.

Keywords Open-vocabulary generalization · Geometric features · Embodied AI tasks

M. Anand (✉)
Department of Mathematics, M. S. Ramaiah Institute of Technology, Bangalore, India

S. Aggarwal
Department of Physics, M. S. Ramaiah Institute of Technology, Bangalore, India

K. Anand
Georgia Institute of Technology, Atlanta, GA, USA

A. Omar
Emory University, Atlanta, GA, USA

S. Selventhiran
R. V. College of Engineering, Bangalore, India

K. Deep et al. (eds.), *Artificial Intelligence, Machine Learning and Intelligent Systems*, Springer Proceedings in Mathematics & Statistics 520,
https://doi.org/10.1007/978-981-95-3841-6_10

1 Introduction

A recent survey that polled around 1500 people about their desired functionality for a service robot revealed that 'organizing and cleaning' is the third highest desired capability [2]. Such tasks of household cleaning and tidying can be formalized for mobile robot agents as an embodied AI rearrangement problem [3], where the environment is transformed from an initial or untidy state to a desired or tidy state by moving objects. Specifically, tidying is a rearrangement task wherein the goal may not be explicitly and exhaustively specified by the user. Instead, users might expect the robot to understand the goal based on prior observations of tidy spaces, and potentially incorporate verbal feedback. The task of tidying a household needs to incorporate personal preferences as well as generalize to novel and potentially unrecognizable objects. Personalization in tidying is associated with both object instances and classes. For instance, one person might keep kitchen knives in a drawer, while another might keep then on a knife block on the counter, and yet another might place a specific cutlery set in a glass case. This sort of specificity requires object instances to be recognized and treated individually [2]. On the other hand, novel objects might be encountered, requiring the robot to extrapolate its knowledge to novel objects based on semantic similarities with the familiar objects. Furthermore, households are composed of a variety of objects and object detection or segmentation might not be able to detect and classify all these objects. This requires the robot to be able to rearrange objects that might not be recognizable by object detectors. In this work, we propose a framework that allows the agent to autonomously navigate a previously unseen room configuration, identify objects misplaced in the room and move those objects into the appropriate receptacle. We validate our model on the Housekeep [4] dataset built on Habitat Sim [5], against the benchmarks provided in by Housekeep. Our main contributions through this work are: Open-set generalization to novel objects in object rearrangement and personalized learning of tidying preferences.

2 Prior Work

Several versions of object rearrangement have been proposed as embodied AI benchmarks. The RVSU scene change detection (SCD) challenge [6] requires identifying and mapping objects that had changed between mapping episodes. The AI2-THOR Visual Room Rearrangement (RoomR) task [7] is a two-phase process, allowing the robot to observe and map in phase 1, and requiring the robot to move objects in phase 2 to achieve the configuration observed in phase 1. Finally, the TDW-Transport Challenge [8] requires the robot to find a specified set of objects and transport them to a specified desired location. Of the proposed solutions to these rearrangement tasks, two-stage pipelines decoupling the perception/mapping from planning have shown better results [9]. Note that traditionally the rearrangement problem is specified over object instances of common objects, with goals being specified as inputs or through observations. This neither tests generalization to novel objects nor preference learning. Recent works [4, 10] address the aspect of generalization to novel objects in object rearrangement. These methods address the problem of tidying up objects in the household by identifying misplaced objects and arranging them in the correct place. However, they assume the agents have access to perfect perception, including instance and semantic segmentation masks, and object-receptacle relation sensing. They use external commonsense knowledge in the form of language models over the ground truth object classes to learn desired object locations. In addition to making the perfect perception assumption, this limits the learned preferences to object categories, and cannot encode instance-level arrangements. Also, they learn a generic object-receptacle mapping from the dataset representing the entire population without considering user-specific placement preferences. User preferences have been incorporated in object rearrangement for determining desirable spatial object layouts [11, 12]. However this relies on object class labels to derive the semantic features, which the preferred arrangements depend on. Another work, performs language-conditioned table-top rearrangement of objects directly from raw point cloud observations [13], but it does not learn from experience and rather uses a language prompt from the user. Beyond tabletop rearrangement, Wei et al. [14] present a framework to rearrange furniture in a messy room using an iterative denoising approach, but assumes full observability of the environment along with class labels of individual objects. In our work, we aim to combine these agent capabilities to enable robot agents to perform rearrangement in more realistic scenarios [3], while learning rearrangement preferences from observations. Pretrained foundational models like CLIP [15] have been used in embodied applications towards open vocabulary generalization of perceived spaces [16, 17] as well as to ground language commands in manipulation tasks [18]. It has also been shown to excel zero-shot at embodied AI navigation tasks, and build world models which can be queried by downstream planners for various robotic tasks. We utilize this open-vocabulary generalization of pretrained CLIP models to enable our model to generalize to unseen, and potentially unclassifiable, objects.

3 Method

In this work, we create a framework which allows an agent to build a map of the environment from egocentric RGB-D observations, and determine a set of object movements to transition the household to a tidy state. This framework, outlined in Fig. 1 comprises of two main modules: an open-vocabulary semantic perception module and a personalizable preference learning module.

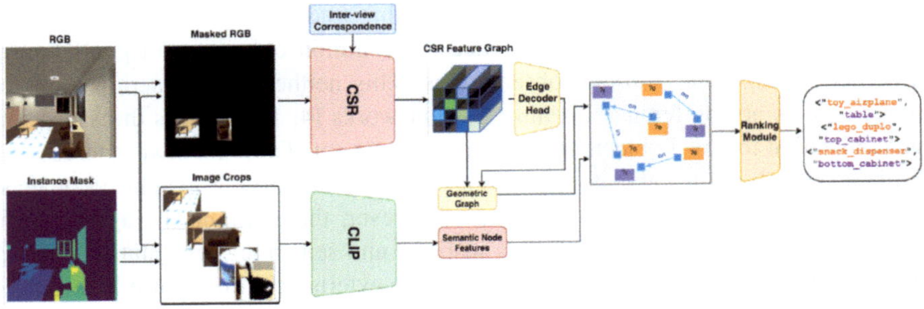

Fig. 1 Pipeline for the proposed solution, including the perception module consisting of CSR and CLIP models, and ranking module which identifies misplaced objects and suggests their desired locations

3.1 Problem Formulation

Given a household with a set of objects O, and receptacles R, an embodied agent must perform a sequence of object movement actions $A = \{(o_i, l_j)\}$ required to transition the household to a tidy state. A tidy state of the house is defined as every object being located at one of its preferred locations $P_o = \{l_i\}$. The agent is given observations of correctly and incorrectly placed objects in the household as prior experience, which it can leverage to learn the user's preferred object-receptacle mappings.

For training, the agent has access to egocentric RGB observations I_t^{RGB}, and the object-receptacle pairs are marked as tidy or untidy for each user, which it must use to extract the state of the environment and learn a personalized model of object location preferences. At test time, the agent must use the RGB-D observations to determine the set of object movement actions $A = \{(o_i, l_j)\}$ required to transition the household to a tidy state for each object. Note that the agent may encounter previously unseen objects at test time.

3.2 Perception Module

The objective of the perception module is twofold: to build a map of observed objects and their current locations, and to learn comprehensive object features to aid downstream object-receptacle ranking model learning. The object features must contain both instance information, to help learn instance-specific preferences, and semantic information, to aid generalization to novel unseen objects. To create a perception module with these characteristics, we utilize ideas from prior works on Continuous-Scene Representations (CSR) [8] which learns to map an environment with latent representations of objects and their relationships, and CLIP [10] which helps augment the object features with requisite semantic information, while maintaining open-vocabulary generalization.

We train a CSR model to generate a map of the environment based on RGB observations and interview point correspondences. The CSR module is based on a momentum contrastive architecture, outlined in Fig. 2, and learns to encode every observed object or receptacle, together with entities, and relationships between them as continuous vectors in latent space. It learns to capture geometric information in this latent space using a viewpoint invariance prior [4]. The model is trained to push the features of the same node or edge obtained from different viewpoints closer together in latent space. In order to normalize, it then pushes distinct node-pair-features away from one another. Specifically, this amounts to minimizing the InfoNCE loss [5] on the feature representations generated from node-pairs. The loss is defined as:

$$L_{CSR} = -\log \frac{\exp(\text{CosSim}(q, k^{+})/\tau)}{\sum_{i=0}^{K} \exp(\text{CosSim}(q, k_i)/\tau)}$$

Here q is the feature representation of a single node or node-pair from the CSR encoder, k^{+} is the encoder representation for the second view of the same relationship, each k_i is a negative drawn from a bank of K views of other relationships, τ is the temperature term, and CosSim is the dot product between $L2$-normalized features.

CSR can be trained in an unsupervised manner, given RGB observations and segmentation masks of elements in a scene, as well as object correspondences between observations obtained from different viewpoints. Prior work has produced several segmentation and region proposal methods which can be used to generate bounding boxes of salient regions in an image, and several association methods which robustly provide point-to-point correspondences between images [3, 8, 14].

Therefore, we assume that these segments and correspondences are available, and train CSR on a set of images with objects viewed from several viewpoints obtained from Habitat Sim. Given the bounding boxes and correspondences for different objects and receptacles (collectively entities), we generate a dataset of entity-entity relations, characterized by an RGB image and bounding boxes for each entity. We obtain a set of positive samples containing a pair of viewpoints of every node-node relationship. In addition to training for node-node relations, we also train for multiple views of a single node. This occurs as a special case of the above framework when both nodes in the node-pair extracted in an image are identical, resulting in two identical crops.

On top of the self-supervised CSR model, we train an edge decoding head with a smaller dataset (1000 examples) of labelled object-receptacle relations. This is a linear feature decoder head which decodes an edge into a probability of support relation indicating whether node A is located on node B. By inferring the most likely receptacle node that every object node is located at, we obtain an object-receptacle map of the perceived environment. Note that in order to enforce validity of object locations being receptacles, we assume the receptacles to be known or recognizable,

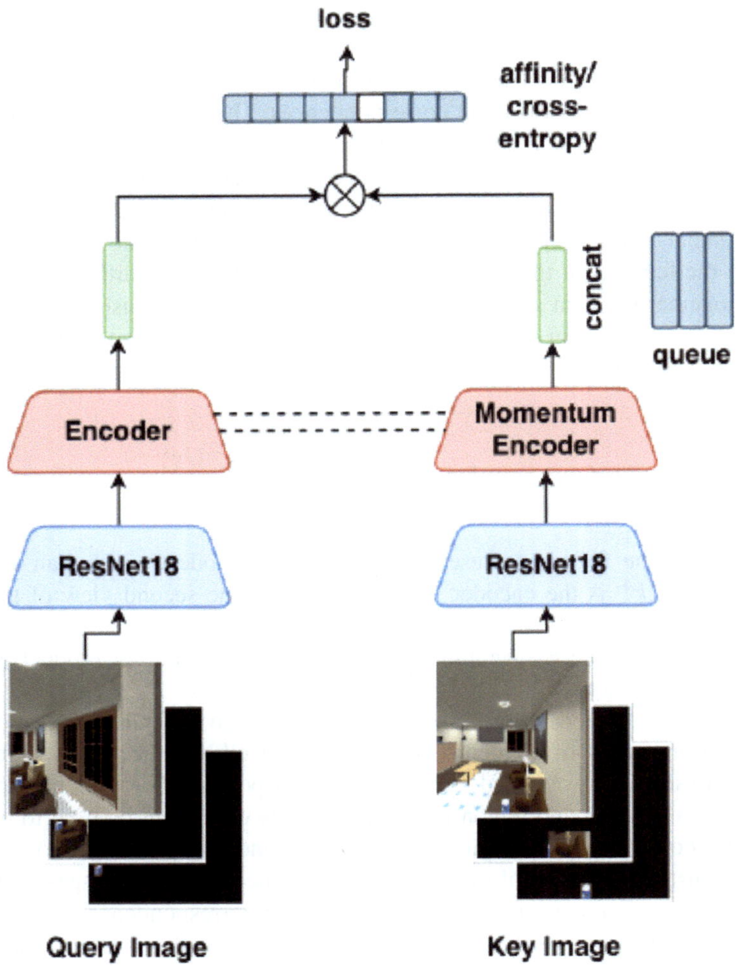

Fig. 2 CSR generation module using momentum contrastive architecture over pretrained ResNet-18 features

since objects like furniture in a house are typically static and known. However, we make no assumptions on knowing anything about the dynamic objects.

At inference time, CSR is used to generate latent vectors for observed entities and relationships from egocentric observations, and features corresponding to each object or relationship are mean pooled to obtain a single latent feature. The result is an $n \times n \times f$ matrix, with n entities and latent feature size of f. The diagonal elements represent node features, which we use as a geometric embedding for downstream ranking module [5]. The off-diagonal elements represent entity-entity relations. The edge decoding head then translates each edge feature into a probability of whether it represents a support relation, i.e. whether one entity is on top of another.

To enable open-vocabulary generalization to unseen, and potentially uncategorizable objects, we use CLIP [14] to generate semantic embeddings, to complement the geometric information encoded in the CSR embeddings. We aggregate the embeddings belonging to the same object by average pooling and concatenate them with the learned CSR features to complete the semantic-geometric representation. Ultimately, this module generates a semantic-geometric scene graph with nodes representing objects and receptacles described by CSR+CLIP latent vectors, and edges representing locations of all observed objects.

3.3 Personalizable Preference Learning

The objective of the preference learning module, outlined in Fig. 3, is to determine the preferred receptacle(s) for every object in a scene, for the specific user [6]. We model this as an object-receptacle-pair compatibility function. Our model is a feed forward network that accepts as input the features for the object and location obtained from the perception network, as well as a latent embedding of the user, and outputs a scalar score. We pick the receptacle with the highest compatibility score with a given object as its tidy location. The latent embedding of the user is obtained by passing a user-specific learned scalar z through a latent embedding generator, which is also modeled as a feed-forward network. At train time, we obtain examples of 'tidy' and 'untidy' scenes, to train our preference model in a supervised manner. We train a universal model across the whole dataset, as well as user-conditioned models for each split of our dataset [7]. These examples contain the perception module's object features for all objects and receptacles, as well as the user ID and tidy/untidy label.

4 Evaluations

We evaluate our CLIP+CSR perception system with the ranking modules on the rearrangement task [8]. In this section, we first describe our setup and training details, and followed by metrics and results.

4.1 Experimental Setup

In this section, we discuss the details of data generation of observations using Housekeep with Habitat2.0 [9], user-specific preferences, and training of our CSR model.

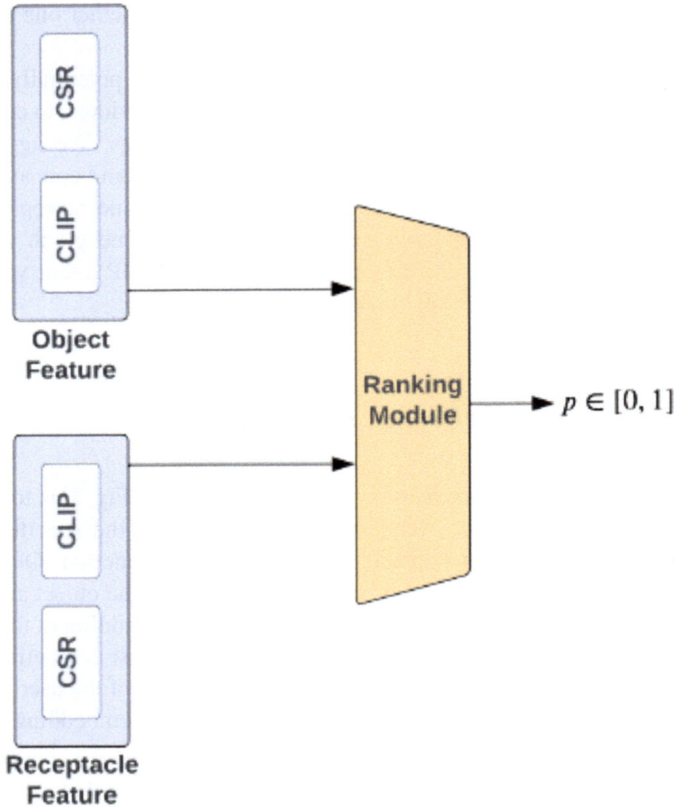

Fig. 3 Object-receptacle ranking module to predict the probability of a receptacle being a 'tidy' place for an object

Housekeep Data Collection Our code is based on the official repositories for Housekeep [13] which employs Habitat Lab as its underlying framework. To facilitate the collection of observations, we modified their Hierarchical Policy Runner to solely perform exploration. Specifically, we changed the image observation size to 512×512, and set the maximum frontier exploration steps to 1000. During exploration, we collected images every 10 steps and ran the exploration policy for 50 episodes. We only stored observations that had at least one object/receptacle present. We use the ihlen 1 int scene for our experiments [10]. We performed the observation collection on a single Nvidia A40 GPU [27].

CSR Training For CSR, we refer to the code for [1]. The code uses PyTorch Lightning as their framework. In our experiments, the backbone for encoders is a frozen ResNet18 pretrained on ImageNet. The learning rate is 0.001, weight decay is 0.0001, and we use Adam optimization. The batch size is set to 16 and the queue size is set to 256. We used a single Nvidia A40 GPU to train CSR encoders. The Housekeep benchmark [4] includes a set of objects annotated with crowdsourced

Fig. 4 Example visual observation collection by the exploration agent wherein the instance mask does not contain any class-level information of entities

(a) Sofa Chair (rec) (b) Bottom Cabinet (rec) (c) Lime Squeezer (obj)

Fig. 5 Example cropped and resized objects and receptacles from the observations

receptacle mappings. It supports rendering of episodes containing these objects in various scenes on the Habitat 2.0 simulator [27]. We use the frontier based exploration agent available in Housekeep to explore environments and obtain RGB-D observations with instance masks as shown in Fig. 4. We use the instance masks to generate bounding boxes, allowing a margin (1.5 times the original dimension) in order to have some background/context for the images. An example of the cropped objects/receptacles is shown in Fig. 5. We generate our train set using 50 episodes and 4300 RGB-D images containing 106 objects and 33 receptacles. Of these, we reserve 10 episodes for evaluation. We also only use 95 objects for training and validation [26], by hiding them from the seen episodes, and reserve the remaining 11 objects as unseen objects for testing. The object crops are used to train the CSR model, and the features resulting from the trained CSR and pretrained CLIP models with the tidy/untidy labels are used to train the object-receptacle preference model.

User-Specific Data Generation The Housekeep benchmark by Kant et al. also comes with a crowd-sourced dataset of annotated tidy room-receptacle pairs for individual object categories [25], with a total of 268 object categories and 128 room-receptacle pairs. The data is collected by giving each object-room pair to 10 different annotators and obtaining one of three labels for each receptacle in the room: correct, implausible, and misplaced. To evaluate the downstream rearrangement metrics via a learned object-receptacle mapping model, we first choose a set of 95 seen and 11 unseen objects, same as the splits for training CSR [11]. We then split the correct receptacle labels for each seen and unseen object-room pair by annotator ID, giving us 10 clusters of labels per object-room pair. Each cluster of a seen object-room pair is randomly assigned to a different user persona, resulting in 10 user personas overall [24]. To maintain similarity between the seen and unseen object preferences for a single user persona, we perform a linear sum assignment to map the label clusters of unseen object-room pairs labels to user personas using the Hungarian algorithm [12].

The following similarity measure was maximized by the linear sum assignment. Given a set of labels C_i^{unseen} for unseen object o_i^{unseen}, the similarity measure between C_i^{unseen} and user persona P_k is:

$$S_{\text{assignment}}(C_i^{\text{unseen}}, P_k) = \sum_{(o_j^{\text{seen}}, C_j^{k,\text{seen}}) \in P_k} \Big[f_{\text{semantic}}(o_i^{\text{unseen}}, o_j^{\text{seen}}) \times N(C_i^{\text{unseen}} \cap C_j^{k,\text{seen}}) \Big]$$

where C_i^{unseen}, $C_j^{k,\text{seen}}$ are sets (clusters) of receptacle labels for objects o_i^{unseen}, o_j^{seen} respectively [13], f_{semantic} is the measure of similarity between the semantic categories of objects (calculated from the cosine similarity of their BERT text embeddings), and N is the length of a set. The intuition behind this metric is that, for a given persona, semantically similar objects should be placed in the same kind of receptacles [14].

4.2 Evaluation Metrics and Results

We evaluate the effectiveness of our perception module on two downstream tasks, each constructed by training separate MLP models:

- **Predicting Object-Receptacle Spatial Relations**: The first downstream task is predicting the existence of a spatial relation between objects and receptacles in an environment from CSR embeddings. The CSR embeddings are collected from observations collected by an exploratory agent. This task evaluates the quality of geometric information available in CSR embeddings [15].
- **Learning Tidy Object-Receptacle Mapping**: The second downstream task is learning an object-receptacle mapping from the Housekeep Household dataset. For the scope of this course, we do not learn user-specific preferences and instead consider user-agnostic placement preferences. Note that the difference between this evaluation and the Housekeep benchmark is that the feature

representations of objects and receptacles in our evaluation are obtained from raw observations via CSR and CLIP, while Housekeep generates object and receptacle representations from ground truth semantic classes [16]. This task evaluates the ability of our perception model to distinguish between objects using instance-level and class-level features without additional supervision [17].

More concretely, we define evaluation metrics for each task:

1. **Edge Prediction Accuracy (EPA)**: Calculated as an accuracy metric of the spatial relation prediction model when evaluated on an excluded dataset of scenes containing seen and unseen object categories. This metric measures the fraction of objects for which their current locations are correctly identified.
2. **Preference Classification Score (PCS)**: Calculated as an accuracy metric of the object-receptacle mapping model when evaluated on an excluded dataset of scenes containing seen and unseen object categories. This metric measures the fraction of objects observed in the scene, for which the model correctly identified a target receptacle.

For the object-receptacle mapping task, we compare against a baseline approach that relies solely on CLIP embeddings [23]. This evaluation helps us understand the importance of instance-level features captured by CSR. Specifically, we train an MLP model that learns the object-receptacle mapping from CLIP visual embeddings generated from image crops of the object and receptacle [18].

Our test dataset consists of a held out set of 10 episodes. Tables 1 and 2 present the results obtained on either downstream task.

5 Conclusion

In this study, we explored the use of pretrained foundational models and unsupervised learning with viewpoint-invariant priors to develop an object-centric representation from egocentric observations. These representations capture instance-level

Table 1 Edge Prediction Accuracy (EPA) calculated after training an MLP model to predict the existence of a spatial relation between objects and receptacles from CSR node and edge embeddings

	Total	Seen Obj	Unseen Obj
EPA	27.1%	27.5%	25.0%

Table 2 Preference Classification Score (PCS) calculated after training an MLP model to predict tidy object-receptacle mapping scores from CSR node and CLIP embeddings

Method	Total	Seen Obj	Unseen Obj
Our system (CSR + CLIP)	49.0%	46.5%	62.5%
CLIP (vision only)	19.6%	18.6%	25.0%

information and demonstrate the ability to generalize to open-vocabulary objects [19]. Our evaluations focused on object rearrangement tasks, specifically identifying spatial relations between objects and receptacles and learning object-receptacle preferences. The results indicated that these representations could generalize to objects not encountered during training.

Our method for tidying tasks in personalized and open environments leverages geometric features learned through viewpoint invariance and visual features obtained using a frozen CLIP model. This approach, which consists of a perception module to recreate a map of the environment and a ranking module to predict rearrangement sequences, achieves notable accuracy in predicting correct object locations (27.1%) and identifying desired receptacles for objects (49.0%). However, several challenges emerged. Obtaining diverse training examples, crucial for effective learning, is feasible through simulation but requires careful consideration of factors like diverse viewpoints, object proximity, and object variety to ensure good resolution and generalization. Additionally, the training process needs careful hyperparameter tuning and coordinated training and finetuning of different system components. Despite our efforts [20], we couldn't fully optimize the system, indicating the need for further exploration in these areas.

Future work should improve the training setup and explore the system's full capabilities. While we tested generalization to unseen objects, evaluations should extend to objects not detected by object detectors. Testing sim2real or sim2sim generalization could provide insights into the transfer performance of the combined model, given the individual strengths of CSR and CLIP. Moreover, expanding evaluations beyond the object-receptacle preferences in the housekeep dataset to instance-specific preferences could reveal if geometric embeddings encoded by CSR capture necessary features for learning such preferences [21].

Rearrangement, being a complex problem, could benefit from incorporating more context [22], including verbal feedback and other objects. For instance, identical drawers might require different object placements based on their contents. User feedback in natural language could be integrated to refine and enhance the model's performance, leveraging the alignment of CLIP embeddings with text. This direction presents an exciting avenue for future research to improve object-rearrangement models.

References

1. S. Y. Gadre, K. Ehsani, S. Song, and R. Mottaghi, "Continuous scene representations for embodied ai," in *Proceedings of the IEEE/CVF Conference on Computer Vision and Pattern Recognition*, pp. 14849–14859, 2022.
2. C. Li, R. Zhang, J. Wong, C. Gokmen, S. Srivastava, R. Martín-Martín, C. Wang, G. Levine, M. Lingelbach, J. Sun, et al., "Behavior-1k: A benchmark for embodied ai with 1,000 everyday activities and realistic simulation," in *Conference on Robot Learning*, pp. 80–93, PMLR, 2023.
3. D. Batra, A. X. Chang, S. Chernova, A. J. Davison, J. Deng, V. Koltun, S. Levine, J. Malik, I. Mordatch, R. Mottaghi, et al., "Rearrangement: A challenge for embodied ai," arXiv preprint arXiv:2011.01975, 2020.

4. Y. Kant, A. Ramachandran, S. Yenamandra, I. Gilitschenski, D. Batra, A. Szot, and H. Agrawal, "Housekeep: Tidying virtual households using commonsense reasoning," in *Computer Vision–ECCV 2022: 17th European Conference, Tel Aviv, Israel, October 23–27, 2022, Proceedings, Part XXXIX*, pp. 355–373, Springer, 2022.
5. A. Szot, A. Clegg, E. Undersander, E. Wijmans, Y. Zhao, J. Turner, N. Maestre, M. Mukadam, D. Chaplot, O. Maksymets, A. Gokaslan, V. Vondrus, S. Dharur, F. Meier, W. Galuba, A. Chang, Z. Kira, V. Koltun, J. Malik, M. Savva, and D. Batra, "Habitat 2.0: Training home assistants to rearrange their habitat," in *Advances in Neural Information Processing Systems (NeurIPS)*, 2021.
6. D. Hall, B. Talbot, S. R. Bista, H. Zhang, R. Smith, F. Dayoub, and N. Sünderhauf, "The robotic vision scene understanding challenge," arXiv preprint arXiv:2009.05246, 2020.
7. L. Weihs, M. Deitke, A. Kembhavi, and R. Mottaghi, "Visual room rearrangement," in *Proceedings of the IEEE/CVF conference on computer vision and pattern recognition*, pp. 5922–5931, 2021.
8. C. Gan, S. Zhou, J. Schwartz, S. Alter, A. Bhandwaldar, D. Gutfreund, D. L. Yamins, J. J. DiCarlo, J. McDermott, A. Torralba, et al., "The threedworld transport challenge: A visually guided task-and-motion planning benchmark towards physically realistic embodied ai," in *2022 International Conference on Robotics and Automation (ICRA)*, pp. 8847–8854, IEEE, 2022.
9. M. Deitke, D. Batra, Y. Bisk, T. Campari, A. X. Chang, D. S. Chaplot, C. Chen, C. P. D'Arpino, K. Ehsani, A. Farhadi, et al., "Retrospectives on the embodied ai workshop," arXiv preprint arXiv:2210.06849, 2022.
10. G. Sarch, Z. Fang, A. W. Harley, P. Schydlo, M. J. Tarr, S. Gupta, and K. Fragkiadaki, "Tidee: Tidying up novel rooms using visuo-semantic commonsense priors," in *Computer Vision–ECCV 2022: 17th European Conference, Tel Aviv, Israel, October 23–27, 2022, Proceedings, Part XXXIX*, pp. 480–496, Springer, 2022.
11. I. Kapelyukh and E. Johns, "My house, my rules: Learning tidying preferences with graph neural networks," in *Conference on Robot Learning*, pp. 740–749, PMLR, 2022.
12. Y. Jiang, M. Lim, and A. Saxena, "Learning object arrangements in 3d scenes using human context," arXiv preprint arXiv:1206.6462, 2012.
13. W. Liu, T. Hermans, S. Chernova, and C. Paxton, "Structdiffusion: Object-centric diffusion for semantic rearrangement of novel objects," arXiv preprint arXiv:2211.04604, 2022.
14. Q. A. Wei, S. Ding, J. J. Park, R. Sajnani, A. Poulenard, S. Sridhar, and L. Guibas, "Lego-net: Learning regular rearrangements of objects in rooms," arXiv preprint arXiv:2301.09629, 2023.
15. A. Radford, J. W. Kim, C. Hallacy, A. Ramesh, G. Goh, S. Agarwal, G. Sastry, A. Askell, P. Mishkin, J. Clark, G. Krueger, and I. Sutskever, "Learning transferable visual models from natural language supervision," 2021. URL https://arxiv.org/abs/2103.00020.
16. H. Ha and S. Song, "Semantic abstraction: Open-world 3d scene understanding from 2d vision-language models," in *Conference on Robot Learning*, 2022.
17. K. M. Jatavallabhula, A. Kuwajerwala, Q. Gu, M. Omama, T. Chen, S. Li, G. Iyer, S. Saryazdi, N. Keetha, A. Tewari, et al., "Conceptfusion: Open-set multimodal 3d mapping," arXiv preprint arXiv:2302.07241, 2023.
18. M. Shridhar, L. Manuelli, and D. Fox, "Cliport: What and where pathways for robotic manipulation," in *Conference on Robot Learning*, pp. 894–906, PMLR, 2022.
19. S. Y. Gadre, M. Wortsman, G. Ilharco, L. Schmidt, and S. Song, "Clip on wheels: Zero-shot object navigation as object localization and exploration," arXiv preprint arXiv:2203.10421, 2022.
20. B. Chen, F. Xia, B. Ichter, K. Rao, K. Gopalakrishnan, M. S. Ryoo, A. Stone, and D. Kappler, "Open-vocabulary queryable scene representations for real world planning," arXiv preprint arXiv:2209.09874, 2022.
21. A. v. d. Oord, Y. Li, and O. Vinyals, "Representation learning with contrastive predictive coding," 2018. URL https://arxiv.org/abs/1807.03748.
22. A. Kirillov, E. Mintun, N. Ravi, H. Mao, C. Rolland, L. Gustafson, T. Xiao, S. Whitehead, A. C. Berg, W.-Y. Lo, et al., "Segment anything," arXiv preprint arXiv:2304.02643, 2023.

23. S. Ren, K. He, R. Girshick, and J. Sun, "Faster r-cnn: Towards real-time object detection with region proposal networks," *Advances in neural information processing systems*, vol. 28, 2015.
24. Z. Zhong, L. Sun, and Q. Huo, "An anchor-free region proposal network for faster r-cnn-based text detection approaches," *International Journal on Document Analysis and Recognition (IJDAR)*, vol. 22, pp. 315–327, 2019.
25. W. Jeong and K. M. Lee, "Cv-slam: A new ceiling vision-based slam technique," in *2005 IEEE/RSJ International Conference on Intelligent Robots and Systems*, pp. 3195–3200. IEEE, 2005.
26. P.-E. Sarlin, D. DeTone, T. Malisiewicz, and A. Rabinovich, "Superglue: Learning feature matching with graph neural networks," in *Proceedings of the IEEE/CVF conference on computer vision and pattern recognition*, pp. 4938–4947, 2020.
27. M. Savva, A. Kadian, O. Maksymets, Y. Zhao, E. Wijmans, B. Jain, J. Straub, J. Liu, V. Koltun, J. Malik, D. Parikh, and D. Batra, "Habitat: A Platform for Embodied AI Research," in *Proceedings of the IEEE/CVF International Conference on Computer Vision (ICCV)*, 2019.

Sentiment Analysis of Sarcastic Hindi Sentences: Analysis of SVM and XGBoost Models

Aagam Bharatkumar Modi, Nikita P. Desai, and Manav Jayesh Patel

Abstract Detecting sarcasm in text presents a significant challenge due to its tendency to subvert literal sentiment. This study proposes a novel approach for classifying sarcastic sentences as either positive or negative. We developed a machine learning model that integrates contextual and syntactic features to enhance sarcasm detection. A major contribution of this research is the creation of a specialized dataset of sarcastic sentences derived from movie reviews, specifically curated for this study. Our model, trained on this dataset, demonstrates superior performance over traditional sentiment analysis techniques in accurately identifying the underlying sentiment of sarcastic remarks. The enhanced precision of our approach holds substantial implications for applications in movie review analytics, social media analytics, customer feedback systems, and automated content evaluation.

Keywords Hindi language · Sarcastic movie reviews · ML models · SVM · XGBoost

1 Introduction

Sarcasm detection in Hindi text is complex, often requiring a deeper understanding of context and cultural nuances. Traditional models used for sentiment analysis often misclassify sarcastic content, leading to inaccuracies. This research focuses on building an effective model to accurately classify sarcastic Hindi sentences into positive and negative sentiments.

A. B. Modi (✉) · N. P. Desai · M. J. Patel
Department of Information Technology, Dharmsinh Desai University, Nadiad, Gujarat, India
e-mail: nikitadesai.it@ddu.ac.in

K. Deep et al. (eds.), *Artificial Intelligence, Machine Learning and Intelligent Systems*, Springer Proceedings in Mathematics & Statistics 520,
https://doi.org/10.1007/978-981-95-3841-6_11

2 Motivation for the Research Work

Recognition of sentiment in sarcasm is profoundly challenging because of the complexity of the Hindi language, which is rich in sound patterns. Traditional sentiment analysis models fail to interpret the subtle meanings and details present in Hindi. Improving the reliability of sentiment analysis systems therefore becomes much more imperative in applications that span areas like sarcastic movie review sentiment classification, customer feedback analysis or social media monitoring applications if the detectability of sarcasm is enhanced. However, cultural and context-specific features unique to the Hindi language would greatly improve the study, thus increasing the accuracy of the models. Moreover, it appears that this problem does not get enough address as most work on sarcasm detection has been focused on languages like English, while very little work has been done in Hindi. The current research addresses this gap and attempts to fulfill this urgent need to overcome the challenges specific to Hindi sarcasm. The main contributions of this research are:

- Manual curation of Hindi Movie Reviews Dataset [1] data to include sentences that contain strong clues of sarcasm and thereby let the focused analysis come out to train on a more fitting representation of the sarcastic expressions.
- Implementation of machine learning techniques such as Support Vector Machines (SVM) and XGBoost, and comparing their performance on Hindi text.

3 Literature Review

In 2023, Rahma, Azab, and Mohammed presented a comprehensive survey on sarcasm detection in Arabic texts, covering both machine learning and deep learning techniques. Their study highlighted key challenges, such as imbalanced datasets, lack of context, and the complexity of detecting implicit sarcasm, which remain unresolved issues in this field [2]. Work on sarcasm detection in Urdu tweets in 2024 utilized an mBERT-BiLSTM-MHA model, achieving effective results but requiring significant computational resources and a large dataset to function optimally [3].

Phukan, Pal, and Ekbal [4] introduced a Hybrid Quantum-Classical Neural Network (HQNN) that used a Variational Quantum Circuit (VQC) for multitask sarcasm, emotion, and sentiment analysis [4]. This quantum-classical approach showed promise, though scalability and implementation were limited by the early stage of quantum technology.

Zhang's [5] review on sarcasm detection methods highlighted various machine learning and deep learning approaches, noting that these methods demand high-end hardware and balanced datasets to be effective. It was suggested that future research could benefit from employing RNN-LSTM models for non-literal emotion detection and utilizing multimodal inputs to enhance sarcasm detection accuracy [5].

In 2022, Abdullah Amer and Siddiqui proposed a novel sarcasm detection algorithm using supervised machine learning. Although effective, they reported high computational demands in feature engineering, indicating a need for optimized feature selection for greater efficiency in sarcasm detection [6]. A hybrid model for sarcasm detection on social media combined Word2Vec, GloVe, and BERT embeddings with fuzzy logic. This approach encountered challenges with informal language, including slang, emojis, and subtle sarcasm, which affected the accuracy of sarcasm-based sentiment analysis [7].

Pandey et al. [8] leveraged a GRU-based framework with an attention mechanism and GloVe embeddings for sarcasm classification across multilingual social media posts [8]. However, the limited dataset size restricted the model's effectiveness. Li and colleagues [9] combined BERT with commonsense knowledge from COMET to improve sarcasm detection accuracy. They recommended enhancing knowledge selection and expanding dataset diversity to cover different types of sarcasm [9].

Chen et al. [10] implemented a Bi-GRU architecture with self-supervised learning to jointly detect sentimental clues and context incongruity in sarcasm detection. This model's limitations included a restricted sentiment lexicon and a simplified approach to detecting context incongruity, suggesting future improvements through semantic chunking and transformer integration [10].

In Hindi sarcasm detection research from 2016, an SVM model achieved an accuracy of 83% when clear sarcasm indicators like hashtags, emoticons, and punctuation were present [11], which served as explicit cues in identifying sarcasm accurately. A recent 2024 evaluation by Desai assessed GPT's effectiveness in detecting sarcasm within Hindi text, with the model achieving only 40% accuracy. This low success rate highlights current limitations in large language models for handling nuanced sarcasm in Hindi [12].

4 Evaluation Setup

We ran our experiments on Google Cloud infrastructure, optimizing each task according to a specific resource requirements. General preprocessing tasks were run with the CPU runtime that appeared to be reasonable in all standard data handling cases. For word embedding tasks involving IndicNLP embeddings, we utilized TPUs, which greatly benefited from massive memory and parallel processing capabilities, thus greatly accelerating computation.

The initial dataset utilized in this research is the Hindi Movie Reviews Dataset available on Kaggle [1]. The dataset comprises approximately 900 Hindi movie reviews obtained from Hindi news websites and is mainly used for sentiment analysis in Hindi. The reviews are categorized into three sentiment classes: positive, neutral, and negative. The experience column of the original dataset denotes the sentiment of each review with values [0,1,2] representing negative (0), neutral (1), and positive (2) sentiments, respectively.

The first dataset consisted of a collection of movie reviews presented in paragraph format. This is a useful source, but within this dataset, there were no sentence-level labels for sarcasm, so the task proved challenging to approach using the model as it stands. Given that sarcasm typically exists at the sentence level, not the paragraph level, the use of full review datasets may have detracted from the effectiveness of applying the model to more sentences than is necessary. Therefore, we addressed the limitation by manually curating a subset: isolating only those sentences that had clear indicators of sarcasm, thus enabling us to carry out a more focused analysis, where performance would improve by training on data that is closer in representation to the actual sarcastic expressions.

It thus generated a customized subset of Hindi Movie Reviews Dataset, strictly focusing on sarcastic sentences from Hindi movie reviews. Every sentence was re-evaluated, retaining those with all obvious elements of sarcasm and labeling them either positive or negative. The final dataset will amount to 217 sarcastic sentences—115 of them were labeled as negative and 102 as positive—and provides a targeted collection that can be used to more effectively do sarcasm detection in Hindi text by focusing solely on sarcastic content.

The dataset was divided into 70% training and 30% testing subsets. Model performance was evaluated using accuracy, precision, recall, and F1-score, allowing for a comprehensive assessment of each feature extraction and classification method's effectiveness in sarcasm detection for Hindi text (Fig. 1).

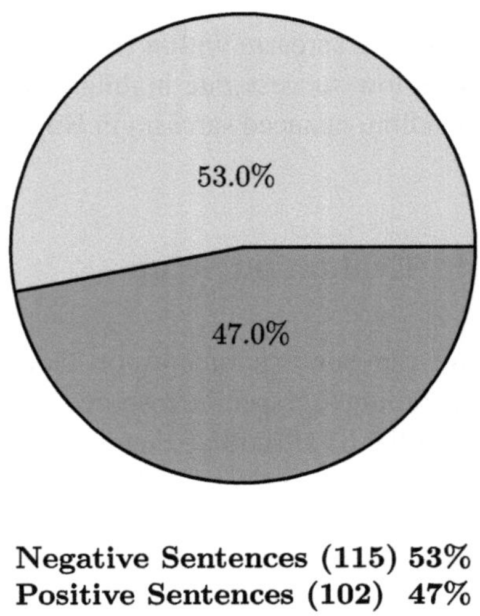

Fig. 1 Weighted distribution of sentences in the curated dataset

5 Methodology

This research focused on the detection of sentiment, whether it is positive or negative, within sarcastic sentences in Hindi using two models: Support Vector Machine(SVM)[1] and XGBoost.[2] The choice of SVM was primarily because of its efficiency with high-dimensional spaces. XGBoost was selected because complex patterns are handled well [13] for subtle cues about sentiment that lie within sarcastic content.

The methodology involved the following steps:

5.1 Data Preprocessing

- **Stopword Removal:** Common stopwords in the Hindi language were removed from the dataset. Stopwords generally hold very little informational meaning, and their removal gave more importance to words of significance both sentimentally as well as contextually.
- **Punctuation Handling:** All punctuation from the datasets was removed except for the question marks (?) and the exclamation marks (!), as these punctuation marks give an emotional emphasis or rhetorical clues that are very crucially important in the context of sarcasm and its intent.

5.2 Feature Vectorization

To capture various textual characteristics, we applied four different vectorization techniques:

- **One-Hot Encoding (OHE):** This basic encoding method represents words as binary vectors, indicating their presence or absence in a sentence.
- **TF-IDF (Term Frequency-Inverse Document Frequency):** By weighting words based on their frequency across the dataset, TF-IDF provided insights into the importance of words relevant to sentiment.
- **IndicNLP Embeddings**:[3] Custom embeddings for Indian languages, providing rich contextual representation specific to Hindi.
- **AI4BH Embeddings**:[4] A specialized embedding approach that captures semantic nuances and contextual depth in Hindi text.

[1] https://scikit-learn.org/stable/modules/generated/sklearn.svm.SVC.html.

[2] https://xgboost.readthedocs.io/en/stable/.

[3] https://indic-nlp-library.readthedocs.io/en/latest/.

[4] https://huggingface.co/ai4bharat/indic-bert.

5.3 Model Training and Prediction

Preprocessed and vectorized sentences were used to train SVM and XGBoost models for classifying positive/negative sentiment within sarcastic sentences. This cross-validation with different feature extraction techniques helps to determine effective configurations for accurately identifying sentiment within sarcastic content.

Through the adoption of this structured method, we aimed to devise an optimized sentiment detection pipeline for sarcastic Hindi sentences using various vectorization techniques and model architectures in a bid to improve the detection accuracy and to capture specific nuances of sarcasm during the analysis of sentiment.

6 Model Configurations

For convenience, we use the following notations to refer to different model configurations in our study:

$\mathtt{M}_1$ represents the Support Vector Machine (SVM), and $\mathtt{M}_2$ represents XGBoost. For word embedding techniques, we define $\mathtt{WE}_1$ as One-Hot Encoding, $\mathtt{WE}_2$ as TF-IDF, $\mathtt{WE}_3$ as IndicNLP embeddings, and $\mathtt{WE}_4$ as AI4BH embeddings.

Similarly, for text encoding techniques, $\mathtt{TE}_1$ refers to Text Encoding, and $\mathtt{TE}_2$ refers to TextWords Encoding, which are explained in the next section.

Using this notation, configurations such as $\mathtt{M}_1 + \mathtt{WE}_2 + \mathtt{TE}_1$ indicate the SVM model with TF-IDF word embedding and Text Encoding.

7 Implementation and Results

In our experiments, we utilized two distinct configurations for encoding the data: **text** and **text words**.

7.1 Text Configuration

With this approach, we first performed various embeddings on the preprocessed data and then trained the $\mathtt{M}_1$ and $\mathtt{M}_2$ models. Embedding techniques included $\mathtt{WE}_1$, $\mathtt{WE}_2$, $\mathtt{WE}_3$, and $\mathtt{WE}_4$. These embedding techniques turned the text into numerical vectors, thereby capturing the words' semantic meanings in a continuous vector space. This allowed us to evaluate contextual relationships between words, which was essential for effective sentiment analysis. Thus, a vectorized representation of the modified data was created, which was then fed into the model for further learning and evaluation, enabling an analysis of the sentence structure in our dataset.

7.2 Text Words Configuration

This encoding strategy was more detailed because the sentiment in the data needed to be represented. Thus, we came up with two different lists, a list of positive Hindi words/phrases and a list of negative Hindi words/phrases. For a better standardization of the representation of the sentiment, we wrote a Python script that replaced all positive words/phrases with one particular placeholder word, such as "positive" while all the negative words/phrases would be replaced with another placeholder, such as "negative". The neutral words remained unchanged so that the structure of the dataset was still maintained with emphasis on those elements that carry sentiment.

The transformed dataset had, in addition to the words that were unchanged as being neutral, two sentiment labels: "positive" and "negative." This way of encoding simplified the complexity of text data by targeting sentiment for analysis. Lastly, we apply word embedding to this newly structured data to allow sentiment models to assess sentiment much more easily using simplified representations. We trained and tested our models on this encoded data to check how well they detect sentiment in sarcastic sentences.

We employed these two settings and attempted to compare the performance of various encoding schemes for sentiment analysis, especially in the challenging domain of sarcastic Hindi text. The key findings from our experiments are summarized in the following points, with relevant graphs and tables presented on the next page:

- $\mathtt{M}_1$ with $\mathtt{WE}_4$ features achieved the highest accuracy (up to 87.8%), outperforming other combinations across all metrics.
- $\mathtt{M}_2$ with $\mathtt{WE}_4$ features performed best within $\mathtt{M}_2$ configurations, with a maximum accuracy of 78.8%.
- $\mathtt{WE}_2$ as a feature yielded the lowest performance in both $\mathtt{M}_1$ and $\mathtt{M}_2$ models, suggesting limitations in handling sarcasm detection in Hindi.
- $\mathtt{WE}_1$ showed the lowest accuracy when paired with $\mathtt{M}_2$, implying it may lack sufficient detail for sarcasm detection.
- **$\mathbf{TE}_1$ versus $\mathbf{TE}_2$**: Both $\mathtt{M}_1$ and $\mathtt{M}_2$ performed variably across feature sets, with $\mathtt{TE}_2$ representations generally boosting accuracy compared to simple $\mathtt{TE}_1$ representations.
- Across all feature configurations, $\mathtt{M}_1$ consistently outperformed $\mathtt{M}_2$, with up to a 9% advantage in accuracy when using $\mathtt{WE}_4$ and $\mathtt{WE}_3$ features (Figs. 2 and 3). Bold entries highlight peak accuracy and significant performance gains (Tables 1 and 2). Single bold values indicate the highest accuracy per model, with AI4BH Embeddings ($\mathtt{WE}_4$) achieving the top scores for both SVM ($\mathtt{M}_1$, 0.8780)

and XGBoost ($\mathtt{M}_2$, 0.7878). Crucially, the paired bold values demonstrate the superior efficacy of TextWords Encoding ($\mathtt{TE}_2$), which drove significant accuracy increments over $\mathtt{TE}_1$ – specifically boosting IndicNLP ($\mathtt{WE}_3$) by +0.06 in the SVM model and One-Hot Encoding ($\mathtt{WE}_1$) by +0.10 in the XGBoost model.

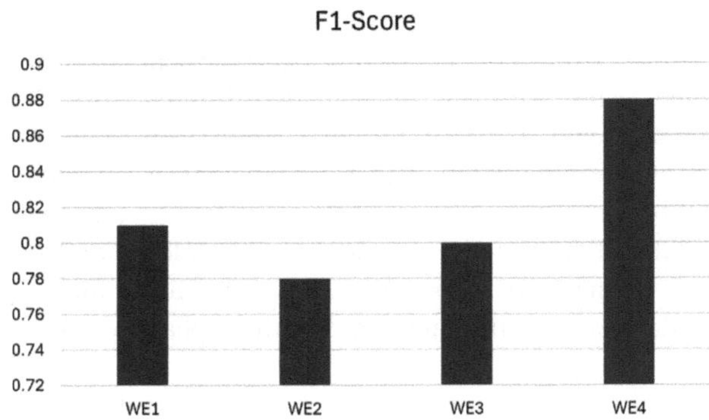

Fig. 2 F1-Score results for $\mathtt{M}_1$ based on different word embedding techniques

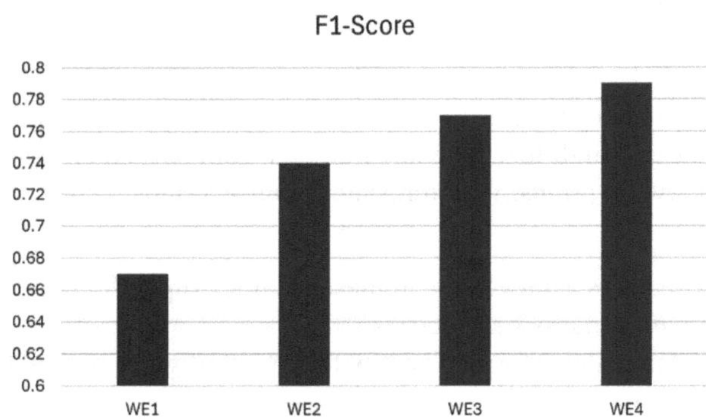

Fig. 3 F1-Score results for $\mathtt{M}_2$ based on different word embedding techniques

Table 1 $\mathtt{M}_1$ analysis using various word embedding techniques

Word embedding	$\mathtt{TE}_1$ accuracy	$\mathtt{TE}_2$ accuracy
$\mathtt{WE}_1$	0.8181	0.8030
$\mathtt{WE}_2$	0.7878	0.7121
$\mathtt{WE}_3$	**0.8030**	**0.8636**
$\mathtt{WE}_4$	**0.8780**	0.8330

Table 2 M_2 analysis using various word embedding techniques

Word embedding	TE_1 accuracy	TE_2 accuracy
WE_1	**0.6666**	**0.7727**
WE_2	0.7424	0.7424
WE_3	0.7727	0.7727
WE_4	0.7878	**0.7878**

8 Conclusion

In this research, we analyzed the performance of M_1 and M_2 algorithms for sarcasm detection in Hindi text. We compared the performance of these algorithms over different feature extraction techniques, specifically WE_4, WE_3, WE_2, and WE_1, analyzing how each feature set impacted accuracy, precision, recall, and F1-score in sarcasm detection. The evaluation was conducted using both TE_1 and TE_2 representations, allowing us to understand the behavior of each model.

We conducted a series of experiments to observe the performance of M_1 and M_2 models across various feature sets. The M_1 model, when combined with WE_4 features, was the top-performing configuration, achieving accuracy between 83.3% and 87.8%. It also demonstrated the highest precision, recall, and F1 score for both TE_1 and TE_2 representations. WE_3 features also yielded strong results with M_1, achieving 80.3% accuracy with TE_1 and 86.3% with TE_2. In contrast, M_1 with WE_2 resulted in the lowest performance, suggesting that WE_2 may be less effective for detecting sarcasm in Hindi. M_2, when paired with WE_4 features, achieved its best performance, reaching a precision, recall, and F1-score of 0.79 with an accuracy of 78.8%. A moderate performance was observed with M_2 combined with WE_2 (accuracy of 74.2%), while the lowest accuracy was recorded with M_2, when combined with WE_1, indicating that WE_1 may not effectively capture nuances of Hindi sarcasm. Overall, M_1 consistently outperformed M_2 across all feature sets. M_1 achieved a maximum accuracy of 87.8% with WE_4 features, while the highest accuracy for M_2 was 78.8%. M_1 demonstrated a performance advantage of up to 9% in accuracy over M_2 in configurations using WE_4 and WE_3 features.

Future research could explore various directions to enhance sarcasm detection in Hindi, such as expanding the feature set to include more sophisticated linguistic features like sentiment scores, part-of-speech tagging, and dependency parsing, which could improve model performance by better capturing the nuances of sarcasm in Hindi. Additionally, combining traditional machine learning models, such as M_1 and M_2, with deep learning methods like transformers and LSTM networks may improve context capture. Since sarcasm on social media frequently includes images or emojis, future studies could also incorporate multimodal data analysis that combines text with images or emoticons to improve sarcasm detection accuracy.

References

1. Gaurav, "Hindi Movie Reviews Dataset." [Online]. Available: https://www.kaggle.com/datasets/disisbig/hindi-movie-reviews-dataset
2. A. Rahma, S. S. Azab, and A. Mohammed, "A comprehensive survey on Arabic sarcasm detection: approaches, challenges and future trends," *IEEE Access*, vol. 11, pp. 18 261–18 280, 2023.
3. M. E. Hassan, M. Hussain, I. Maab, U. Habib, M. A. Khan, and A. Masood, "Detection of sarcasm in Urdu tweets using deep learning and transformer based hybrid approaches," *IEEE Access*, 2024.
4. A. Phukan, S. Pal, and A. Ekbal, "Hybrid quantum-classical neural network for multimodal multitask sarcasm, emotion, and sentiment analysis," *IEEE Transactions on Computational Social Systems*, 2024.
5. Z. Zhang, "Sarcasm detection methods based on machine learning and deep learning," Applied and Computational Engineering, vol. 36, pp. 119–127, 2024.
6. A. Y. Abdullah Amer and T. Siddiqu, "A novel algorithm for sarcasm detection using supervised machine learning approach." *AIMS Electronics & Electrical Engineering*, vol. 6, no. 4, 2022.
7. D. K. Sharma, B. Singh, S. Agarwal, N. Pachauri, A. A. Alhussan, and H. A. Abdallah, "Sarcasm detection over social media platforms using hybrid ensemble model with fuzzy logic," *Electronics*, vol. 12, no. 4, p. 937, 2023.
8. R. Pandey, A. Kumar, J. P. Singh, and S. Tripathi, "A hybrid convolutional neural network for sarcasm detection from multilingual social media posts," *Multimedia Tools and Applications*, pp. 1–29, 2024.
9. J. Li, H. Pan, Z. Lin, P. Fu, and W. Wang, "Sarcasm detection with commonsense knowledge," *IEEE/ACM Transactions on Audio, Speech, and Language Processing*, vol. 29, pp. 3192–3201, 2021.
10. W. Chen, F. Lin, X. Zhang, G. Li, and B. Liu, "Jointly learning sentimental clues and context incongruity for sarcasm detection," *IEEE Access*, vol. 10, pp. 48 292–48 300, 2022.
11. N. Desai and A. D. Dave, "Sarcasm detection in Hindi sentences using support vector machine," *International Journal*, vol. 4, no. 7, pp. 8–15, 2016.
12. N. Desai, "Performance evaluation of GPT for sentiment analysis of movie reviews in Indian Languages," Jun. 2024. [Online]. Available: https://doi.org/10.5281/zenodo.11408108
13. T. Chen and C. Guestrin, "Xgboost: A scalable tree boosting system," in *Proceedings of the 22nd ACM SIGKDD International Conference on Knowledge Discovery and Data Mining. New York, NY, USA: Association for Computing Machinery*, 2016, pp. 785–794.

Enhanced Intrusion Detection for Cloud-Integrated IoT Networks Using Autoencoder-Driven Unsupervised Deep Learning

K. Giri Babu and J. Bhavana

Abstract With the increasing integration of IoT devices into cloud environments, robust Network Intrusion Detection (NID) systems are critical for ensuring data security and network resilience within these interconnected infrastructures against sophisticated cyber threats. Conventional NID methods often struggle to keep pace with the dynamic and complex traffic patterns in cloud-connected IoT networks, necessitating advanced deep learning techniques for effective threat detection and anomaly identification. This research provides a comprehensive performance and convergence analysis of unsupervised autoencoder models, including dense autoencoder, variational autoencoder (VAE), long short-term memory autoencoder (LSTM-AE), and convolutional neural network autoencoder (CNN-AE), for NID tasks. We experiment with a standard dataset to analyze the convergence properties by evaluating the behavior of the reconstruction loss function $L(\theta)$. We evaluate the anomaly detection efficacy of each model utilizing F1 scores and accuracy. The results show that the VAE is consistently better than other models for numerical stability. It has the smoothest, most monotonic de-crease in $L(\theta)$, the least variation in the gradient norm $\nabla L(\theta)$, and the fastest convergence. Furthermore, the VAE attains the highest F1 score of 89.64%, closely followed by the LSTM-AE at 86.50%. The CNN-AE and dense autoencoder models, albeit successful, have slower convergence and marginally reduced detection accuracy. This thorough research highlights the trade-offs between convergence behavior and detection performance, establishing the VAE as the most resilient and effective model for real-time NID applications, especially in settings requiring strong numerical stability and swift model convergence.

Keywords Network intrusion detection (NID) · Autoencoder · Anomaly detection · Convergence stability · Deep learning

---

K. Giri Babu (✉) · J. Bhavana
Department of Computer Science and Artificial Intelligence, SR University, Warangal, Telangana, India
e-mail: 2303C50118@sru.edu.in; j.bhavana@sru.edu.in

K. Deep et al. (eds.), *Artificial Intelligence, Machine Learning and Intelligent Systems*, Springer Proceedings in Mathematics & Statistics 520,
https://doi.org/10.1007/978-981-95-3841-6_12

1 Introduction

Security is very crucial in the fast-evolving world of networked systems, especially as networks become increasingly complex and vulnerable to advanced cyberattacks. These networks handle large amounts of diverse data at high speeds, making them vulnerable to multiple attacks such as DDoS, malware infections, and unauthorized access [1]. Traditional NID systems, especially signature-based ones, use attack signatures to detect malicious activities. Traditional systems can recognize prior attacks, but they struggle to react to new threats, such as zero-day vulnerabilities and advanced persistent threats (APTs), which use novel attack vectors [2].

Anomaly-based detection approaches that focus on network activity anomalies, are more flexible and proactive [3]. For indicating an intrusion, we need modern algorithms that can learn complex representations and detect minor changes. This will effectively model complex, high-dimensional network data for anomaly detection [4]. Unsupervised learning algorithms, especially deep learning ones are powerful for this. Although promising, NID has not properly examined the performance and convergence dynamics of various autoencoder configurations [5]. The performance and convergence of unsupervised autoencoder architectures for NID tasks are examined in this paper. Dense, variational, long short-term memory, and CNN-AE autoencoders are tested for their ability to discover network anomalies.

Numerical stability and convergence characteristics are our main concerns. Training autoencoders minimizes a reconstruction loss function, usually measured by mean squared error (MSE) between input and reconstruction [6]. The practicality of these models in real-time intrusion detection depends on numerical stability and loss function convergence during training. We analyze the smoothness of the loss reduction trajectory, gradient magnitude variance, and iterations needed to attain a loss threshold to evaluate convergence. These measurements reveal each autoencoder model's computational efficiency and dependability [7]. This research compares deep autoencoder designs for different NID aspects.

Autoencoders are trained on ordinary traffic data and only identify anomalies using reconstruction mistakes. The models' convergence and detection skills are thoroughly assessed to understand their performance in diverse network scenarios. This study emphasizes convergence stability and processing efficiency, which are necessary for implementing these models in operational NID systems. This study also illuminates the trade-offs between anomaly detection and convergence. It also provides crucial information on selecting and using autoencoder models in real-world NID applications.

2 Literature Review

This section provides an in-depth analysis of deep learning models for Network Intrusion Detection (NID). Table 1 presents a tabular overview, highlighting key findings, advantages, and limitations of recent relevant approaches.

Table 1 Comparative evaluation of various network intrusion detection studies

Study	Techniques	Key metrics	Dataset	Description
[8]	VAE + WGAN, stacked LSTM, MSCNN	Acc: 83.45% (NSL-KDD), >98.9% (AWID)	NSL-KDD, AWID	Hybrid model for balanced dataset generation and multi-scale feature extraction
[9]	VAWGAN + Gazelle optimization	Recall +17.58% to +23.18%, AUC +21.63%	NSL-KDD	Cloud intrusion detection using feature selection and data preprocessing
[10]	Convolutional autoencoder + DNN	Acc: 79.18%	NSL-KDD, NSL-KDD Test+	Integrates attention for feature selection, addressing class imbalances
[11]	Optimal feature selection, semi-supervised learning	Acc: 99% (MQTT-IoT-IDS2020), 97% (CIC-IDS2017)	MQTT-IoT-IDS2020, CIC-IDS2017	Lightweight model optimized for IoT systems, effective under resource constraints
[12]	Lightweight VAE + autoregression + permutation entropy	F1-score: 84.81%	Industrial Control Systems (ICS)	Efficient anomaly detection for ICS with low resource consumption
[13]	DCNN	Acc: 99.79–100%	ISCX-IDS 2012, DDoS, CICIDS2017, CICIDS2018	High accuracy and low false positives via deep learning and GPU acceleration
[14]	Autoencoder + modified PSO	Acc: 98.8% (UNSW-NB15), 99.22% (BoT-IoT)	UNSW-NB15, BoT-IoT	Combines feature selection with DNN for robust IoT cyber threat detection
[15]	Supervised sparse autoencoder	Acc: 90.11% (NSL-KDDTest+), 91.21% (CI-CIDS2017)	NSL-KDD, KDDCUP'99, CICIDS2017	Improves detection rates through enhanced feature extraction in supervised classification
[16]	Unsupervised anomaly detection with GAN	ROC-AUC: 0.99, ROC-EER: 0.035	MQTTset	Identifies anomalies in IoT data using adversarial training
[17]	Class-wise focal loss VAE	Acc: 88.08%, FPR: 3.77%	NSL-KDD	Addresses class imbalance by generating novel samples for minority attack classes

A novel intrusion detection system by Li et al. [8] combines a VAE with a Wasserstein Generative Adversarial Network (WGAN) to generate balanced traffic data. The model outperforms current techniques in accuracy, precision, recall, and F1-score tests on the NSL-KDD and AWID datasets, with an accuracy of 83.45% on NSL-KDD and reaching 98.9% on AWID. Senthilkumar et al.'s [9] proposed network detects cloud incursions using Variational Autoencoder Wasserstein Generative Adversarial Networks and Gazelle Optimization Algorithm. The Archerfish Hunting Optimizer selects NSL-KDD features after data pre-processing to reduce redundancy and handle missing values. The proposed method outperforms exsisting methods in recall (17.58-23.18%), AUC (21.63%), and calculation times.

Sivasubramanian et al. [10] introduce CAAE-DNN, a network intrusion detection model that uses a Convolutional Auto-Encoder, attention mechanism, and Deep Neural Network (DNN) for feature extraction and threat categorization. Using a 90:10 train-test split and the NSL-KDD Test+ dataset, CAAE-DNN achieves 79.18% classification accuracy and ROC-AUC analysis reveals class-specific efficacy. Otokwala et al. [11] introduce OCFSDA, a lightweight intrusion detection model for IoT embedded devices with memory constraints and cyber threats. To improve performance, OCFSDA uses optimal common feature selection, data compression, pruning, and deparameterization. The results show high computational efficiency (99% and 97% accuracy) with low memory consumption (2KB) and fast inference speeds (0.30 and 0.12 s).

The lightweight, unsupervised intrusion detection model LVA-SP by Ren et al. [12] is designed for industrial control systems (ICSs) vulnerable to public network threats. The model uses Spectral Residual (SR) for preliminary data processing, a lightweight variational autoencoder (LVA) with autoregression for data reconstruction, and permutation entropy (PE) for anomaly identification to reduce ICS resource usage. LVA-SP is designed for efficient detection accuracy and reduced system overhead, with an F1-score of 84.81% and reduced time and memory usage, making it ideal for resource-constrained ICS environments.

Hnamte et al. [13] propose a deep CNN (DCNN)-based intrusion detection system (IDS) to detect complex cyber threats. With GPU acceleration, the model achieves detection accuracies of 99.79% to 100%, surpassing conventional methods. HAEMPSO, a hybrid intrusion detection system by Saheed et al. [14], uses Autoencoder with Modified Particle Swarm Optimization (PSO) for feature selection and a Deep Neural Network (DNN) for classification to protect IoT against cyberattacks. On the UNSW-NB15 and BoT-IoT datasets, HAEMPSO performed well, achieving 98.8% accuracy and 99.9%. The strategy works against advanced IoT cyber threats.

For 5G IoT Network Intrusion Detection Systems (NIDS), Tu et al. [18] offer a pseudo-siamese stacked autoencoder (PSSAE). Using KDDTest+ and KDDTest-21 datasets, the PSSAE outperformed current methods with accuracy gains of 2-15%, detection rate improvements of 1-14%, and F1-score improvements of up to 11.55%. Idrissi et al. [16] offer "EdgeIDS" a generative adversarial network-trained unsupervised anomaly detection method for IoT devices. With ROC-AUC of 0.99 and ROC-EER of 0.035 on the MQTTset dataset of five attack types, EdgeIDS performed well. EdgeIDS can detect IoT cyber anomalies, according to the results.

Khanam et al. [17] introduce the Class-wise Focal Loss Variational AutoEncoder (CFLVAE) to balance IoT intrusion detection.

Novel minority attack class samples and a Class-wise Focal Loss (CFL) target boost training efficacy in this model. In studies using the imbalanced NSL-KDD dataset, the CFLVAE-DNN model outperforms other methods in detecting low-frequency attacks (79.25% for U2R and 67.5% for R2L) with an overall detection accuracy of 88.08% and a false positive rate of 3.77%.

3 Data and Methods

3.1 Dataset

The UNSW-NB15 dataset [19], used for network intrusion detection, consists of 257,673 samples across 10 classes (9 attack types and 1 benign class) with 49 features per instance, including flow-based, content-based, and time-based attributes. The dataset is divided into a training set of 175,341 samples and a test set of 82,332 samples, allowing for balanced model evaluation across diverse network behaviors. The distribution includes 30% test traffic and 70% train samples, with attack classes such as DoS, reconnaissance, and exploit representing the majority which are clubbed into a single class.

3.2 Dense Autoencoder Models

Autoencoders in general are unsupervised neural networks that compress and reconstruct data, comprising an encoder and a decoder [20]. The encoder creates reduced-dimensional latent representations, while the decoder reconstructs the input, as seen in Fig. 1.

This approach effectively captures essential data features and reduces noise, making autoencoders valuable for Network Intrusion Detection (NID). By training on benign traffic, the autoencoder minimizes reconstruction error. If the error

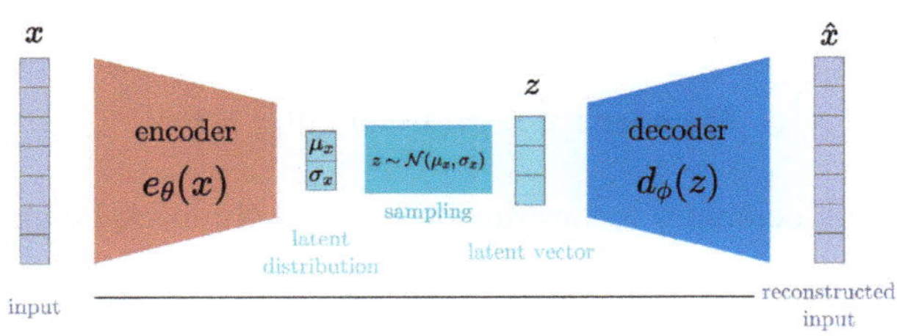

Fig. 1 Architecture of Autoencoder Neural Network with encoder and decoder sections. [21]

exceeds a threshold, it indicates a potential anomaly. The effectiveness of an NID system hinges on the autoencoder architecture, with variations including Dense, Variational, LSTM, and CNN autoencoders. A Dense Autoencoder is a feedforward neural network with fully connected layers, featuring:

$$\textbf{Encoder}: z = f(x) = \sigma\left(W_{1x} + b_1\right) \tag{1}$$

$$\textbf{Decoder}: \hat{x} = g(z) = \sigma\left(W_{2z} + b_2\right) \tag{2}$$

where:

- $x \in R^n$ is the input vector of network traffic features.
- $z \in R^m$ is the latent representation ($m < n$).
- $W_1 \in R^{m \times n}$, $W_2 \in R^{n \times m}$ are weight matrices.
- b_1, b_2 are bias vectors.
- $\sigma(\cdot)$ is an activation function (e.g., ReLU or sigmoid).

The training objective is to minimize reconstruction loss:

$$L\left(\vartheta\right) = \frac{1}{N}\sum_{i=1}^{N}\left\|x_i - \hat{x}_i\right\|^2 \tag{3}$$

Anomalies are detected when:

$$\left\|x_i - \hat{x}_i\right\|^2 > \epsilon \tag{4}$$

3.3 VAE

The VAE extends the standard autoencoder using a probabilistic framework. It learns parameters of a distribution over the latent space:

$$\textbf{Encoder}: \mu = W_\mu x + b_\mu, \;\; \sigma = W_\sigma x + b_\sigma \tag{5}$$

$$\textbf{Sampling}: z = \mu + \sigma \odot \epsilon, \; \epsilon \sim \mathrm{N}\;(0, I) \tag{6}$$

$$\textbf{Decoder}: \hat{x} = g(z) = \sigma\left(W_2 z + b_2\right) \tag{7}$$

The loss function combines reconstruction loss and Kullback-Leibler (KL) divergence:

$$L\left(\vartheta\right) = E_{q(z|x)}\left[\left\|x - \hat{x}\right\|^2\right] + D_{KL}\left(q\left(z|x\right)\left\|p(z)\right.\right) \tag{8}$$

with

$$D_{KL}\left(q\left(z|x\right)\Big\|p(z)\right) = \int q\left(z|x\right)\log\frac{q\left(z|x\right)}{p(z)}dz \tag{9}$$

3.4 LSTM-AE

LSTM-AE is tailored for sequential data analysis, utilizing LSTM units to capture long-range dependencies:

$$\textbf{Encoder}: h_t, c_t = \text{LSTM}\left(x_t, h_{t-1}, c_{t-1}\right) \tag{10}$$

$$\textbf{Decoder}: \hat{x}_t = g\left(h_T\right) \tag{11}$$

The training objective is:

$$L\left(\vartheta\right) = \frac{1}{T}\sum_{t=1}^{T}\left\|x_t - \hat{x}_t\right\|^2 \tag{12}$$

Anomalies are identified based on reconstruction error.

3.5 CNN-AE

CNN-AE employs convolutional layers to extract spatial features, suitable for high-dimensional data:

$$\textbf{Encoder}: z = f(x) = \text{Conv}(x) \rightarrow \text{Activation} \rightarrow \text{Pooling} \tag{13}$$

$$\textbf{Decoder}: \hat{x} = g(z) = \text{TransposeConv}(z) \rightarrow \text{Activation} \tag{14}$$

The loss function mirrors that of the dense autoencoder:

$$L\left(\vartheta\right) = \frac{1}{N}\sum_{i=1}^{N}\left\|x_i - \hat{x}_i\right\|^2 \tag{15}$$

Anomalies are detected via the reconstruction error calculated using the L_2 norm.

Autoencoders serve as powerful tools for unsupervised NID, each model presenting unique advantages. The dense autoencoder provides a straightforward implementation, while the VAE enhances robustness through probabilistic modeling [22]. The LSTM-AE excels in capturing temporal patterns, and the CNN-AE is adept at identifying spatial features [23]. Each model's performance in anomaly detection is significantly influenced by its architectural choices and the mathematical formulations guiding its training process.

3.6 Trade-off Analysis

This analysis reveals a trade-off between convergence behavior and anomaly detection performance. The VAE's superior stability and rapid convergence correspond with its leading performance metrics, making it optimal for real-time NID applications. The LSTM-AE, while effective, may require further optimization for faster convergence. In contrast, the CNN-AE and dense autoencoder may not be the best choices in scenarios demanding quick and accurate detection.

4 Result and Discussion

All experiments were conducted on Google Colab, utilizing an NVIDIA Tesla T4 GPU, 12GB RAM, and Intel Xeon CPU, providing a stable environment for training and testing. The models were trained using holdout training set randomly sampled with 70% of the data and 30% for testing. We use key metrics such as F1 Score, Precision, Recall, Accuracy, Training Time, and Convergence Ratio (Training Time per Epoch) to evaluate each model's anomaly detection efficacy and efficiency.

In terms of training performance, the VAE model demonstrated superior stability and efficiency, achieving the highest F1 Score of 90.82%, closely followed by the LSTM-AE at 88.19%, CNN-AE at 87.57%, and Dense AE at 85.43%. Additionally, the VAE attained the highest Accuracy (91.50%) and demonstrated smooth, monotonic convergence of the reconstruction loss $L(\theta)$ with minimal gradient variation, indicating robust learning dynamics. The VAE's convergence ratio was 13.06 s/epoch, reflecting fast model convergence with minimal computational cost, whereas the Dense AE model, with the lowest convergence ratio of 12.41 s/epoch, required significantly more epochs and exhibited slower progress in reducing $L(\theta)$. Test performance metrics further reinforced the VAE's effectiveness. Achieving an F1 Score of 89.64% and an Accuracy of 90.30%, the VAE outperformed other models in generalization capabilities, with stable Precision and Recall of 89.20% and 90.10%, respectively. The LSTM-AE and CNN-AE models, while competitive, yielded F1 Scores of 86.50% and 85.72%, respectively, with a marginally lower accuracy than the VAE. Dense AE displayed the lowest test performance, with an F1 Score of 83.21% and an Accuracy of 84.50%, suggesting limited efficacy in capturing the dataset's complex patterns. These results establish VAE as the most reliable and effective model, showcasing a balance between high detection accuracy, efficient convergence, and robust stability. The training and testing results along with convergence statistics can be summarised in Tables 2 and 3.

Table 2 Performance and convergence metrics of various autoencoder models for network intrusion detection

Model	F1 score (%)	Accuracy (%)	Precision (%)	Recall (%)	Training time (s)	Epochs	Convergence ratio (s/Epoch)
Dense AE	85.43	86.21	84.20	86.80	720	58	12.41
LSTM AE	88.19	89.02	87.00	89.40	830	52	15.96
CNN-AE	87.57	88.71	86.90	88.20	765	46	16.63
VAE	90.82	91.50	90.10	91.70	640	49	13.06

Table 3 Test performance metrics of various autoencoder models for network intrusion detection

Model	F1 score (%)	Accuracy (%)	Precision (%)	Recall (%)
Dense AE	83.21	84.50	82.10	84.35
LSTM AE	86.50	87.62	85.70	87.30
CNN-AE	85.72	86.90	85.10	86.35
VAE	89.64	90.30	89.20	90.10

5 Conclusion

This research underscores the VAE's superiority in the unsupervised autoencoder-based models for network intrusion detection, proving its worth in both detection accuracy and convergence efficiency. By leveraging a probabilistic latent space, the VAE effectively models complex and nonlinear dependencies in network traffic data, resulting in enhanced feature representation and improved anomaly detection performance. This probabilistic framework not only ensures smoother gradient trajectories, minimizing oscillations in the optimization process, but also accelerates convergence with a reduced computational footprint, achieving an F1 score of 89.64% and a convergence ratio of 13.06 s/epoch.

In contrast with it, deterministic approaches such as dense and CNN autoencoders, which rely on rigid latent representations, exhibited slower convergence and struggled to generalize across diverse traffic patterns. Furthermore, the LSTM-AE, while effective, showed higher computational demands, limiting its scalability in real-time scenarios. The VAE's ability to adapt to varying traffic types highlight its robustness, scalability, and suitability for modern intrusion detection systems. Particularly in cloud-connected IoT environments where numerical stability and rapid convergence are critical. These findings position the VAE as an important model for advancing the field of intrusion detection by bridging the gap between computational efficiency and high detection fidelity.

6 Future Work

As a future research work, we can explore the integration of hybrid autoencoder architectures, combining the strengths of models such as VAEs and LSTM-AEs. This will further enhance anomaly detection in dynamic and high-dimensional

IoT network environments. Additionally, expanding the scope to include semi-supervised or self-supervised learning paradigms could improve detection accuracy by leveraging partially labeled datasets. A another promising direction would be to investigate the impact of transfer learning to adapt models across heterogeneous network datasets and real-world traffic scenarios. Finally, the development of lightweight VAE-based models optimized for deployment on edge devices could significantly advance intrusion detection systems, enabling real-time analysis with minimal resource consumption in decentralized IoT ecosystems.

Acknowledgements The authors acknowledge the support from SR University for the completion of this study and experiments.

References

1. Yin, S., Li, H., Laghari, A.A., Gadekallu, T.R., Sampedro, G.A., Almadhor, A.: An anomaly detection model based on deep auto-encoder and capsule graph convolution via sparrow search algorithm in 6g internet-of-everything. IEEE Internet of Things Journal (2024)
2. Andresini, G., Appice, A., Malerba, D.: Autoencoder-based deep metric learning for network intrusion detection. Information Sciences **569**, 706–727 (2021)
3. Verma, A., Amin, S.B., Naeem, M., Saha, M.: Detecting covid-19 from chest computed tomography scans using ai-driven android application. Computers in biology and medicine **143**, 105298 (2022)
4. Verma, A., Singh, N., Khanna, V., Singh, B.P., Singh, N.P.: Automated tongue contour extraction from ultrasound sequences using signal enhancing neural network and energy minimized spline. Multimedia Tools and Applications **83**(19), 57511–57530 (2024)
5. Mekemte, L.A.K., Chalhoub, G.: On the use of autoencoders in unsupervised learning for intrusion detection systems. In: International Symposium on Ubiquitous Networking. pp. 54–69. Springer (2023)
6. Talaei Khoei, T., Kaabouch, N.: A comparative analysis of supervised and unsupervised models for detecting attacks on the intrusion detection systems. Information **14**(2), 103 (2023)
7. Thiyam, B., Dey, S.: An improved deep autoencoder-based network intrusion detection system with enhanced performance. International Journal of Internet Technology and Secured Transactions **13**(3), 270–290 (2024)
8. Li, Z., Huang, C., Qiu, W.: An intrusion detection method combining variational auto-encoder and generative adversarial networks. Computer Networks **253**, 110724 (2024)
9. Senthilkumar, G., Tamilarasi, K., Periasamy, J.: Cloud intrusion detection framework using variational auto encoder wasserstein generative adversarial network optimized with archerfish hunting optimization algorithm. Wireless Networks **30**(3), 1383–1400 (2024)
10. Sivasubramanian, A., Devisetty, M., Bhavukam, P.: Feature extraction and anomaly detection using different autoencoders for modeling intrusion detection systems. Arabian Journal for Science and Engineering pp. 1–13 (2024)
11. Otokwala, U., Petrovski, A., Kalutarage, H.: Optimized common features selection and deep-autoencoder (ocfsda) for lightweight intrusion detection in internet of things. International Journal of Information Security pp. 1–23 (2024)
12. Ren, Y., Feng, K., Hu, F., Chen, L., Chen, Y.: A lightweight unsupervised intrusion detection model based on variational auto-encoder. Sensors **23**(20), 8407 (2023)
13. Hnamte, V., Hussain, J.: Dependable intrusion detection system using deep convolutional neural network: A novel framework and performance evaluation approach. Telematics and Informatics Reports **11**, 100077 (2023)

14. Saheed, Y.K., Usman, A.A., Sukat, F.D., Abdulrahman, M.: A novel hybrid autoencoder and modified particle swarm optimization feature selection for intrusion detection in the internet of things network. Frontiers in Computer Science **5**, 997159 (2023)
15. Ghorbani, A., Fakhrahmad, S.M.: A deep learning approach to network intrusion detection using a proposed supervised sparse auto-encoder and svm. Iranian Journal of Science and Technology, Transactions of Electrical Engineering **46**(3), 829–846 (2022)
16. Idrissi, I., Azizi, M., Moussaoui, O.: An unsupervised generative adversarial network based-host intrusion detection system for internet of things devices. Indones. J. Electr. Eng. Comput. Sci **25**(2), 1140–1150 (2022)
17. Khanam, S., Ahmedy, I., Idris, M.Y.I., Jaward, M.H.: Towards an effective intrusion detection model using focal loss variational autoencoder for internet of things (IOT). Sensors **22**(15), 5822 (2022)
18. Tu, S., Waqas, M., Badshah, A., Yin, M., Abbas, G.: Network intrusion detection system (nids) based on pseudo-siamese stacked autoencoders in fog computing. IEEE Transactions on Services Computing (2023)
19. Nour Moustafa, Jill Slay: Unsw-nb15 (2024). https://doi.org/10.34740/KAGGLE/DSV/9350725, https://www.kaggle.com/dsv/9350725
20. Ortega-Fernandez, I., Sestelo, M., Burguillo, J.C., Piñón-Blanco, C.: Network intrusion detection system for ddos attacks in ics using deep autoencoders. Wireless Networks **30**(6), 5059–5075 (2024)
21. Verma, S., Verma, P.K., Singh, N.P.: Analysis on machine learning strategies for carcinoma detection biomarker. In: Proceedings of the 5th International Conference on Information Management & Machine Intelligence. pp. 1–7 (2023)
22. Verma, A., Mansoori, S., Srivastava, A., Rathee, P., Singh, N.P.: Deep learning driven automated malaria parasite detection in thin blood smears. In: Artificial Intelligence, Blockchain, Computing and Security Volume 1, pp. 375–380. CRC Press (2023)
23. Verma, A., Mallick, R., Harursampath, D., Sahay, P., Mishra, K.K.: Physics- informed neural networks with application in computational structural mechanics

Segmentation-Based Word Spotting in Handwritten Ayurveda Manuscripts Using Siamese Convolutional Network

Muskan Agarwal, S. Indu, and N. Jayanthi

Abstract Word spotting is typically formulated as an image-based retrieval problem, wherein one or more query word images are systematically compared against a repository of candidate word images to identify instances exhibiting visual or semantic similarity. This task follows the query-by-example paradigm, where the goal is to retrieve instances from the database that match the query based on similarity. This paper presents a segmentation-based word spotting framework specifically designed for Ayurveda manuscripts, which frequently exhibit ambiguous or indistinct word boundaries. A new dataset tailored for word spotting is also introduced. The system spots words using proposed Siamese convolutional network. Our experimental results, evaluated on a complex multi-writer corpus, demonstrate that the proposed framework consistently outperforms traditional matching approaches. The system achieved promising results in word spotting for handwritten Ayurveda manuscripts when tested on this new dataset.

Keywords Ayurveda manuscripts · Word spotting · Segmentation-based · Siamese convolutional network

1 Introduction

Ancient Ayurvedic documents kept in libraries and museums have been digitized in large quantities recently, and users can access this digital data via online portals. The Significant advancements in document digitization have established a robust and efficient method for the preservation of ancient handwritten manuscripts, enabling historians and researchers to access these invaluable resources with greater ease and convenience. Traditionally, searching through these large collections of digitized documents has relied on indexation methods based on Optical Character

M. Agarwal (✉) · S. Indu · N. Jayanthi
Department of Electronics and Communication Engineering, Delhi Technological University, New Delhi, Delhi, India

K. Deep et al. (eds.), *Artificial Intelligence, Machine Learning and Intelligent Systems*, Springer Proceedings in Mathematics & Statistics 520,
https://doi.org/10.1007/978-981-95-3841-6_13

Recognition (OCR) or manual transcription [1]. However, the performance of current OCR systems on historical documents is often suboptimal due to variability in handwriting and font styles, linguistic and script differences, and degraded document quality resulting from extensive deterioration. Manual transcription of handwritten texts is also a labor-intensive and costly process. In response to these challenges, word-spotting techniques have emerged as a viable and increasingly explored alternative, garnering significant research attention.

Word spotting has been extensively applied to Latin-script documents, but there has been no significant development of word spotting systems for handwritten Ayurvedic manuscripts. These manuscripts present unique challenges due to the lack of clear word boundaries, primarily caused by document degradation. In this paper, we introduce a segmentation-based word spotting framework designed for Ayurvedic manuscripts, utilizing a Siamese convolutional network architecture. Siamese convolutional networks use a learned distance metric, making them more robust to variations in handwriting or font styles. The framework is scalable to multi-writer scenarios, addressing the issue of unclear word boundaries. Its objective is to facilitate word spotting in this context, offering potential applications such as document retrieval and online searches in cultural heritage collections housed in libraries worldwide. Notably, we focus on the segmentation-based word spotting problem where no labeled training data is available, which is a common issue given the significant costs and time associated with accurately labeling data.

The remainder of this article is organized as follows: Section 2 reviews the related literature. In Sect. 3, we introduce our proposed method, detailing the dataset, candidate image selection, and spotting framework. Section 4 addresses the system's complexity, while Sect. 5 presents the experimental evaluation.

2 Related Work

Konidaris, Thomas, et al. [2] suggest a novel method for identifying keywords in historical manuscripts by combining artificial intelligence data and human feedback. Zagoris, Konstantinos, and colleagues [3] offer a new method for word spotting in handwritten texts that is based on document-specific local characteristics that take into account textual information surrounding representative keypoints. The suggested word spotting method is assessed on Washington dataset. Zhong, Zhuoyao, et al. [4] provide a convolutional neural network-based word spotting in historical documents. The GW dataset is used for experiments, and no recognition techniques or previous word classification knowledge is used.

Stauffer, Michael, et al. [5] describe an innovative, reliable method for template-based word spotting in historical documents. Specifically, their approach employs various graph representations for segmented word pictures, as well as a complex matching algorithm. Furthermore, they apply our approach to a detecting ensemble. They conducted experiments on two standard datasets namely George Washington (GW) and Parzival (PAR). Malakar, Samir, et al. [6] introduced a two-stage word searching methodology. This approach is tested using the publicly available QUWI English database. Barakat, Berat Kurar at el [7] describe a solution for word spotting based on a convolutional Siamese network.

Serdouk, Yasmine, et al. [8] provide a novel Siamese deep network with three inputs that can retrieve the most similar terms to a given query. The proposed system uses a query-by-example strategy based on segmentation to learn appropriate representations of word images, which may be matched using a Euclidean distance. They obtained the experimental results on the GW dataset for the spotting words. Zagoris et al. [9] employed local features designed for word spotting in handwritten manuscripts. As a result, regardless of the document's scaling, it can identify significant local points or keypoints of the characters. The suggested method for word spotting system is tested on George Washington dataset.

A method for attribute-based word spotting with CNNs is presented by Sudholt, Sebastian, et al. [10]. They theoretically develop loss functions that can be used to train models with both binary and real-valued attribute representations. They further demonstrate that normalizing the output and label vectors results in equivalent outcomes for both Euclidean Loss and Binary Cross Entropy Loss. They assessed their architectures on different data sets. Their research was limited to word spotting based on segmentation. Retsinas, George, et al. [15] proposed a method for segmentation-based word spotting on handwritten documents. The method involves three main steps: preprocessing, feature extraction, and feature matching, which address significant changes in text images (for example, translation, skew and different writing styles). They conducted experiments on seven publicly available datasets that demonstrates the efficiency of the proposed method.

3 Proposed Method

The ancient Ayurveda manuscripts suffer from various problems such as degradations, typesetting errors and irregular spacing between words which prevent the implementation of any segmentation task. Figure 1 shows pipeline of segmentation based word spotting. First, GNU Image Manipulation Program (GIMP) tool is used to manually segment word candidates. Then, the extracted candidate images are characterized as sequences of features. Finally, by using a similarity measure based on Siamese convolutional network, candidate images are matched to the query image and subsequently ranked based on their similarity scores.

3.1 Dataset Discription

We have collected images of handwritten Ayurveda documents from open source websites. Total 4512 scanned images are taken from various Ayurveda sources like Charak Samhita, Madhav Nidana, Sushrut Samhita, Madan Nighantu etc. These

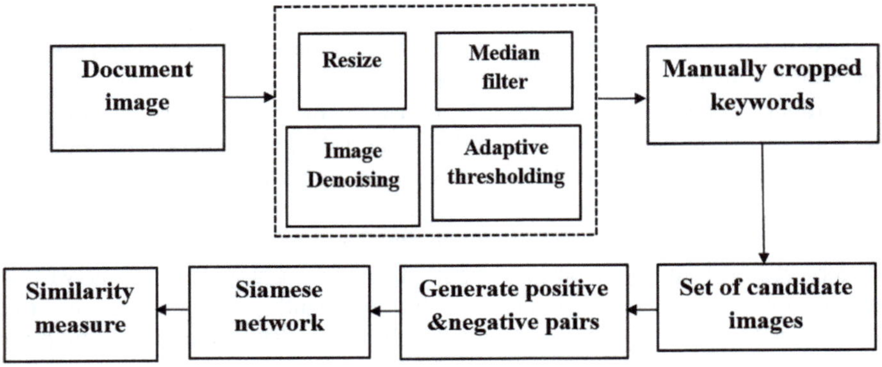

Fig. 1 Pipeline of segmentation based word spotting in Ayurveda documents

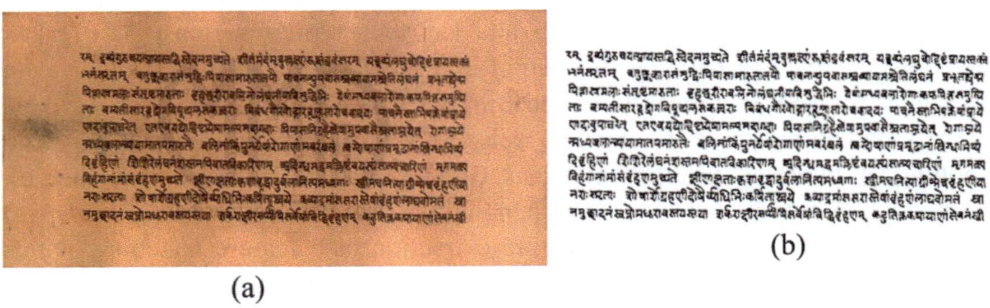

Fig. 2 (**a**) Sample images of Ancient Ayurvedic documents (**b**) Ayurveda document after preprocessing

ancient Ayurveda documents consist of several diseases such as jwar, kaph, pandu, vaman etc. which acts as keywords. Figure 2a shows sample images of ancient Ayurvedic documents. Figure 2b shows Ayurveda document after applying some image preprocessing technique noise removal, image denoising and thresholding

3.2 Selection of Candidate Words

We selected 4512 scanned images and manually marked all diseases as keywords. Some Diseases like Grahni, gulm, hridayarog, jwar, kaph, kas, Kushta, pandu, Prameh, Vaman, sannipat, Udavart and vishamjwar are denoted as candidate words. These keywords are semantically significant and commonly appear in the Ayurveda manuscripts. Figure 3 demonstrates Examples of some candidate word images (diseases) for training our segmentation-based word spotting model. To train our segmentation-based model, we developed a manual training dataset consisting of 13 separate folders, each containing more than 50 copies of a candidate word. The

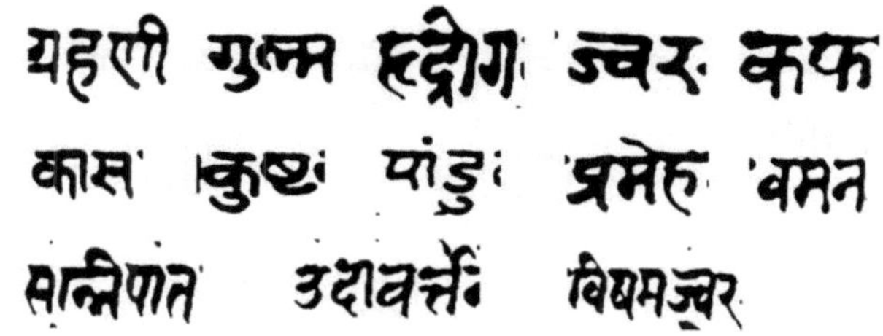

Fig. 3 Examples of some candidate word images (diseases) for training our segmentation-based word spotting model

Fig. 4 Example of positive pairs (first row) and Negative pairs (second row) for training Siamese convolutional network

dataset comprises total 719 candidate words from different writer. We assigned a class to each of these folders in order to distinguish them.

3.3 Positive and Negative Pairs

We have a total of 719 word images, and we are randomly creating pairs and labelling them. If an image pair belongs to the different class, labelled them as 0 else as 1. These image pairs and corresponding labels serve as training data for our Siamese convolutional network. To generate positive-pairs, we search for each word image (one disease) in the manuscript and its equivalent occurrences (same disease) in the other manuscripts and generate all possible pairs. To generate the Negative-pairs, we look up for each word image (one disease) in the first manuscript and randomly chosen other word image (other disease) taken from the any other manuscripts. These pairs serve as our training dataset such that the Siamese convolutional network can learn image similarity. Positive-pairing and Negative-pairing examples can be shown in Fig. 4.

3.4 Siamese Convolutional Network

Siamese convolutional network computes similarity between two word images using two similar convolutional networks. Siamese convolutional network consists of two identical Convolutional Neural Networks (CNNs) with equivalent architecture,

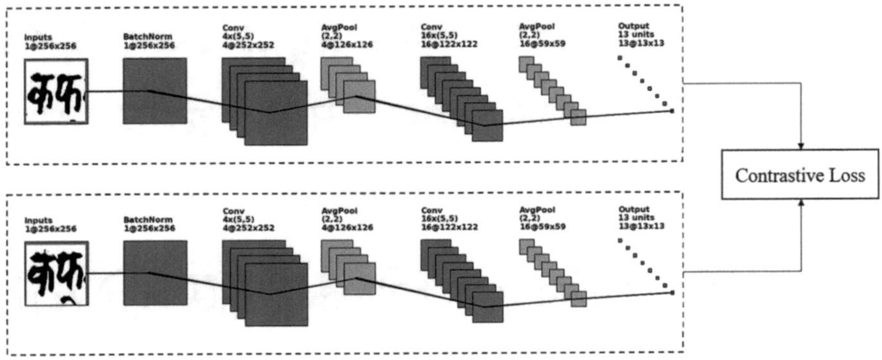

Fig. 5 Architecture of proposed Siamese Convolutional network for sementation-based word spotting in Ayurveda documents

parameters, and shared weights [10–14]. The parameter updates are synchronized across both subnetworks. The input to siamese convolutional network is the image pairs and the output represents a normalized Euclidean distance, constrained within the interval [0, 1]. Each subnetwork applies a set of batch normalization layer, convolutional layer (4, 16 filters with kernel 5, 5), ReLU and 2*2 avg-pooling layers and dropout layer (dropout of 0.2) to one of the image pairs. Figure 5 illustrates the architecture of the proposed Siamese convolutional network. These modifications aim to improve performance metrics and to overcome the over-fitting problem of proposed model. Each subnetwork produces a descriptor for the matching input image. The two subnetworks are linked through a loss function referred to as the contrastive loss function.

4 Experimental Results

In this section, we report the experimental findings from the query-by-example spotting approach on the ancient Ayurveda document images. The Ayurveda document images was collected from the website. It includes 5662 scanned pages with 4894 word instances. We divided the dataset into three sets: training, validation, and testing. There are total 1480 pairs in which 740 are positive pairs and 740 are negative pairs. The training set consists of 888 pairs, the validation set consists of 296 pairs, and the testing set consists of 296 pairs. As a performance metric, we provide accuracy, precision, recall and the mean average precision (mAP) that is computed by comparing the similarity of each word image in the test set to the query image. Experiments show that our proposed word spotting model achieved accuracy of 98.53% on training dataset and 96.09% on testing dataset as shown in Table 1 and Fig. 6.

Table 1 Performance metrics for Siamese convolutional network (%)

Performance metric	Proposed model (%)
Training accuracy	98.53
Testing accuracy	96.09
Precision	98.03
Recall	95.47
Mean average precision (mAP)	97.68

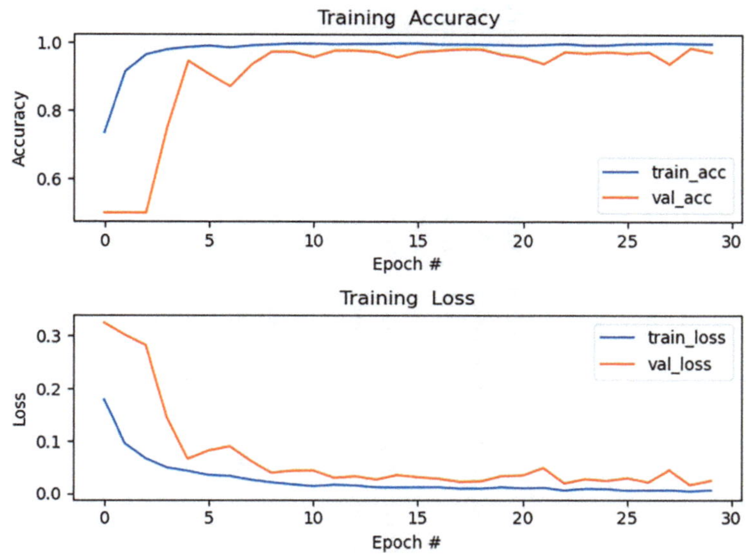

Fig. 6 Training/Validation accuracy for siamese convolutional network

4.1 Comparison with Segmentation Based Word Spotting Methods

The performance of the proposed approach is compared with few cutting-edge techniques. Table 1 presents a comparison between our approach and the cutting-edge query-by-example segmentation based word spotting technique. We borrow the baseline results from segmentation-based word spotting model for comparison to our method. The lack of a standard benchmark dataset of ancient Ayurveda document for word spotting approaches make it clear that it's difficult to compare methods directly. Our proposed word spotting model based on Siamese convolutional network gives mAP 0f 0.97 on ancient Ayurveda dataset which is better than other segmentation based models. Table 2 depicts Comparison with cutting-edge Segmentation Based Methods.

Table 2 Comparison with cutting-edge segmentation based methods

S.No.	Author	Dataset	Type of segmentation	Method	Performance metrics
1	Zhong, Zhuoyao et al. [4]	GeorgeWashington (GW) dataset	Segmentation-based	CNN	mAP = 80.03%
2	Malakar, Samir, et al. [6]	QUWI English database	Segmentation-based	Two-stage word searching schema	mAP = 0.72
3	Barakat, Berat Kurar et al. [7]	GeorgeWashington (GW) dataset	Segmentation-based	Siamese network	mAP = 0.49
4	Serdouk, Yasmine et al. [8]	GeorgeWashington (GW) dataset	Segmentation-based	Siamese convolution-nal network	mAP = 91.63
5	Zagoris et al. [9]	George Washington dataset	Segmentation-based	KeyPoints Detection	mAP = 0.692
6	Proposed	Ancient Ayurveda dataset	Segmentation-based	Siamese convolutional Network	mAP = 97.68

Fig. 7 Six top-most retrieved images for some queries in ancient Ayurveda documents

4.2 Evaluation of Segmentation-Based Model

We are evaluating our model on the test dataset. Each word image in the test dataset serves as a query to perform retrieval of similar images from the whole test dataset. The system then generates a ranked list of retrieved word images, ordered based on their Euclidean distance from the query image. Once our network is trained, cropped word images, together with query word images from the test dataset, are fed through the siamese convolutional network in pairs. We next compute the Euclidean distance between the candidate images and the query image in the test dataset, and in the output layer, we utilize sigmoid activation to predict 1/0, indicating whether the pair is identical or not. Finally, Fig. 7 illustrates some qualitative outcomes highlighting the effectiveness of the word spotting approach on the ancient Ayurveda dataset. In Fig. 7, we depicts the first sixth retrieved words that are similar to the query. For instance, retrieving the word images “jwar” for the query image “jwar” is correct.

5 Conclusion

We present an advanced handwritten word-spotting system specifically designed for Ayurveda manuscripts, utilizing a Siamese convolutional network. A notable characteristic of these manuscripts is the high variability in handwriting styles, contributing to the complexity of the task. Our approach demonstrates superior accuracy compared to traditional query-by-example (QBE) methods, while maintaining efficiency and requiring only a moderate amount of labeled training data. We propose a novel dataset for handwritten Ayurveda manuscripts. Experiments on our self-made dataset indicate that our method achieves state-of-the-art performance in QBE word spotting with respect to existing approaches. Moreover, our system offers high accuracy when compared to other approaches tailored to this problem.

References

1. Khurshid, Khurram, Claudie Faure, and Nicole Vincent. "Word spotting in historical printed documents using shape and sequence comparisons." *Pattern Recognition* 45.7 (2012): 2598-2609.
2. Konidaris, Thomas, et al. "Keyword-guided word spotting in historical printed documents using synthetic data and user feedback." *International Journal of Document Analysis and Recognition (IJDAR)* 9 (2007): 167-177.
3. Zagoris, Konstantinos, Ioannis Pratikakis, and Basilis Gatos. "Segmentation-based historical handwritten word spotting using document-specific local features." *2014 14th International Conference on Frontiers in Handwriting Recognition*. IEEE, 2014.
4. Zhong, Zhuoyao, et al. "Spottingnet: Learning the similarity of word images with convolutional neural network for word spotting in handwritten historical documents." *2016 15th International Conference on Frontiers in Handwriting Recognition (ICFHR)*. IEEE, 2016.
5. Stauffer, Michael, Andreas Fischer, and Kaspar Riesen. "Keyword spotting in historical handwritten documents based on graph matching." *Pattern Recognition* 81 (2018): 240-253.
6. Malakar, Samir, et al. "Development of a two-stage segmentation-based word searching method for handwritten document images." *Journal of Intelligent Systems* 29.1 (2019): 719-735.
7. Barakat, Berat Kurar, Reem Alasam, and Jihad El-Sana. "Word spotting using convolutional siamese convolutionalnetwork." *2018 13th IAPR International Workshop on Document Analysis Systems (DAS)*. IEEE, 2018.
8. Serdouk, Yasmine, et al. "Keyword spotting using siamese convolutionaltriplet deep neural networks." *2019 International Conference on Document Analysis and Recognition (ICDAR)*. IEEE, 2019.
9. Zagoris, Konstantinos, Ioannis Pratikakis, and Basilis Gatos. "Unsupervised word spotting in historical handwritten document images using document-oriented local features." *IEEE Transactions on Image Processing* 26.8 (2017): 4032-4041.
10. Sudholt, Sebastian, and Gernot A. Fink. "Attribute CNNs for word spotting in handwritten documents." *International journal on document analysis and recognition (ijdar)* 21 (2018): 199-218.
11. J. Bromley, I. Guyon, Y. LeCun, E. Sackinger, and R. Shah,""Signature verification using a" siamese" time delay neural network," in Advances in Neural Information Processing Systems, 1994, pp. 737–744.

12. Dey, Sounak, et al. "Signet: Convolutional siamese convolutionalnetwork for writer independent offline signature verification." *arXiv preprint arXiv:1707.02131* (2017).
13. Kassis, Majeed, Jumana Nassour, and Jihad El-Sana. "Alignment of historical handwritten manuscripts using siamese convolutionalneural network." *2017 14th IAPR International Conference on Document Analysis and Recognition (ICDAR)*. Vol. 1. IEEE, 2017.
14. Koch, Gregory, Richard Zemel, and Ruslan Salakhutdinov. "Siamese convolutionalneural networks for one-shot image recognition." *ICML deep learning workshop*. Vol. 2. No. 1. 2015.
15. Retsinas, George, Georgios Louloudis, Nikolaos Stamatopoulos, and Basilis Gatos. "Keyword spotting in handwritten documents using projections of oriented gradients". In *2016 12th IAPR Workshop on Document Analysis Systems (DAS)*, pp. 411–416. IEEE, 2016.

AI-Enhanced Pediatric Pneumonia Classification

Anisha Jadhav

Abstract Pneumonia remains a leading cause of mortality among children under 5 years old, with a significant impact in low-income regions. This study explores the application of transfer learning with convolutional neural networks (CNNs) to improve the classification of pediatric chest X-rays into normal, bacterial, and viral Pneumonia. Utilizing pre-trained models on the ImageNet dataset, specifically InceptionV3, these models were repurposed for classification tasks. The framework included both binary and multi-class classification approaches, with the best binary classifier achieving an F1-score of 0.941 and an accuracy of 91.44%. The multi-class classifier reached an F1-score of 0.597 and an accuracy of 83.78%. These results demonstrate the potential of advanced AI techniques to enhance diagnostic accuracy and address challenges in Pneumonia classification, particularly in resource-limited settings.

Keywords Pediatric Pneumonia · Transfer learning · Chest X-rays · InceptionV3 · Convolutional neural networks · Artificial intelligence

1 Introduction

According to a study done by UNICEF, Pneumonia remains the leading cause of death among children under 5 years old [1]. In 2015, Pneumonia was responsible for 922,000 child deaths, making up almost 16% of all child deaths [2]. More than one million people in the US alone are hospitalized from Pneumonia every year [2]. Despite there being effective methods to diagnose and treat Pneumonia, high-poverty areas with suboptimal health conditions (malnutrition, contaminated water, and air pollution) are greatly affected by the disease [1].

Pneumonia arises when inflammation occurs in the alveoli of the lungs due to the presence of bacteria or viruses. These contaminants can be introduced via water

A. Jadhav (✉)
California State University, Monterey Bay, Seaside, CA, USA

K. Deep et al. (eds.), *Artificial Intelligence, Machine Learning and Intelligent Systems*, Springer Proceedings in Mathematics & Statistics 520,
https://doi.org/10.1007/978-981-95-3841-6_14

droplets or food particles that enter the respiratory tract [3]. While healthy adults can recover quickly, for younger and older patients with weakened immune systems, Pneumonia may be life-threatening [4].

Methods of diagnosing Pneumonia may range depending on the resources available. Diagnoses in low-income countries primarily rely on identifying symptoms early with simple clinical procedures, such as observing the breathing quality and rate of the patient using a stethoscope [5]. Patients with access to more precise methods may have an x-ray of their lungs taken to identify which part of the lung tissue is inflamed [3].

Several issues remain with the status quo of radiology exams. A combination of low resources and high volume may lead to misclassifications and consequent misdiagnoses by WHO protocol-trained radiologists. In fact, studies have shown that there are discrepancies in radiograph evaluation between radiologists (with concordance rates between two reviewers to be as low as 48%) [6]. In addition, the proper treatment of Pneumonia depends on the correct identification of the pathogen. Antibiotics, such as amoxicillin, are administered as a blanket drug to children diagnosed with Pneumonia in low-income regions, which may not be effective against viral Pneumonia or other forms of bacteria with different site specificities [7].

Artificial intelligence has shown promise in surpassing the performance of medical professionals for classification tasks. Traditional methods of image analysis involved a triad approach of manually segmenting objects of interest, developing classifiers for each class of objects, and classifying the image [8]. Now, convolutional neural networks have become a standard architecture for processing and classifying images. Deep neural networks have shown an ability to quantify features that are indicative of diseases like diabetes mellitus (from retinopathy images) [9] and skin cancer (from images of skin) with high accuracy [10].

For this study, a transfer learning approach was adopted, which involves reusing a base network trained on a base dataset and repurposing the learned features for a dataset of interest. This is made possible because deep neural networks trained on images tend to learn similar features. The weights of the model learned on extremely large datasets, like the Stanford ImageNet dataset, can be leveraged and re-trained on the last layer using backpropagation to recognize shared features in biomedical images such as chest X-rays. In this research work, a transfer learning framework using Google's model is explored to effectively classify normal, viral, and bacterial Pneumonia in pediatric chest X-ray images.

2 Materials and Methods

2.1 Problem Formulation

For this experiment, two problems were considered. Given the nature of the dataset, which was divided into normal and abnormal (bacterial and viral separately labeled) Pneumonia images, the performance of the network was assessed as both, a binary classification (normal vs. abnormal) and a multi-class classification (normal vs. bacterial vs. viral) problem. Input X is the provided chest X-ray images, and the output Y is an output class label $y \in \{0, 1\}$ for the binary classifier and $y \in \{0, 1, 2\}$ for the multi-class classifier. A binary cross entropy loss function was optimized for the binary classifier, and a sparse softmax cross entropy (categorical cross-entropy loss) for the multi-class classifier.

To combat the potentially dangerous misdiagnoses, precision and recall performance metrics can be used to assign higher costs to false negatives than false positives. The F1 score accounts for both precision and recall scores and is used to gauge the network's performance. Accuracy is also recorded.

2.2 Model Architecture and Training

Large convolutional neural networks pre-trained on the ImageNet dataset were utilized, with their last classification layers replaced to align with the specific output requirements of the task. The experiments were conducted using both Xception and InceptionV3 models, which were also previously used in studies [9, 11], but better results were found with InceptionV3 model. After the pretrained model, a global average pooling layer was added, followed by a variable number of dense layers. The output from the global average pooling layer was extracted and fed into a separate model composed solely of dense layers, which was then trained. This technique significantly sped up training times without sacrificing the performance. The Adam Optimizer was used for training, and the best number of epochs, batch size, and training steps were determined empirically. The training was conducted on a *Google Cloud Deep Learning VM* with an *NVIDIA Tesla V100 GPU*, and all code was written in *TensorFlow*. Figures 4 and 5 summarize the results of all the hyperparameters tested on both classifiers.

The models were trained using the Adam optimizer with a learning rate of 0.001. For binary classification, the model used a batch size of 64 and was trained for 50 epochs. For multi-class classification, a batch size of 1000 and 400 epochs were used. The loss functions used were binary cross-entropy for binary classification and sparse categorical cross-entropy for multi-class classification. Early stopping and dropout regularization were employed to reduce overfitting. The input image size was standardized to 300×300 pixels with RGB channels.

2.3 Data

The Chest X-Ray Images (Pneumonia) dataset made publicly available by Paul Mooney on Kaggle is used for this study. The dataset contains 5863 front chest X-ray images of pediatric patients classified into Normal, Bacterial Pneumonia and Viral Pneumonia categories. Refer to Fig. 1 (taken from [11] for sample

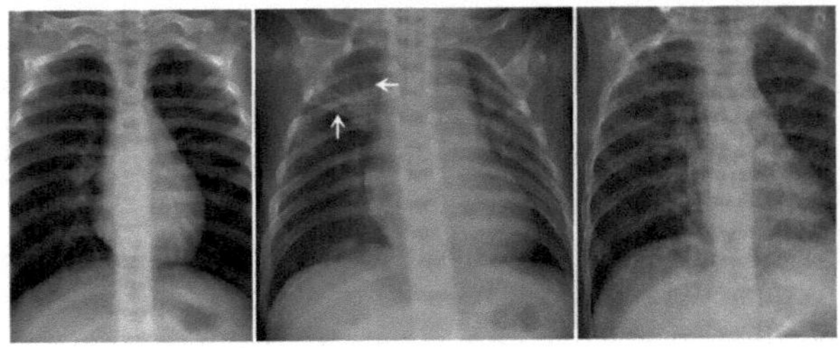

Fig. 1 Sample chest X-rays for each class [11]

Table 1 Number of images for training, validation, and test sets per label, before and after preprocessing

	Before balancing			After balancing		
	Normal	Bacterial	Viral	Normal	Bacterial	Viral
Train	1342	2530	1345	1342	1342	1342
Validate	9	8	0	74	74	74
Test	234	242	148	74	74	74

X-rays). There was no patient overlap between training, validation, and testing datasets. The dataset can be accessed publicly here: https://www.kaggle.com/datasets/paultimothymooney/chest-xray-pneumonia.

Label analysis showed that the number of images in each folder was inherently unbalanced. Specifically, the training dataset contained 1342 normal, 2530 bacterial and 1345 viral Pneumonia images, while the validation dataset had only nine normal, eight bacterial, and zero viral Pneumonia images. The test dataset included 234, 242 and 148 images for normal, bacterial and viral Pneumonia cases, respectively.

To address this imbalance, two preprocessing strategies were applied. First, the validation and test datasets were merged, shuffled, and then evenly re-distributed to ensure an equal class distribution in the new validation and test sets. Second, a random under-sampling approach was adopted to equalize the number of training examples per class to the minimum available across the categories. Table 1 presents the number of images per class before and after balancing.

All the images from training, validation, and testing sets were resized to 300×300 pixels and converted into RGB format to resolve the shape inconsistencies between the images and to match the input dimensions required by the InceptionV3 and Xception models. Additionally, data augmentation was performed to enrich the training set by applying horizontal flipping, random brightness adjustments, and rotations of up to 20 degrees.

As shown in Table 1, the dataset was rebalanced to ensure equal class representation across all splits.

3 Results

It is found that the best model for multi-class classification had three additional dense layers, a batch size of 1000 images, a learning rate of 0.001, and it was trained on 400 epochs. This model had a F1 score of 0.597 and an accuracy of 83.78% for 3-class classifications. Additionally, the model predicted labels with above a 0.8 true positive rate. Refer to Fig. 2 for the normalized confusion matrix and Fig. 3 for the train and validation loss curves. These parameters were chosen to match previous work, but there were better models in binary classification.

A model with three dense layers, a batch size of 64, a learning rate of 0.001, and 50 epochs reached the F1 score of 0.941 and an accuracy of 91.44% for binary classification.

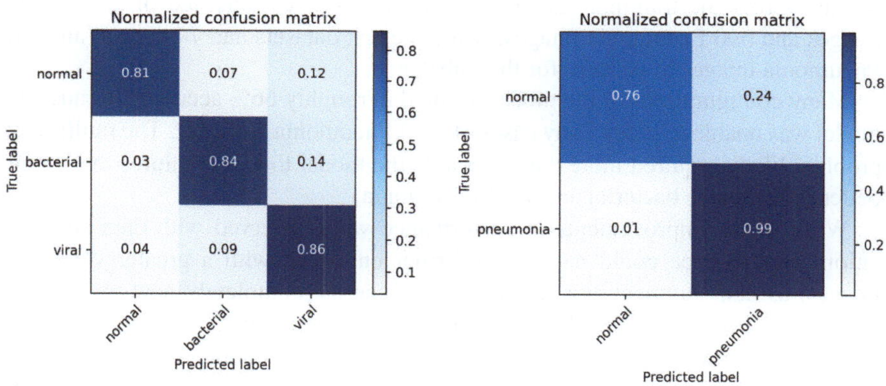

Fig. 2 Normalized confusion matrix for (left) multi-class label and (right) binary-class model

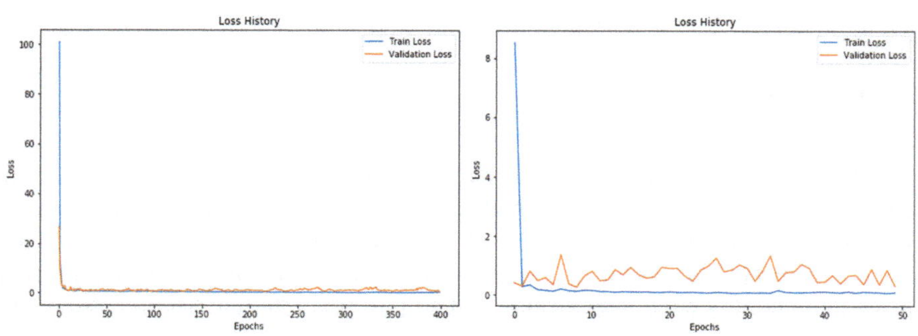

Fig. 3 Train and validation loss for (left) multi-class model and (right) binary-class model

4 Discussion

As expected, the binary classifier performed better than the multi-class classifier model because the differences in features between the viral and bacterial Pneumonia images were very subtle. The binary model performed similarly to Kermany et al., who used a similar kind of dataset and achieved 92.8% with a sensitivity of 93.2% and a specificity of 90.1% for their Pneumonia versus normal model [11]. For a bacterial versus viral Pneumonia model, which, in the study by Kermany et al. [11], they achieved a test accuracy of 90.7%, with a sensitivity of 88.6% and a specificity of 90.9%. The final results for the first binary model were slightly lower, and exact replication was difficult due to the large batch size of 1000, which increased randomness across runs. Another factor contributing to the discrepancy was the data imbalance; this was addressed by under-sampling to match the minimum number of images across the training, validation of Kermany's set consisted of 234 normal images and 390 Pneumonia images, while the test datasets had 74 normal and 148 Pneumonia images to account for the imbalance.

Many configurations of the model resulted in roughly 66% accuracy because the model was unable to predict any cases of viral Pneumonia correctly. The multi-class problem likely required more data to enable the model to learn features that would better differentiate bacterial and viral Pneumonia.

While small improvements in performance were observed with data augmentation, performance could have been further enhanced with a greater variety of images, as data augmentation alone cannot generate completely new photos. For example, radiologists typically use supplemental lateral view X-rays when making a diagnosis; however, the dataset from Kaggle only contained frontal images. Additionally, patient health records are often considered in diagnosis, which could have supplemented the images. Using a more comprehensive dataset for each patient would better reflect clinical practices and may lead to improved performance.

5 Conclusion

It is found that convolutional neural networks are effective at diagnosing normal and abnormal chest X-rays with high accuracy and precision but can struggle with differentiating types of abnormal X-rays. Replication of previous work on the same dataset [11] was achieved. Future work involves looking at class activation maps to determine what features the model thinks are relevant. These maps could inform the model's decisions, a useful feature in medical diagnosis, and could help determine why it confuses viral with bacterial Pneumonia. This study has shown that transfer learning can be used to apply large convolutional neural networks to chest X-rays.

Appendix

See Figs. 4 and 5.

Model #	# of hidden layers	# of nodes	batch size	epochs	learning rate	train_loss	test_loss	train_acc	test_acc	train_F1	test_F1
1	3	1024, 512, 3	64	50	0.01	0.081	0.914	97.201%	84.234%	0.980	0.896
2	3	1024, 512, 3	32	50	0.01	0.637	0.637	66.667%	66.667%	0.797	0.798
3	3	1024, 512, 3	64	50	0.1	0.645	0.621	66.667%	66.667%	0.795	0.812
4	3	1024, 512, 3	64	50	0.001	0.084	0.321	97.241%	91.441%	0.979	0.941
5	3	1024, 512, 3	64	50	0.0001	0.137	1.044	96.172%	81.532%	0.970	0.881
6	3	1024, 512, 3	64	100	0.01	0.061	0.728	98.011%	83.784%	0.985	0.895
7	3	1024, 512, 3	64	200	0.01	0.026	0.82	99.304%	81.982%	0.995	0.894
8	3	1024, 512, 3	1000	200	0.01	0.041	0.864	98.260%	84.685%	0.986	0.896
9	3	1024, 512, 3	1000	200	0.001	0.051	0.714	97.962%	83.784%	0.982	0.891
10	3	1024, 512, 3	1000	400	0.01	0.034	1.139	98.509%	76.577%	0.991	0.851
11	3	1024, 512, 3	1000	400	0.001	0.037	1.242	98.136%	73.874%	0.989	0.836
12	6	1024, 512, 256, 128, 64, 3	64	50	0.01	0.097	0.811	96.868%	81.081%	0.976	0.875
13	6	1024, 512, 256, 128, 64, 3	32	50	0.01	0.637	0.638	66.667%	66.667%	0.798	0.798
14	6	1024, 512, 256, 128, 64, 3	64	50	0.1	0.64	0.652	66.667%	66.667%	0.799	0.792
15	6	1024, 512, 256, 128, 64, 3	64	50	0.001	0.056	3.375	98.111%	82.432%	0.985	0.887
16	6	1024, 512, 256, 128, 64, 3	64	50	0.0001	0.034	1.131	99.130%	81.982%	0.993	0.880
17	6	1024, 512, 256, 128, 64, 3	64	100	0.01	0.637	0.634	66.667%	66.667%	0.798	0.803
18	6	1024, 512, 256, 128, 64, 3	64	200	0.01	0.637	0.62	66.667%	66.667%	0.798	0.812
19	6	1024, 512, 256, 128, 64, 3	1000	200	0.01	0.074	1.556	96.093%	74.324%	0.981	0.839
20	6	1024, 512, 256, 128, 64, 3	1000	200	0.001	0.086	1.407	96.520%	72.072%	0.978	0.827
21	6	1024, 512, 256, 128, 64, 3	1000	400	0.01	0.168	0.338	93.060%	86.486%	0.958	0.907
22	6	1024, 512, 256, 128, 64, 3	1000	400	0.001	0.029	0.877	98.733%	86.486%	0.992	0.907
23	9	1024, 512, 256, 128, 64, 32, 16, 8, 3	64	50	0.01	0.087	1.396	96.992%	80.631%	0.977	0.864
24	9	1024, 512, 256, 128, 64, 32, 16, 8, 3	32	50	0.01	0.637	0.637	66.667%	66.667%	0.797	0.798
25	9	1024, 512, 256, 128, 64, 32, 16, 8, 3	64	50	0.1	0.64	0.634	66.667%	66.667%	0.799	0.807
26	9	1024, 512, 256, 128, 64, 32, 16, 8, 3	64	50	0.001	0.096	1.191	96.445%	74.324%	0.973	0.834
27	9	1024, 512, 256, 128, 64, 32, 16, 8, 3	64	50	0.0001	0.05	0.477	98.732%	87.838%	0.990	0.914
28	9	1024, 512, 256, 128, 64, 32, 16, 8, 3	64	100	0.01	0.637	0.637	66.667%	66.667%	0.798	0.795
29	9	1024, 512, 256, 128, 64, 32, 16, 8, 3	64	200	0.01	0.637	0.64	66.667%	66.667%	0.799	0.796
30	9	1024, 512, 256, 128, 64, 32, 16, 8, 3	1000	200	0.01	0.08	0.572	97.042%	78.378%	0.977	0.860
31	9	1024, 512, 256, 128, 64, 32, 16, 8, 3	1000	200	0.001	0.068	0.899	97.166%	77.477%	0.983	0.855
32	9	1024, 512, 256, 128, 64, 32, 16, 8, 3	1000	400	0.01	0.639	0.637	66.667%	66.667%	0.798	0.800
33	9	1024, 512, 256, 128, 64, 32, 16, 8, 3	1000	400	0.001	0.626	0.637	66.667%	66.667%	0.810	0.806

Fig. 4 Hyperparameters tested for binary classifier (normal vs. Pneumonia). The model with the highest accuracy is highlighted

Model #	# of hidden layers	# of nodes	batch size	epochs	learning rate	train_loss	test_loss	train_acc	test_acc	train_F1	test_F1
1	3	1024, 512, 3	64	50	0.01	0.599	0.832	68.854%	56.757%	0.652	0.665
2	3	1024, 512, 3	32	50	0.01	1.1	1.099	32.960%	33.333%	nan	0.666
3	3	1024, 512, 3	64	50	0.1	1.103	1.106	35.706%	33.333%	nan	0.658
4	3	1024, 512, 3	64	50	0.001	0.418	1.197	81.854%	60.018%	0.665	0.587
5	3	1024, 512, 3	64	50	0.0001	0.268	1.218	90.455%	69.820%	0.665	0.635
6	3	1024, 512, 3	64	100	0.01	0.56	1.31	70.296%	49.550%	0.704	0.681
7	3	1024, 512, 3	64	200	0.01	1.099	1.099	32.911%	33.333%	nan	0.666
8	3	1024, 512, 3	1000	200	0.01	1.1	1.099	33.333%	33.333%	0.666	0.667
9	3	1024, 512, 3	1000	200	0.001	0.413	0.976	80.661%	70.721%	0.643	0.627
10	3	1024, 512, 3	1000	400	0.01	0.432	1.144	79.021%	65.315%	0.680	0.577
11	3	1024, 512, 3	1000	400	0.001	0.193	1.232	91.897%	63.784%	0.670	0.597
12	6	1024, 512, 256, 128, 64, 3	64	50	0.01	0.522	1.142	67.810%	54.955%	0.716	0.703
13	6	1024, 512, 256, 128, 64, 3	32	50	0.01	0.535	1.107	73.080%	61.712%	0.675	0.614
14	6	1024, 512, 256, 128, 64, 3	64	50	0.1	1.102	1.103	33.656%	33.333%	nan	0.665
15	6	1024, 512, 256, 128, 64, 3	64	50	0.001	0.407	0.971	82.240%	72.523%	0.685	0.636
16	6	1024, 512, 256, 128, 64, 3	64	50	0.0001	0.276	1.079	90.157%	76.775%	0.668	0.624
17	6	1024, 512, 256, 128, 64, 3	64	100	0.01	0.414	1.008	81.979%	77.027%	0.662	0.619
18	6	1024, 512, 256, 128, 64, 3	64	200	0.01	0.373	1.269	83.669%	68.468%	0.670	0.598
19	6	1024, 512, 256, 128, 64, 3	1000	200	0.01	0.469	1.045	79.442%	68.468%	0.646	0.612
20	6	1024, 512, 256, 128, 64, 3	1000	200	0.001	0.360	1.088	82.351%	68.468%	0.661	0.598
21	6	1024, 512, 256, 128, 64, 3	1000	400	0.01	0.333	1.004	84.439%	67.117%	0.690	0.641
22	6	1024, 512, 256, 128, 64, 3	1000	400	0.001	0.411	1.493	80.686%	56.306%	0.648	0.628
23	9	1024, 512, 256, 128, 64, 32, 16, 8, 3	64	50	0.01	1.099	1.099	32.563%	33.333%	nan	0.665
24	9	1024, 512, 256, 128, 64, 32, 16, 8, 3	32	50	0.01	1.099	1.099	33.209%	33.333%	nan	0.659
25	9	1024, 512, 256, 128, 64, 32, 16, 8, 3	64	50	0.1	1.105	1.107	35.905%	33.333%	nan	0.494
26	9	1024, 512, 256, 128, 64, 32, 16, 8, 3	64	50	0.001	1.099	1.099	32.214%	33.333%	nan	0.654
27	9	1024, 512, 256, 128, 64, 32, 16, 8, 3	64	50	0.0001	0.376	0.987	84.638%	74.324%	0.662	0.601
28	9	1024, 512, 256, 128, 64, 32, 16, 8, 3	64	100	0.01	1.099	1.099	32.438%	33.333%	nan	0.663
29	9	1024, 512, 256, 128, 64, 32, 16, 8, 3	64	200	0.01	1.099	1.099	32.627%	33.333%	nan	0.670
30	9	1024, 512, 256, 128, 64, 32, 16, 8, 3	1000	200	0.01	0.574	1.344	68.841%	40.090%	0.745	0.687
31	9	1024, 512, 256, 128, 64, 32, 16, 8, 3	1000	200	0.001	1.099	1.099	33.333%	33.333%	0.664	0.667
32	9	1024, 512, 256, 128, 64, 32, 16, 8, 3	1000	400	0.01	0.466	0.792	78.672%	76.577%	0.681	0.613
33	9	1024, 512, 256, 128, 64, 32, 16, 8, 3	1000	400	0.001	1.098	1.099	33.333%	33.333%	0.674	0.667

Fig. 5 Hyperparameters tested for multi-class classifier (normal vs. bacterial vs. viral Pneumonia). The model with the highest accuracy is highlighted

References

1. Ending child deaths from Pneumonia and diarrhea, https://data.unicef.org/resources/one-many-ending-child-deaths-pneumonia-diarrhoea/
2. Pneumonia can be prevented - vaccines can help, https://www.cdc.gov/pneumonia/prevention/
3. Pneumonia: Overview (Aug 2018), https://www.ncbi.nlm.nih.gov/books/NBK525774/
4. Zhang, Y., Yao, Q., Chen, L.: Efficient, simple and automated negative sampling for knowledge graph embedding. arXiv preprint arXiv:2010.14227 (2020)

5. Chen, X., Zhang, W., Yao, Z., Chen, M., Tang, S.: Negative sampling with adaptive denoising mixup for knowledge graph embedding. In: International Semantic Web Conference. pp. 253–270. Springer (2023)
6. Elemraid, M.A., Muller, M., Spencer, D.A., Rushton, S.P., Gorton, R., Thomas, M.F., Eastham, K.M., Hampton, F., Gennery, A.R., Clark, J.E., et al.: Accuracy of the interpretation of chest radiographs for the diagnosis of pediatric pneumonia. Plos One 9(8) (2014). https://doi.org/10.1371/journal.pone.0106051
7. Patel, A.: Optimizing knowledge graph embeddings for enhanced question answering performance. In: 2024 International Conference on Intelligent Computing, Communication, Networking and Services (ICCNS). pp. 191–199 (2024). https://doi.org/10.1109/ICCNS62192.2024.10776440
8. Goldbaum, M., Moezzi, S., Taylor, A., Chatterjee, S., Boyd, J., Hunter, E., Jain, R.: Automated diagnosis and image understanding with object extraction, object classification, and inferencing in retinal images. Proceedings of 3rd IEEE International Conference on Image Processing (1996). https://doi.org/10.1109/ICIP.1996.560760
9. Srinivasan, S., Shetty, S., Natarajan, V., Sharma, T., Raman, R.: Development and validation of a diabetic retinopathy referral algorithm based on single-field fundus photography. Plos One 11(9) (2016). https://journals.plos.org/plosone/article?id=10.1371/journal.pone.0163108
10. Esteva, A., Kuprel, B., Novoa, R.A., Ko, J., Swetter, S.M., Blau, H.M., Thrun, S.: Dermatologist-level classification of skin cancer with deep neural networks. Nature 542(7639), 115–118 (2017). https://doi.org/10.1038/nature21056
11. Kermany, D.S., Goldbaum, M., Cai, W., Valentim, C.C., Liang, H., Baxter, S.L., McKeown, A., Yang, G., Wu, X., Yan, F., et al.: Identifying medical diagnoses and treatable diseases by image-based deep learning. Cell 172(5), 1122–1131 (2018)

Anomaly Detection in Cryptocurrency Prices

Anushka Pawar, Shruti Choudhari, Sachin Naik, and Rajeshree Khande

Abstract This Paper focuses on creating and utilizing ML algorithms that can detect anomalies in the pricing of key cryptocurrencies, including Bitcoin, Ethereum, Avalanche, Tether, and Litecoin. The volatility of the cryptocurrency market highlights the significance of identifying abnormal price movements that might indicate market scams or other discrepancies. Through this research, we employ Random Forest, Isolation Forest, and XGBoost models, all offering distinct advantages: Random Forest merges different educational approaches, Isolation Forest is skilled at pinpointing anomalies, and XGBoost applies enhancement techniques. By utilizing these models, we enhance the detection of abnormal trading behaviors, promoting a more secure and open environment in the cryptocurrency market. Additionally, we present the findings of the anomaly detection through PowerBI and Python tools to clearly communicate insights to stakeholders. The outcomes show the effectiveness of these machine learning approaches in detecting anomalies, aiding both investors and regulatory bodies in making well-informed decisions about the direction of the cryptocurrency market.

Keywords Cryptocurrency · Anomaly detection · Machine learning · Random forest · Isolation forest · XGBoost · Data visualization · Market manipulation · Cryptocurrency prices · Financial security

1 Introduction

The main objective of this study is to detect discrepancies in the price fluctuations of major digital assets, such as Bitcoin, Ethereum, Avalanche, Tether, and Litecoin. Given the volatile nature of the digital asset markets, it is crucial to recognize

---

A. Pawar · S. Choudhari (✉) · S. Naik · R. Khande
Department of Computer Science and Computer Applications, Dr. Vishwanath Karad MIT World Peace University, Pune, India
e-mail: anushka.pawar@mitwpu.edu.in; shruti.Choudhari@mitwpu.edu.in

K. Deep et al. (eds.), *Artificial Intelligence, Machine Learning and Intelligent Systems*, Springer Proceedings in Mathematics & Statistics 520,
https://doi.org/10.1007/978-981-95-3841-6_15

unusual price trends that might indicate market fraud or unlawful activities. To achieve this, we are employing machine learning methods, specifically Random Forest, Isolation Forest, and XGBoost, to detect these discrepancies. Random Forest is chosen for its ability to merge different models, Isolation Forest is effective in pinpointing anomalies, and XGBoost is selected for its speed and accuracy in enhancing performance.

The datasets utilized in this research encompass the price histories from 2014 to 2023 for Bitcoin and Litecoin, and from 2017 to 2023 for Ethereum. These datasets include various price-related metrics, such as the initial and final prices, the peak and trough values, the adjusted final prices, and the trading volumes. Before starting the analysis, we address any missing data and handle outliers. The machine learning models are then trained and assessed to find any irregularities in the cryptocurrency price trends. The detected irregularities are displayed using sophisticated tools like PowerBI and Python libraries (Matplotlib, Seaborn, Plotly) to offer stakeholders detailed and interactive visualizations that can guide their decision-making process.

The anticipated results are the discovery of irregular trading behaviors, which will aid in enhancing the decision-making processes for investors, regulators, and other participants in the market. This research highlights the role of machine learning and data visualization in enhancing the security and transparency of the cryptocurrency markets.

2 Previous Work

Numerous studies have explored various machine learning approaches to detect anomalies in cryptocurrency transactions and the broader financial systems, with a particular emphasis on methods such as Convolutional Neural Networks (CNNs), Support Vector Machines (SVM), and deep learning models. Below is a summary of the key studies relevant to the detection of anomalies and the development of trading strategies in cryptocurrency markets.

Anomaly detection in cryptocurrency transactions has garnered significant attention in recent research, leveraging various machine learning approaches to enhance detection accuracy and computational efficiency. This section provides an overview of some key studies that have contributed to the field.

Leandro L. Cunha et al. (2023) proposed a method utilizing Convolutional Neural Networks (CNN) with active learning strategies to detect anomalies in cryptocurrency transactions. Their study highlighted the effectiveness of focusing the model's learning on the most informative samples, resulting in an improvement in detection accuracy by 5–10% over standard models [6]. The study suggests future integration of more sophisticated deep learning techniques to further enhance model performance.

Michel Rwibasira et al. (2022) explored the application of Support Vector Machines (SVM) for anomaly detection in blockchain platforms. Using a dataset of historical transactions, the authors fine-tuned SVM parameters through grid search and combined it with unsupervised learning methods for improved accuracy. Their approach achieved approximately 95% detection accuracy, underscoring the robustness of SVM in this context. The authors recommended exploring alternative machine learning algorithms for enhanced detection capabilities [5].

Liu et al. [2] introduced a deep reinforcement learning framework for analyzing and predicting Bitcoin transaction patterns. The study employed a multi-layered CNN to extract features from transaction data, achieving a prediction accuracy of 92%. This approach demonstrates the potential of deep learning for financial forecasting, with a focus on expanding the model's applicability to other cryptocurrencies [2].

Lin Liu et al. (2022) presented a method for detecting fraudulent activities in Ethereum smart contracts using Heterogeneous Graph Transformer Networks (HGTNs) combined with CNNs. Their model processed heterogeneous information networks, achieving an F1-score of 0.87 for normal contracts and 0.78 for fraudulent ones. This study demonstrated the benefits of utilizing graph-based learning for anomaly detection, especially in complex blockchain environments [3].

ElBahrawy et al. [1] conducted a cluster analysis of the Bitcoin transaction network, using community detection algorithms to identify structural patterns and anomalies. Their study revealed the presence of a "rich club" phenomenon where a few nodes dominated transaction volumes. This finding underscores the importance of understanding network structures in detecting potential manipulations within blockchain transactions [1].

Shin Morishima [4] explored scalable anomaly detection using CNNs with GPU acceleration, applied to Bitcoin transaction data. The proposed method achieved an 11.onefold speed improvement over traditional methods without compromising accuracy. This study highlights the potential of leveraging parallel processing capabilities for real-time anomaly detection in large-scale blockchain datasets [4].

Other research, such as that by Rudkiewicz and Hernes [7], utilized autoencoders and Long Short-Term Memory (LSTM) neural networks for detecting anomalies in cryptocurrency time series. Their results demonstrated the practical application of anomaly detection for trading decisions, emphasizing the role of anomalies as indicators of significant market movements [7].

Across these studies, various methodologies such as CNNs, SVMs, HGTNs, and reinforcement learning have been applied to anomaly detection, reflecting a broad exploration of machine learning techniques in the domain. While CNN-based approaches excel in capturing complex patterns from transaction data, SVMs and graph-based methods provide robust solutions for anomaly classification. Future research is recommended to integrate these approaches with larger datasets and refine the models for improved real-time detection and broader applicability across different blockchain platforms.

3 The Proposed Approach

This section outlines the methodology and system architecture for detecting anomalies in cryptocurrency price data. The process is divided into six distinct phases: data collection, data preprocessing, data analysis, model selection and training, anomaly detection, and visualization (Fig. 1).

3.1 Data Collection

The dataset used in this research includes historical data for eight major cryptocurrencies: Bitcoin, Ethereum, Litecoin, Avalanche, Dogecoin, Polkadot, Solana, and Tether. Each cryptocurrency dataset contains seven columns representing key trading metrics (Table 1)

Bitcoin data spans from 2015 to 2023, with a total of 3413 rows. Ethereum data covers the period from 2018 to 2023, comprising 2264 rows. Litecoin's dataset

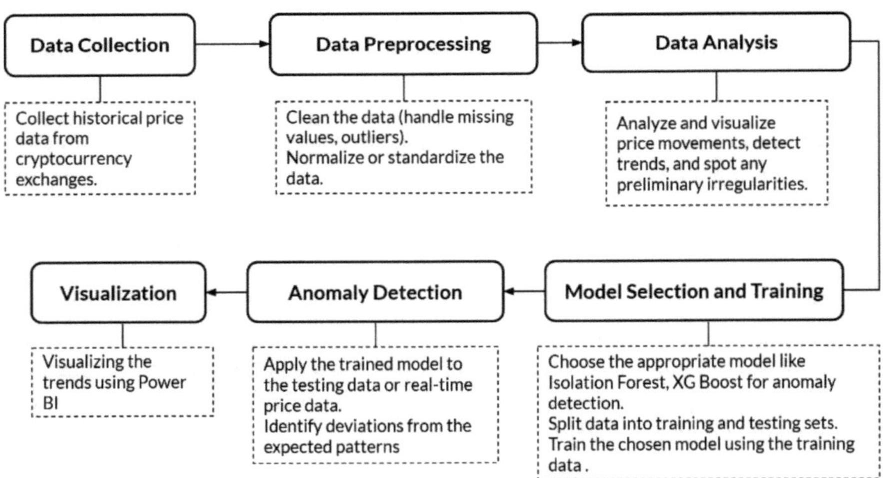

Fig. 1 Cryptocurrency price anomaly detection workflow

Table 1 Features in cryptocurrencies data

Column name	Description
Date	The specific date of the data point
Open	The opening price on that date
High	The highest price recorded on that date
Low	The lowest price recorded on that date
Close	The closing price on that date
Adj close	The adjusted closing price on that date
Volume	The trading volume on that date

ranges from 2015 to 2022, with 3128 rows. Avalanche, with data from 2021 to 2022, includes 1002 rows. Dogecoin, Polkadot, and Solana data range from 2018 to 2022, 2021 to 2022, and 2021 to 2022, containing 1979, 964, and 1096 rows, respectively. Tether data also spans from 2018 to 2022, with 1979 rows. The total number of columns for each cryptocurrency is consistent at 7. This dataset provides a comprehensive overview of the price movements and trading volume for these significant digital assets over their respective time periods, making it suitable for analyzing market trends and cryptocurrency behavior.

3.2 Data Preprocessing

Once the raw data is collected, it undergoes preprocessing to ensure accuracy, consistency, and usability for analysis. Preprocessing is critical to improving the quality of the dataset and involves several key steps:

1. **Handling Missing Data:** Gaps in the dataset can distort the analysis, so any missing values must be addressed. Techniques such as imputation (filling missing values using statistical methods) and removing incomplete rows are applied.
2. **Outlier Detection and Treatment:** Outliers, which are data points significantly different from others, may skew the results. Methods like z-score analysis are used for this purpose.

3.3 Data Analysis and Visualizing Data for Outlier Detection

The third phase, data analysis, involves exploratory data analysis (EDA) to better understand the structure and dynamics of the collected data. The objective is to uncover preliminary patterns, trends, and correlations that can guide model selection and training. During this stage, various statistical tools are used to measure central tendencies (mean, median), dispersion (variance, standard deviation), and relationships between variables (correlation analysis) (Table 2).

Table 2 Mean, median, variance, and standard deviation of price metrics

Metric	Mean	Median	Variance	Standard deviation
Open	3394	51	935,024,360	9670
High	3475	52	979,870,090	9899
Low	3307	49	886,693,360	9416
Close	3397	51	935,670,540	9673
Adj close	3397	51	935,670,540	9673
Volume	1,106,340,555,300	187,520,859,000	4.38×10^{20}	20,939,909,668

3.4 Model Selection and Training

With a well-prepared dataset and initial insights from EDA, the next step is to develop and train machine learning models capable of detecting anomalies in cryptocurrency prices. The process for model selection and training includes the following steps:

1. **Model Selection:** The system relies on established machine learning techniques suitable for anomaly detection. Two key models are considered:
   (a) *Isolation Forest:* An unsupervised learning algorithm that isolates anomalies by randomly selecting features and partitioning data. This model is particularly effective in identifying outliers based on the isolation principle.
   (b) *Random Forest:* An ensemble learning method that constructs multiple decision trees and combines their outputs for improved accuracy and robustness. In anomaly detection, Random Forest can be effective by measuring the average path length for each data point across the forest, with shorter paths indicating potential anomalies.
   (c) *XGBoost:* A powerful gradient boosting algorithm often used in structured data tasks. XGBoost is chosen for its capacity to model complex relationships and detect anomalies by identifying unusual data patterns.
2. **Data Splitting:** The dataset is divided into training and testing subsets to evaluate model performance. This split ensures that the model can generalize to unseen data.
3. **Model Training:** Using the training subset, the models are trained with techniques such as cross-validation to avoid overfitting and optimize their predictive performance. The models learn to recognize normal behavior patterns in the data.
4. **Hyperparameter Tuning:** To further improve accuracy, hyperparameters of the models are fine-tuned. This process involves adjusting key parameters (e.g., learning rate, tree depth for XGBoost) to achieve optimal performance in anomaly detection.

3.5 Anomaly Detection

Once the models are trained, they are applied to new or unseen data for anomaly detection. The objective is to identify deviations from expected behavior in cryptocurrency prices. Anomalies may manifest as sudden spikes or drops in price, unexpected trading volumes, or inconsistencies in price movements. The system continuously analyzes incoming data and flags anomalies based on significant deviations from historical patterns.

Detected anomalies are subjected to validation to ensure they are not due to noise, data quality issues, or external market factors unrelated to inherent cryptocurrency market behavior. The accuracy and relevance of the detected anomalies are crucial for informing further decision-making processes or market predictions.

3.6 Visualization

The final phase of the proposed approach focuses on effective communication of the findings through visualization tools. Visual representations of the data, such as charts, graphs, and dashboards, are essential for interpreting and understanding detected anomalies. Using platforms like Power BI, the system generates visual outputs that display trends, patterns, and anomalies over time. These visuals help stakeholders, including traders and analysts, to easily grasp the nature of market disruptions and anomalies. Moreover, visual insights support further exploration and decision-making.

4 Model Training and Anomaly Detection

4.1 Anomaly Detection Using Isolation Forest

The Isolation Forest algorithm was applied to detect anomalies in the price trends of major cryptocurrencies, including Bitcoin, Ethereum, Litecoin, Dogecoin, Solana, Polkadot, Avalanche, and Tether. The algorithm is recognized for its efficiency in identifying outliers in high-dimensional data, making it well-suited for analyzing cryptocurrency markets. The key goal of this experiment was to uncover significant price anomalies, reflecting unusual market activity or volatility, and provide insights into these deviations.

4.1.1 Formal Model of Isolation Forest

The Isolation Forest works by isolating data points (in this case, cryptocurrency price points) through recursive partitioning. The process involves randomly selecting a feature (price data) and a split value within its range to form isolation trees (iTrees). Anomalies, which are rare and deviate from the general trend, tend to be isolated earlier in the tree structure.

Mathematically, the path length $h(x)$ of a data point x is the number of edges required to isolate that point. Anomalies generally have shorter path lengths since they differ significantly from the rest of the data. The anomaly score $s(x, n)$, computed based on the average path length across trees, is given by:

$$s(x, n) = 2 - \frac{E(h(x))}{c(n)} \tag{1}$$

where $c(n)$ is a normalizing factor based on the sample size n, and $E(h(x))$ is the expected path length. An anomaly score close to 1 indicates a high likelihood of being an anomaly, while a score near 0 suggests normal behavior. This formal model underpins how the algorithm detects unusual price behavior in cryptocurrencies.

4.1.2 Key Observations

The Isolation Forest algorithm detected significant anomalies in cryptocurrency price trends, particularly during periods of rapid price increases and market peaks. Across both established cryptocurrencies like Bitcoin, Ethereum, and Litecoin, and newer assets such as Solana, Polkadot, and Avalanche, anomalies were concentrated during volatile market phases, especially the 2021 bull run. These anomalies, which highlight speculative trading and market overextension, were also prominent in Dogecoin's 2021 surge, driven by social media and celebrity endorsements. Tether, as a stablecoin, exhibited anomalies tied to minor deviations from its \$1 peg, indicating market stress or liquidity issues.

Overall, the algorithm effectively identified outliers during extreme market behavior, aligning with broader cryptocurrency cycles like bull runs and corrections, making it a valuable tool for highlighting periods of heightened volatility and investor risk (Fig. 2).

4.1.3 Implications of Anomaly Detection

This makes the Isolation Forest a valuable tool for risk management, as it offers early detection of potential market bubbles or unsustainable price movements, allowing investors to act preemptively. The ability to detect anomalies during extreme market behavior also supports the development of trading strategies, enabling traders to adjust portfolios, mitigate risks, or exploit market inefficiencies, especially during speculative bubbles or corrections.

The distinction between traditional cryptocurrencies like Bitcoin, Ethereum, and Litecoin, and stablecoins like Tether highlights the divergent anomaly patterns of these assets. While traditional cryptocurrencies exhibit anomalies during price surges, stablecoins, such as Tether, show deviations tied to market stress and liquidity issues, detectable even with minor fluctuations. The algorithm is particularly sensitive to rapid price increases, making it effective at identifying speculative bubbles typically followed by corrections.

Furthermore, it has the potential to serve as an early warning system, detecting anomalies in real-time and enabling proactive risk management to anticipate volatility spikes or market corrections.

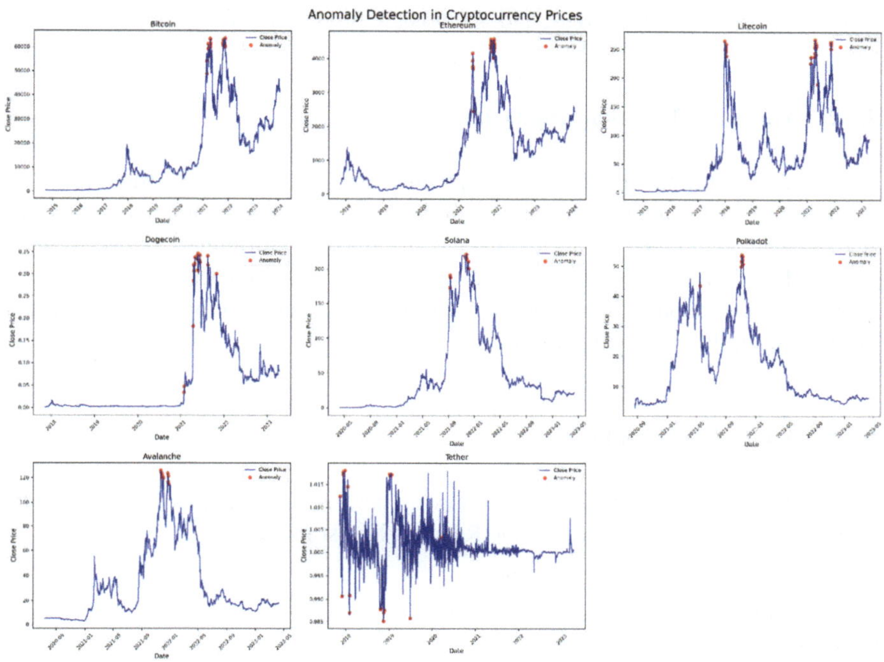

Fig. 2 Anomaly detection in cryptocurrency prices (bitcoin, ethereum, litecoin, dogecoin, solana, polkadot, avalanche, and tether)

4.2 Anomaly Detection Using Random Forest

The Random Forest algorithm was applied to detect anomalies in the price movements of various cryptocurrencies. This ensemble learning method leverages multiple decision trees to identify outliers in complex, high-dimensional data, making it a powerful tool for anomaly detection in financial markets.

4.2.1 Formal Model of Random Forest

Random Forest is an ensemble learning method that operates by constructing multiple decision trees during training and outputting the mean prediction of the individual trees for regression tasks. In the context of anomaly detection, the model functions as follows:

The Random Forest prediction for a given input x is:

$$f(x) = \frac{1}{N} \sum_{t=1}^{N} f_t(x) \tag{2}$$

where N is the number of trees, and $f_t(x)$ is the prediction of the t-th tree.

4.2.2 Anomaly Detection Approach

The model is trained on normal cryptocurrency price patterns. Anomalies are identified when the actual price significantly deviates from the model's predictions. A data point is flagged as an anomaly if the residual (difference between actual and predicted value) exceeds a certain threshold, typically based on the standard deviation of residuals. This approach helps detect unusual price behavior that diverges from the expected patterns.

4.2.3 Analysis of Results

Model Performance Metrics The following performance metrics demonstrate the accuracy and predictive power of the Random Forest model:

- Mean Squared Error (MSE): 0.0000
- Root Mean Squared Error (RMSE): 0.0009
- R^2 Score: 0.9405

These metrics indicate excellent model performance:

- The extremely low MSE and RMSE suggest that the model's predictions closely align with the actual cryptocurrency prices.
- The high R^2 score (0.9405) shows that the model explains 94.05% of the variability in the cryptocurrency prices, demonstrating strong predictive power.

4.2.4 Anomaly Detection Results

The Random Forest model was used to detect anomalies in the price data for several major cryptocurrencies. The table below shows the percentage of anomalies detected for each cryptocurrency (Table 3):

Table 3 Anomalies detected in various cryptocurrencies

Cryptocurrency	Anomalies detected (%)
Bitcoin	2.70
Ethereum	2.22
Litecoin	1.75
Dogecoin	1.95
Solana	2.97
Polkadot	2.08
Avalanche	1.83
Tether	2.60

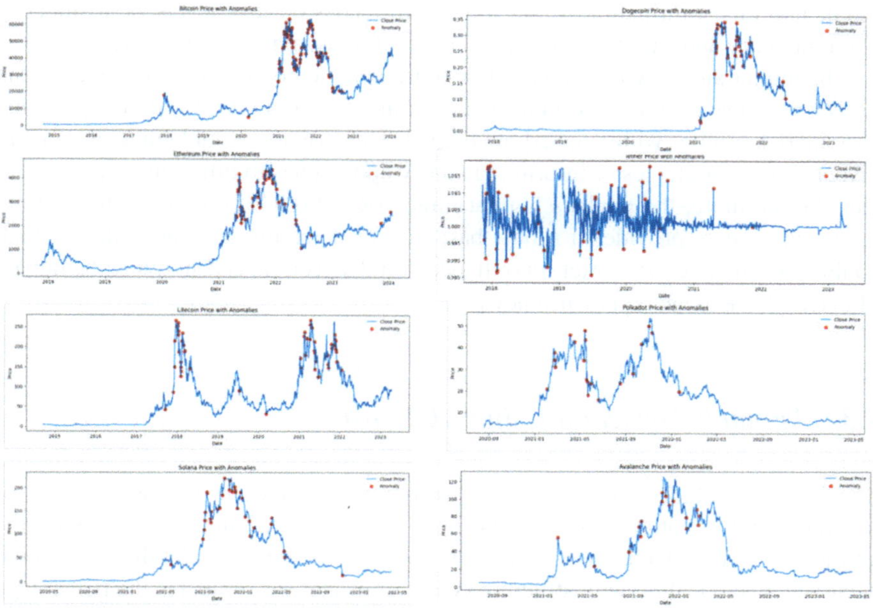

Fig. 3 Detailed cryptocurrency price charts with anomalies

4.2.5 Key Observations

The percentage of anomalies is relatively consistent across different cryptocurrencies, ranging from 1.75% (Litecoin) to 2.97% (Solana). Solana exhibits the highest percentage of anomalies (2.97%), indicating more frequent unusual price movements or higher volatility. Litecoin, with the lowest percentage of anomalies (1.75%), suggests more stable or predictable price behavior.

The graphs accompanying the analysis visualize the price movements and detected anomalies for each cryptocurrency. Red dots indicate detected anomalies, which often correspond to sudden price spikes or drops (Fig. 3).

4.2.6 Implications and Insights

The Random Forest model's anomaly detection results offer several valuable insights for cryptocurrency market participants and researchers. The consistently low percentage of anomalies (ranging from 1.75% to 2.97%) across different cryptocurrencies suggests that extreme price movements, while present, are relatively rare events. This finding has important implications for risk management and trading strategies. For investors, it highlights the importance of being prepared for sudden, anomalous price changes that could significantly impact portfolio values. Traders might use this information to set more informed stop-loss levels or to identify potential entry and exit points during periods of unusual market activity.

The higher anomaly rate in newer or more volatile cryptocurrencies like Solana (2.97%) compared to more established ones like Litecoin (1.75%) underscores the varying risk profiles within the crypto market. This disparity could guide investment diversification strategies or inform regulatory approaches to different cryptocurrencies. Furthermore, the model's high accuracy (R^2 of 0.9405) in predicting normal price movements provides a robust baseline for detecting true anomalies, which could be invaluable for market surveillance, fraud detection, and studying the impact of external events on cryptocurrency prices.

4.3 Anomaly Detection Using XGBoost

The XGBoost algorithm was applied to detect anomalies in the price trends of several major cryptocurrencies. XGBoost, an advanced implementation of the gradient boosting technique, is known for its efficiency, scalability, and high accuracy when dealing with structured data. This analysis evaluates the percentage of data points classified as anomalies for each cryptocurrency, along with detailed model performance metrics.

4.3.1 Formal Model of XGBoost

XGBoost (Extreme Gradient Boosting) is a powerful machine learning algorithm based on the gradient boosting framework. It creates an ensemble of weak prediction models, typically decision trees, to form a strong predictive model. In the context of anomaly detection, XGBoost works by minimizing a loss function (residuals) and iteratively improving the model.

Formula for XGBoost Prediction

At each iteration t, the XGBoost model updates the predictions by adding a new tree's prediction $f_t(x)$, aimed at correcting the previous model's errors:

$$\hat{y}^{(t)} = \hat{y}^{(t-1)} + \eta f_t(x) \tag{3}$$

Where:

$\hat{y}^{(t)}$	is the prediction after iteration t,
$\hat{y}^{(t-1)}$	is the prediction from the previous iteration,
$f_t(x)$	is the prediction from the current decision tree at iteration t,
η	is the learning rate, which controls the contribution of each tree.

The model iterates until the residuals (errors) are minimized, creating a strong final model from multiple weak learners.

Table 4 Anomalies detected in various cryptocurrencies using XGBoost

Cryptocurrency	Data points	Anomalies detected (%)
Bitcoin	3413	2.31
Ethereum	2264	1.86
Litecoin	3128	0.10
Tether	1979	1.72
Avalanche	1002	0.30
Dogecoin	1979	1.16
Solana	1096	1.37
Polkadot	964	1.24

4.3.2 Anomaly Detection Approach

XGBoost was trained on cryptocurrency price data to capture normal price behavior. Data points are flagged as anomalies when their predicted values deviate significantly from the actual prices. The degree of deviation is measured based on residuals, and points exceeding a set threshold are considered anomalous.

4.3.3 Anomaly Detection Results

The XGBoost algorithm was used to detect anomalies in the price data for various cryptocurrencies. The table below summarizes the percentage of data points classified as anomalies (Table 4):

4.3.4 Key Observations

- **Variation in Anomaly Rates:** Anomaly rates vary significantly across different cryptocurrencies, ranging from 0.10% for Litecoin to 2.31% for Bitcoin.
- **Established vs. Newer Cryptocurrencies:** Established cryptocurrencies like Bitcoin (2.31%) and Ethereum (1.86%) show higher anomaly rates, potentially due to their central role in the crypto market. Newer cryptocurrencies like Avalanche (0.30%) and Solana (1.37%) show varied anomaly rates, likely reflecting differences in market volatility and investor behavior.
- **Stablecoin Anomalies:** Tether, designed as a stablecoin, exhibits a relatively high anomaly rate of 1.72%, indicating deviations from its $1 peg, potentially due to market volatility or liquidity issues.
- **Low Anomaly Rate in Litecoin:** Litecoin shows the lowest anomaly rate of 0.10%, which suggests more stable price behavior over the analyzed period, potentially reflecting its long-established market position (Fig. 4).

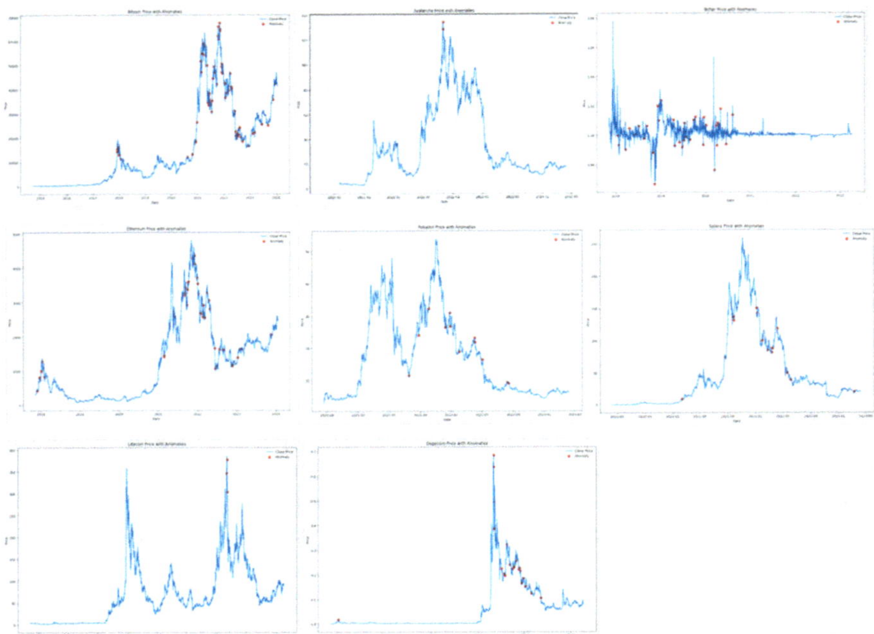

Fig. 4 Extended cryptocurrency price analysis with anomalies

4.3.5 Model Performance Metrics

The XGBoost model demonstrated strong performance in predicting cryptocurrency prices, as shown by the following metrics:

- Mean Squared Error (MSE): 0.0055
- Root Mean Squared Error (RMSE): 0.0741
- R^2 Score: 1.0000
- Average R^2 score across folds (Cross-validation): 0.9978

These performance metrics indicate high accuracy:

- The low MSE (0.0055) and RMSE (0.0741) show that the prediction errors are minimal, indicating that the model's predictions are very close to the actual prices.
- The R^2 score of 1.0000 signifies that the model explains nearly all the variability in cryptocurrency price data.
- The high average R^2 score of 0.9978 across cross-validation folds reinforces the consistency and robustness of the model.

5 Cross-Validation

Cross-validation is a technique used to evaluate the generalizability of a model by splitting the dataset into multiple subsets or "folds" and training the model on different portions of the data. It ensures that the model's performance is consistent across various subsets of the data and prevents overfitting, where the model performs well on the training data but poorly on unseen data.

Purpose in This Context Cross-validation was essential in this study to ensure that the XGBoost model's ability to detect anomalies was not overfitted to a specific segment of cryptocurrency price data. Since cryptocurrency markets are highly volatile and diverse, validating the model's performance across multiple data folds helps guarantee that the model can generalize well to future data and accurately capture unusual price behavior.

5.1 Cross-Validation Results

The high average R^2 score of 0.9978 across cross-validation folds demonstrates that the model is highly reliable and consistent, indicating that it is unlikely to overfit the training data and can accurately identify anomalies across various market conditions.

5.2 Implications and Insights

The XGBoost anomaly detection model offers valuable insights into the behavior of various cryptocurrencies. Market dynamics are reflected in the variation of anomaly rates across different cryptocurrencies, likely due to differences in market maturity, liquidity, and exposure to external factors such as regulatory changes and market sentiment. For instance, Bitcoin and Ethereum volatility is evident in their higher anomaly rates, with Bitcoin at 2.31% and Ethereum at 1.86%, indicating that these cryptocurrencies are more responsive to market-wide events and speculative trading due to their central roles in the crypto ecosystem.

Interestingly, Tether's anomaly rate, which stands at 1.72%, highlights that even stablecoins—designed to maintain price stability—can experience significant deviations during periods of market stress. This finding suggests that further scrutiny of Tether's stability mechanisms is necessary. On the other hand, Litecoin's low anomaly rate (0.10%) suggests a relatively stable price trend, though this could also point to potential refinements needed in the model's sensitivity to detect anomalies in Litecoin.

The reliability of the XGBoost model is underscored by its high R^2 scores and low error rates, demonstrating its effectiveness in capturing underlying patterns in cryptocurrency price movements. This suggests that the model is a reliable tool for detecting anomalies. Furthermore, the model's strong performance indicates its potential for predictive analytics, which could be leveraged to offer insights into future price movements within cryptocurrency markets.

Finally, the model's cross-validation confidence is bolstered by a high score of 0.9978, enhancing trust in its ability to generalize across different data subsets and ensuring robustness for future applications.

6 Conclusion

This study emphasizes the effectiveness of the Random Forest and XGBoost models in detecting anomalies within cryptocurrency price movements. The Random Forest model achieved a strong R^2 score of 0.9405, indicating that it explains over 94% of the variability in cryptocurrency prices. It detected anomalies across several major cryptocurrencies, with anomaly rates ranging from 1.75% for Litecoin to 2.97% for Solana. These rates demonstrate the model's ability to identify abnormal price behavior, particularly in volatile cryptocurrencies like Solana, while detecting fewer anomalies in more stable assets like Litecoin.

The XGBoost model delivered even stronger performance, boasting an R^2 score of 1.0000 and an average cross-validation R^2 score of 0.9978. It identified varying anomaly rates across different cryptocurrencies, with Bitcoin (2.31%) and Ethereum (1.86%) exhibiting higher rates due to their pivotal roles in the crypto market, compared to Avalanche (0.30%) and Litecoin (0.10%), which showed fewer anomalies. Even Tether, designed for price stability, had an anomaly rate of 1.72%, highlighting its vulnerability to deviations during market stress.

Graphical outputs, such as scatter plots and time-series anomaly charts, visualize these findings, illustrating significant price spikes and sudden drops corresponding to detected anomalies. For instance, the Z-score method applied to Bitcoin flagged four outlier days in 2021, with a peak Z-score of 3.241 recorded on November 8, 2021, marking a sharp price surge. Similar trends were observed in Ethereum, Dogecoin, and Solana, particularly during periods of speculative trading and market overextensions.

These models provide a strong foundation for risk management and trading strategies, as they detect anomalies during crucial market events, allowing traders to react quickly. In future work, implementing a real-time dashboard that leverages these models on live data streams would be a significant step forward. This would enable traders and investors to set custom alerts based on anomaly detection, monitor cryptocurrency prices dynamically, and make more informed decisions in real time, ultimately improving risk management in the highly volatile cryptocurrency market.

7 Future Scope

The future scope of this anomaly detection project in cryptocurrency markets could be significantly expanded by creating an interactive dashboard using a web application. This dashboard would serve as a powerful, real-time tool for detecting and visualizing anomalies in cryptocurrency price movements, offering enhanced decision-making capabilities for traders, investors, and market analysts. The implementation of such a system would allow the Random Forest or XGBoost models to work on live data streams, constantly analyzing price changes and flagging any abnormal behavior as it happens. This would provide market participants with timely insights, enabling them to react swiftly to market volatility and adjust their strategies in real time.

The dashboard could feature real-time price monitoring for multiple cryptocurrencies, displaying key metrics such as anomaly rates, historical trends, and live price charts. Users could set customizable alerts for significant anomalies, allowing them to receive notifications when a cryptocurrency exhibits unusual price movements. This could be particularly valuable in periods of market stress, where rapid price changes might trigger anomalies. Additionally, integrating a predictive analytics component would further enhance the tool by offering forecasts of potential future price movements based on the models' outputs, giving users a proactive approach to their trading or investment decisions.

Another key feature of the web-based dashboard could be its user-friendly interface that allows for easy navigation and customization. Market participants could filter cryptocurrencies by volatility, market capitalization, or anomaly rate, enabling them to focus on assets most relevant to their portfolios or strategies. The platform could also offer historical anomaly data for backtesting trading strategies, helping traders refine their approaches based on past market behavior. Moreover, the inclusion of visual analytics like heatmaps, charts, and trend lines would offer a more intuitive understanding of the data, making the dashboard accessible not just to seasoned traders but also to newcomers in the cryptocurrency space.

Overall, the creation of a web application dashboard would transform the anomaly detection models into a highly practical, real-time tool, empowering users to navigate the complexities of the cryptocurrency market with greater confidence and precision.

References

1. ElBahrawy, A., et al. (2019). Cluster analysis on the structure of the cryptocurrency transaction network. *Physica A: Statistical Mechanics and its Applications*, 523, 731–742. https://doi.org/10.1016/j.physa.2019.121339
2. Liu, F., Li, Y., & Xie, H. (2021). Bitcoin transaction strategy construction based on deep reinforcement learning. *Applied Soft Computing*, 113, 107952. https://doi.org/10.1016/j.asoc.2021.107952

3. Liu, L., et al. (2022). Blockchain-enabled fraud discovery through abnormal smart contract detection on Ethereum. *Future Generation Computer Systems*, 128, 158–166. https://doi.org/10.1016/j.future.2021.08.023
4. Morishima, S. (2021). Scalable anomaly detection in blockchain using graphics processing unit. *Computers & Electrical Engineering*, 93, 107087. https://doi.org/10.1016/j.compeleceng.2021.107087
5. Rwibasira, M., & Suchithra, R. (2022). ADOBSVM: Anomaly detection on blockchain using support vector machine. *Measurement: Sensors*, 24, 100503. https://doi.org/10.1016/j.measen.2022.100503
6. Cunha, L. L. (2021). Active learning in the detection of anomalies in cryptocurrency transactions. *Journal of Financial Crime*, 28(4), 1234–1250. https://doi.org/10.3390/make5040084
7. Rudkiewicz, J., & Hernes, M. (2023). Investment strategies based on anomalies detected in the financial time series of cryptocurrencies. *Procedia Computer Science*, 225, 53–62. https://doi.org/10.1016/j.procs.2023.09.091

GEMNet: Gabor Enhanced Multiscale CNN for Breast Tumor Detection Using Biorthogonal Wavelet Transform Features

Shairal Verma and Rajneesh Rani

Abstract Breast cancer remains a prevalent cause of mortality, and accurate early detection through mammography significantly improves patient outcomes. However, current diagnostic techniques often struggle with distinguishing benign from malignant mammographic findings, particularly in dense or subtle presentations. To address these limitations, we propose a novel deep learning framework, the Gabor Enhanced Multi-Scale Neural Network (GEMNet), integrated with biorthogonal wavelet transforms to capture fine-grained spatial and frequency domain features. The wavelet transforms enable multi-resolution analysis, allowing the model to identify localized variations in texture and intensity, which are critical for accurate lesion classification. Our approach leverages Gabor filters for enhanced feature extraction, enabling the model to discern subtle patterns within mammograms. We rigorously benchmarked GEMNet against leading architectures, including EfficientNet B0, ResNet150, VGG16, and MobileNetV2, across comprehensive mammographic datasets. Results indicate that GEMNet outperforms these state-of-the-art models, achieving an F1 score of 97% and an AUC of 99.3%, marking a significant advancement in diagnostic accuracy. This work demonstrates the efficacy of combining multi-scale analysis with wavelet-based transformations, providing a robust framework for high-precision mammographic lesion classification. GEMNet holds promise for integration into clinical workflows, potentially reducing false positives and false negatives, and advancing breast cancer diagnostics through machine learning.

Keywords Breast tumor detection · Wavelet transform · Deep learning · Gabor regularization · Convolutional neural networks

S. Verma (✉) · R. Rani
Department of Computer Science and Engineering, Dr B R Ambedkar National Institute of Technology, Jalandhar, India
e-mail: shairalv.is.23@nitj.ac.in; ranir@nitj.ac.in

K. Deep et al. (eds.), *Artificial Intelligence, Machine Learning and Intelligent Systems*, Springer Proceedings in Mathematics & Statistics 520,
https://doi.org/10.1007/978-981-95-3841-6_16

1 Introduction

Early identification is essential for addressing breast cancer, one of the main causes of mortality among women. Mammography, which detects cancer visually, is the principal screening method. Breast tissue heterogeneity—density and texture—and the minor differences between benign and malignant cancers challenge mammographic interpretation. Diagnostic errors, including false positives that delay vital therapies and false negatives that cause unneeded operations, are widespread due to its complexity [15]. Although useful, conventional computer-aided detection (CAD) approaches generally lack the sensitivity and specificity needed for accurate detection because to their limited ability to capture sensitive spatial and frequency-based mammography information [6]. Thus, new diagnostic instruments that can identify benign and malignant masses and calcifications are needed urgently to eliminate diagnostic ambiguity and facilitate therapeutic decision-making.

Recent years have seen deep learning (DL) improve medical imaging analysis, especially through CNNs that excel at feature extraction and pattern recognition [16]. Some of the known networks such as VGG16, ResNet, and EfficientNet were designed for spatial qualities, which limited their performance on complex mammographic data. Textural alterations and multi-scale characteristics can categorize breast cancer lesions. Traditional CNN models may not collect frequency-domain information essential to discern textural patterns in dense or intricate mammograms [5]. A model that efficiently gathers geographical information and includes frequency-domain analysis would increase diagnosis accuracy. This research aims to develop a deep learning model that diagnoses breast cancer in mammograms more reliably and interpretably than CNN-based methods using multi-scale and frequency-based feature extraction.

Also, conventionally we drive deep learning with multi-modal images directly fed to the models. In this study we introduce the Gabor Enhanced Multi-Scale VGG16 model that uses biorthogonal wavelet transformations to circumvent conventional and deep learning-based mammographic analysis constraints. Biorthogonal wavelet transforms break images into several frequency components, letting the model detect minute textural details while preserving image structure. We use spatial and frequency-domain data in deep learning to enhance mammographic classification.

2 Literature Review

This section provides an in-depth analysis of deep learning models for breast tumor detection. Table 1 presents a tabular overview, highlighting key findings, advantages, and limitations of recent relevant approaches.

John et al. [8] improved Content-Based Image Retrieval using deep learning. It uses variational autoencoder multi-modal features to optimize the search and

Table 1 Comparative evaluation of various studies on breast tumor detection

Author	Techniques	Key metrics	Description	Disadvantage
John et al. [8]	Variational Autoencoders, CSNN	Improved retrieval	Two-stage CBIR using deep learning, achieving significant performance gains on the Oxford dataset.	Model complexity may lead to overfitting.
Razali et al. [12]	CNNs, Wavelet Scattering	98.0% for masses, 99.3% for tissue	CAD system for breast tissue classification, improving representation with WS on a small dataset.	Limited dataset size may affect generalization.
Al Husaini et al. [2]	Inception models	99.748% accuracy	Enhances breast cancer classification using thermal imaging and deep learning, achieving high accuracy.	Relies on specific thermal imaging technology.
Taylor et al. [14]	ResNet34	High sensitivity	Analyzes cancer-associated EVs, effectively distinguishing cancer stages for research and clinical use.	May not generalize to other cancers.
Ahmad et al. [1]	CNNs, SVM	99.16% accuracy	Automated system for mammogram analysis, outperforming traditional methods in breast cancer detection.	Complexity may hinder interpretability.
Kumbhare et al. [10]	Federated Learning, DenseNet	95% accuracy, 91% MCC	Combines FL and deep learning for breast cancer detection while preserving data privacy.	Coordination challenges in federated settings.
Kumar et al. [9]	Stacking Ensemble Learning	99.45% accuracy	Uses ensemble learning to improve early identification of breast cancer accuracy on the Wisconsin dataset.	Potential for overfitting.
Aljuaid et al. [4]	Deep Neural Networks, Transfer Learning	99.7% for binary classification	Utilizes DL for CAD, achieving high accuracy for breast cancer classification on the BreakHis dataset.	Transfer learning effectiveness may vary.

compare with a CSNN. CSNN classify photos by query image similarity using pseudo-labeling. Oxford dataset retrieval is more effective than typical methods.Razali et al. [12] used CNNs and WS to classify breast tissue images in CAD systems. Addressing CNN constraints from faulty training pictures and automated spatial data extraction. WS keeps high-frequency data for visualization. The model classified masses at 98.0% and breast tissue at 99.3% with 112 INbreast mammography images, demonstrating good training with limited data and complexity. Al Husaini et al. [2] discovered that deep learning and real-time thermography enhance breast cancer classification. The modified Inception Mv4 had the highest accuracy (99.748%), topping v4 (99.712%) and v3 (96.8%). By automating dual imaging analysis for single-vesicle molecular profiling of cancer-associated EVs, Taylor et al. [14] can diagnose and track breast cancer. HER2-positive breast cancer biofluid samples contain EpCAM-positive and CD24-positive extracellular vesicles detected by Resnet34. EVs were rare in healthy people and Stage I patients but grew dramatically from Stage II to III.

Ahmad et al. [1] suggest early breast cancer detection for improved outcomes. BreastNet-SVM classifies mammography breast cancer using CNNs. The Digital Database for Screening Mammography trains the nine-layer model using SVM classification. It outperforms other DDSM dataset detection algorithms with 99.16% accuracy, 97.13% sensitivity, and 99.30% specificity.

Federated Learning (FL) is used by Kumbhare et al. [10] to protect sensitive breast cancer data while working on a large dataset Federated and deep learning employ DDSM mammography. Hybrid Dragon-Rider Optimization enhances DenseNet feature extraction and E-RNN classification. With 95% accuracy and 91% MCC, the model beats conventional breast cancer diagnosis approaches.

Kumar et al. [9] found that early breast cancer identification can reduce mortality rates in women in their 40s. On the Wisconsin breast cancer dataset, the OSEL model outperforms AdaBoostM1, gradient boosting, and XGBoost with 99.45% accuracy. This study shows integrated prediction algorithms aid breast cancer diagnosis and prevention.

Mohamed et al. [11] study breast cancer detection and biopsy needs worldwide. The top models, Densenet169, Resnet50, and Resnet101, attained accuracy rates of 62, 68, and 85%, respectively. Data augmentation and segmentation improved accuracy by 20%. Ensemble learning increased accuracy to 92.5. As breast cancer rates rise due to lack of awareness and delayed diagnoses, Aljuaid et al. [4] prioritize detection and categorization. A new computer-aided diagnostic technique for BreakHis uses transfer learning with deep neural networks (ResNet 18, ShuffleNet, and Inception)V3Net). ResNet, InceptionV3Net, and ShuffleNet achieved 99.7, 97.66, and 96.94% binary classification accuracy for benign and malignant occurrences. The multi-class classification strategy improved early breast cancer diagnosis for the same networks, achieving 97.81, 96.07, and 95.79% accuracy.

3 Methodology

3.1 Dataset and Pre-processing

We utilized the CBIS-DDSM (Curated Breast Imaging Subset of the Digital Database for Screening Mammography) [13], an updated, standardized version of the DDSM dataset. CBIS-DDSM includes 6775 mammographic series from 1566 participants, with cases categorized as normal, benign, or malignant based on verified pathology. For our analysis, we grouped normal and benign cases into a single "non-tumorous" class, while malignant cases comprised the "tumorous"

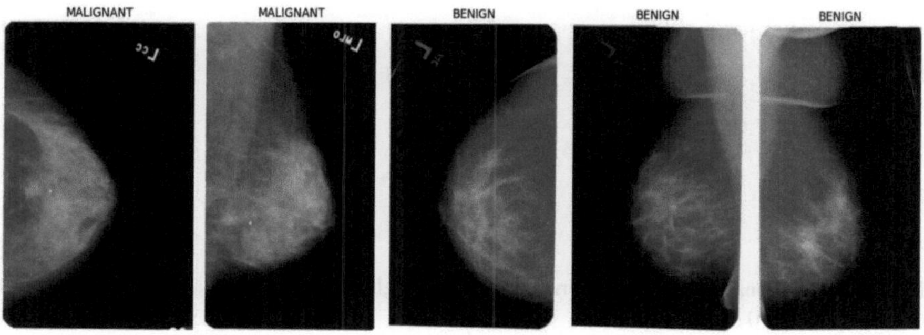

Fig. 1 Full mammogram images of some sampled images from CBIS-DDSM dataset

class, resulting in 1018 benign/non-tumorous and 570 malignant/tumorous images. We focused exclusively on full mammogram images (as shown in Fig. 1), excluding cropped ROIs and ROI masks, to retain complete anatomical context in model training.

Pre-processing Prior to model training, we performed several pre-processing steps to standardize the images. First, we rescaled each image to a fixed resolution of 1024 × 1024 pixels using bilinear interpolation. This resizing process ensures that all input images have consistent dimensions, as described by Eq. (1), where I represents the original image, and $w_{i,j}$ are the interpolation weights.

$$I'_{x',y'} = \sum_{i=0}^{1}\sum_{j=0}^{1} w_{i,j} \cdot I_{x+i,y+j} \tag{1}$$

$$I''_{x,y} = \frac{I'_{x,y} - p_{\min}}{p_{\max} - p_{\min}} \tag{2}$$

The pixel intensities were scaled to the range [0, 1] to ensure consistency across the dataset, using the formula (2), where $p_{\min}$ and $p_{\max}$ represent the minimum and maximum pixel values, correspondingly.

3.2 *Biorthogonal Wavelet Transforms*

Biorthogonal wavelet transforms are widely used for image decomposition, enabling the extraction of spatial and frequency components. Biorthogonal wavelets support symmetric filter functions and near-perfect reconstruction, which are beneficial for image processing [17]. For our study we have considered this method as it will aid the feature extraction process from large sizes of original mammograms. There are following two processes:

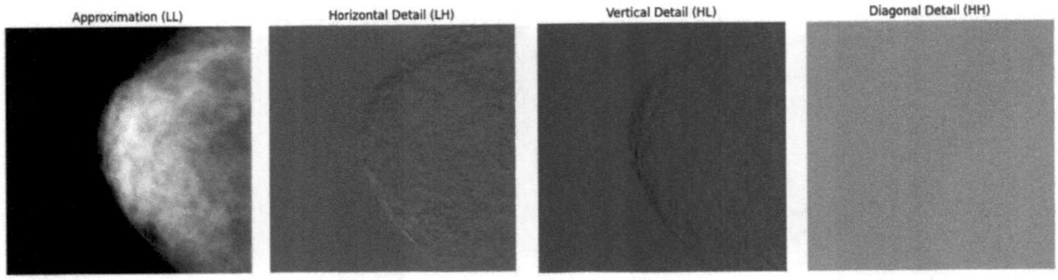

Fig. 2 Biorthogonal Wavelet Transform applied to a single mammogram image (Cropped to cover only breast region)

Decomposition Decompose image I into four subbands: LL, LH, HL, and HH by applying low-pass (g) and high-pass (h) filters both row-wise and column-wise as described in formula 3,4,5, and 6, where g_r and h_r denote row filters, and g_c and h_c denote column filters. The outputs are shown in Fig. 2.

$$LL = (I * g_r) * g_c^T, \tag{3}$$

$$LH = (I * g_r) * h_c^T, \tag{4}$$

$$HL = (I * h_r) * g_c^T, \tag{5}$$

$$HH = (I * h_r) * h_c^T, \tag{6}$$

Reconstruction The original image I is approximated by combining the four sub bands as formula (7). Further, for better de-noising of the filtered features, we apply gaussian de-noising as described by formula (8).

$$I \approx LL * g_r * g_c^T + LH * g_r * h_c^T + HL * h_r * g_c^T + HH * h_r * h_c^T \tag{7}$$

$$G(x, y) = \frac{1}{2\pi\sigma^2} \exp\left(-\frac{x^2 + y^2}{2\sigma^2}\right), \quad I' = I * G \tag{8}$$

Properties Biorthogonal wavelets support symmetric filters, which minimize boundary effects in image processing [7]. Also, the filters meet the perfect reconstruction condition, allowing accurate image reconstruction with minimal loss.

3.3 Gabor Regularization

In this study we have introduced gabor regularization which further puts a constraint on the convolutional filters in CNNs to approximate the characteristics of Gabor filters, which are effective for capturing spatial frequencies and orientations in images. This method enhances model robustness and interpretability by focusing on textures, edges, and other important visual features. The gabor filter is designed to capture both spatial and frequency domain features, making it ideal for tasks involving edge detection and texture analysis [3]. A gabor filter function in a spatial domain (as we are working with images) can be seen in formula (9).

$$G(x, y; \lambda, \theta, \psi, \sigma, \gamma) = \exp\left(-\frac{x'^2 + \gamma^2 y'^2}{2\sigma^2}\right) \cos\left(2\pi \frac{x'}{\lambda} + \psi\right) \tag{9}$$

where $x' = x\cos\theta + y\sin\theta$, $y' = -x\sin\theta + y\cos\theta$, λ is the wavelength of the sinusoidal component, θ is the orientation of the filter, σ is the standard deviation of the Gaussian envelope, γ is the spatial aspect ratio, and ψ is the phase offset.

3.3.1 Gabor Regularization Term

Let W_i be the filters of a convolutional layer in the CNN. The Gabor regularization term $R_{\text{Gabor}}(W)$ ensures that these filters resemble Gabor functions by minimizing the difference between the learned filters and predefined Gabor-like patterns. We can define this regularisation as following formula (10).

$$R_{\text{Gabor}}(W) = \sum_{i=1}^{N} \| W_i - G_{\lambda,\theta,\psi,\sigma,\gamma} \|^2 \tag{10}$$

where W_i are the convolutional filters, $G_{\lambda,\theta,\psi,\sigma,\gamma}$ is the Gabor filter with parameters $\lambda, \theta, \psi, \sigma, \gamma$, and N is the total number of filters in the layer. This regularization term ensures that the filters learn representations similar to Gabor features, which are known to capture key visual patterns in images.

3.3.2 Loss Function with Gabor Regularization

The overall objective function for training the CNN is given by Eq. (11), where, the Cross-Entropy Loss is the standard loss used for classification tasks, and α adjusts the trade-off between standard loss and the Gabor regularization.

$$\text{Loss} = \text{Cross-Entropy Loss} + \alpha R_{\text{Gabor}}(W) \tag{11}$$

3.4 Multi-Scale CNN Network (GEMNet) Architecture

Our model architecture utilizes four parallel streams of VGG-16 networks, each receiving a pre-processed 1024×1024 image input. These images are ouput images from biorthogonal wavelet transform as shown in Fig. 3. Each VGG-16 stream

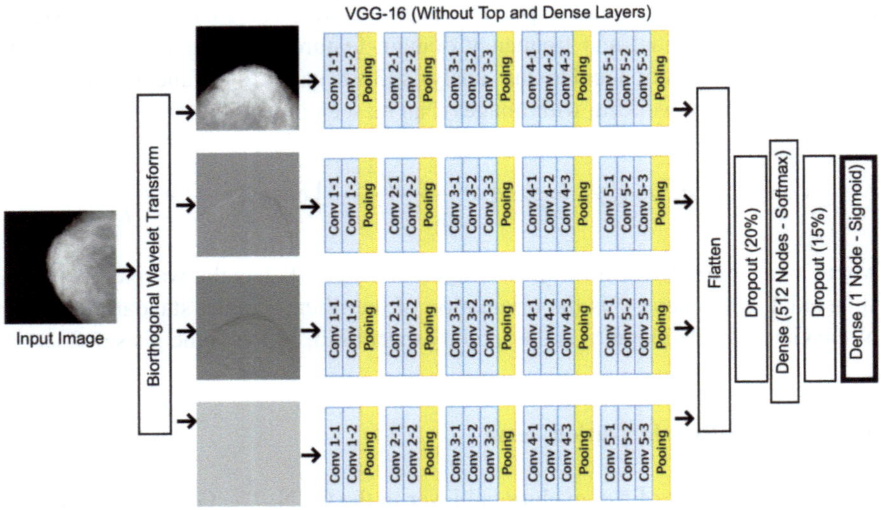

Fig. 3 GEMNet: Gabor enhanced Multi-Scale VGG-16 network architecture with custom top layers and fusion of features

processes the image through a series of convolutional and max-pooling layers, extracting multi-scale spatial features across the depth of the network. These parallel streams operate independently until the final convolutional layer, allowing each to capture complementary features and patterns from the image. Following the last convolutional layer in each stream, the feature maps are flattened and stacked into a single, unified vector representation, effectively consolidating the extracted spatial information from all four streams. The architecture concludes with a final dense layer using a sigmoid activation function, outputting a probability score that classifies the image as either tumorous or non-tumorous as shown in Fig. 3.

Feature Fusion The feature maps F_i from each scale are combined into a unified representation. If using concatenation, the combined feature map F_{combined} is:

$$F_{\text{combined}} = \text{concat}(F_1, F_2, \ldots, F_n) \tag{12}$$

where concat denotes the concatenation of feature maps across the different scales. The combined feature map F_{combined} is then passed through the fully connected layers to obtain the final output:

$$\text{Output} = \text{Fully Connected Layers}(F_{\text{combined}}) \tag{13}$$

The GEMNet leverages input images at multiple resolutions to capture features at varying scales, enhancing performance in tasks such as object detection and image segmentation.

4 Result and Discussion

The trials were performed on Google Colab, employing an NVIDIA Tesla T4 GPU with 16 GB of GPU memory, 12 GB of system RAM, and 100 GB of available disk storage. We applied an 80–20 holdout validation approach, using 80% of the dataset allocated for training and 20% for testing, to ensure robust performance assessment across all evaluation metrics.

Our proposed GEMNet model demonstrates superior performance across all key evaluation metrics on both training and test data, underscoring its potential as a robust tool for mammographic analysis. As shown in Table 2, GEMNet achieved a remarkable F1 score of 0.980, with precision and recall values of 0.979 and 0.981, respectively, on the training set, and an accuracy of 0.980 and AUC of 0.992. This consistent high performance across training metrics reflects GEMNet's ability to effectively capture and generalize complex mammographic features.

On the test set (Table 3), GEMNet maintained its high performance, with an F1 score of 0.968, precision of 0.969, and recall of 0.971. Additionally, it achieved an

Table 2 Summary of results for GEMNet against state-of-the-art models on CBIS-DDSM training data

Model	F1	Precision	Recall	Accuracy	AUC
GEMNet	**0.980**	0.979	0.981	0.980	0.992
EfficientNetB0	0.960	0.958	0.961	0.958	0.971
ResNet-50	0.945	0.946	0.943	0.947	0.965
VGG-16	0.927	0.926	0.928	0.929	0.953
MobileNetV2	0.917	0.915	0.918	0.919	0.947
VGG-19	0.896	0.897	0.894	0.894	0.933

The bold values in this table denote the best F1-score(s) achieved by the proposed model, as described in the Results section of the chapter. These values highlight the top-performing metric(s) for model comparison and indicate the superior performance of the proposed approach relative to the other methods listed.

Table 3 Summary of results for GEMNet against state-of-the-art models on CBIS-DDSM test data

Model	F1	Precision	Recall	Accuracy	AUC
GEMNet	**0.968**	0.966	0.971	0.971	0.990
EfficientNetB0	0.954	0.953	0.955	0.955	0.973
ResNet-50	0.940	0.938	0.942	0.941	0.966
VGG-16	0.923	0.924	0.921	0.922	0.954
MobileNetV2	0.908	0.907	0.909	0.910	0.946
VGG-19	0.885	0.883	0.887	0.884	0.932

The bold values in this table denote the best F1-score(s) achieved by the proposed model, as described in the Results section of the chapter. These values highlight the top-performing metric(s) for model comparison and indicate the superior performance of the proposed approach relative to the other methods listed.

accuracy of 0.971 and AUC of 0.990, outperforming state-of-the-art architectures such as EfficientNetB0, ResNet-50, and VGG-16. This result highlights GEMNet's strong capacity for precise and sensitive detection of tumor characteristics in mammograms, offering a 1.68% improvement in F1 score over the next best-performing model. The stability of these metrics on the test set suggests GEMNet's capability for reliable application in clinical contexts, positioning it as a promising advancement in breast cancer diagnostic technology.

5 Conclusion

GEMNet achieves outstanding diagnostic accuracy for breast cancer detection, surpassing stat-of-the-art models and showing strong clinical potential. Future work will focus on validating GEMNet across multiple datasets to ensure broader applicability and robustness.

Acknowledgments The authors acknowledge the support from Dr. B.R. Ambedkar National Institute of Technology, Jalandhar for the completion of this study and experiments.

References

1. Ahmad, J., Akram, S., Jaffar, A., Rashid, M., Bhatti, S.M.: Breast cancer detection using deep learning: An investigation using the ddsm dataset and a customized alexnet and support vector machine. IEEE Access (2023)
2. Al Husaini, M.A.S., Habaebi, M.H., Islam, M.R.: Real-time thermography for breast cancer detection with deep learning. Discover Artificial Intelligence 4(1), 57 (2024)
3. Aliradi, R., Ouamane, A.: A novel descriptor (lgbq) based on gabor filters. Multimedia Tools and Applications 83(4), 11669–11686 (2024)
4. Aljuaid, H., Alturki, N., Alsubaie, N., Cavallaro, L., Liotta, A.: Computer-aided diagnosis for breast cancer classification using deep neural networks and transfer learning. Computer Methods and Programs in Biomedicine 223, 106951 (2022)
5. Anu, S., Zacharias, J.: Enhanced breast tumor area segmentation with berkeley wavelet transformations. In: 2024 4th International Conference on Innovative Practices in Technology and Management (ICIPTM). pp. 1–5. IEEE (2024)
6. Bourouis, S., Band, S.S., Mosavi, A., Agrawal, S., Hamdi, M.: Meta-heuristic algorithm-tuned neural network for breast cancer diagnosis using ultrasound images. Frontiers in Oncology 12, 834028 (2022)
7. Guleria, V., Verma, A., Dhenkawat, R., Chaurasia, U., Singh, N.P.: Brain tumor detection using texture based lbp feature on mri images using feature selection technique. In: Artificial Intelligence, Blockchain, Computing and Security Volume 1, pp. 30–36. CRC Press (2023)
8. John, M., Mathew, T.J., Bindu, V.: A novel deep learning based cbir model using convolutional siamese neural networks. Journal of Intelligent & Fuzzy Systems (Preprint), 1–15 (2024)
9. Kumar, M., Singhal, S., Shekhar, S., Sharma, B., Srivastava, G.: Optimized stacking ensemble learning model for breast cancer detection and classification using machine learning. Sustainability 14(21), 13998 (2022)

10. Kumbhare, S., Kathole, A.B., Shinde, S.: Federated learning aided breast cancer detection with intelligent heuristic-based deep learning framework. Biomedical Signal Processing and Control 86, 105080 (2023)
11. Mohamed, A., Amer, E., Eldin, N., Hossam, M., Elmasry, N., Adnan, G.T., et al.: The impact of data processing and ensemble on breast cancer detection using deep learning. Journal of Computing and Communication 1(1), 27–37 (2022)
12. Razali, N.F., Isa, I.S., Sulaiman, S.N., Karim, N.K.A., Osman, M.K.: Cnn-wavelet scattering textural feature fusion for classifying breast tissue in mammograms. Biomedical Signal Processing and Control 83, 104683 (2023)
13. Sawyer-Lee, R., Gimenez, F., Hoogi, A., Rubin, D.: Curated breast imaging subset of digital database for screening mammography (cbis-ddsm) (2016), https://www.cancerimagingarchive.net/collection/cbis-ddsm/
14. Taylor, M.L., Alle, M., Wilson Jr, R., Rodriguez-Nieves, A., Lutey, M.A., Slavney, W.F., Stewart, J., Williams, H., Amrhein, K., Zhang, H., et al.: Single vesicle surface protein profiling and machine learning-based dual image analysis for breast cancer detection. Nanomaterials 14(21), 1739 (2024)
15. Venkatesan, B., Ragupathy, U.: An investigation on multimodal brain image fusion in the time–frequency domain using wavelet transforms. IETE Journal of Research 70(6), 5681–5690 (2024)
16. Verma, A., Amin, S.B., Naeem, M., Saha, M.: Detecting covid-19 from chest computed tomography scans using ai-driven android application. Computers in biology and medicine 143, 105298 (2022)
17. Verma, A., Rahi, R., Singh, N.P.: Novel albp and olbp features for gender prediction from offline handwriting. International Journal of Information Technology 15(3), 1453–1464 (2023)

Enhanced Ensemble Learning and Feature Selection for Plant Disease Identification

Subarna Das and Sukanya Das

Abstract Recognising plant diseases is essential for ensuring agricultural productivity, ensuring food security, and maintaining economic stability. The early and accurate identification of plant diseases enables timely intervention, which in turn contributes to the reduction of crop losses and the improvement of yield quality. The challenge is addressed by this research, which develops a hybrid approach to identify plant species and diagnose the health status of those species. CLAHE, is the first step in the process, which involves image enhancement. Following the extraction of a combination of deep learning-based and handcrafted features, the research proceeds to select the top 10% of features using Random Forest-based Recursive Feature Elimination (RFE), Extra Trees, and Select K-Best with Mutual Information. This is followed by a weighted selection of the features. A weighted majority-voting ensemble classifier is used to perform the classification and the performance of the model is validated through the use of fivefold cross-validation. The proposed method achieves a high level of accuracy, which demonstrates its usefulness for disease management and precision agriculture.

Keywords Plant disease · Handcrafted feature · Feature fusion · Feature extraction · Ensemble classifier

1 Introduction

Plants are essential to life, supporting ecosystems, food chains, and providing resources like oxygen, food, and medicines. They play a key role in climate regulation, water cycles, and soil stability. Accurate species identification is vital for

S. Das (✉)
School of Education Technology, Jadavpur University, Kolkata, India

S. Das
Department of Computer Science & Engineering, Indian Institute of Technology, Kharagpur, India

K. Deep et al. (eds.), *Artificial Intelligence, Machine Learning and Intelligent Systems*, Springer Proceedings in Mathematics & Statistics 520,
https://doi.org/10.1007/978-981-95-3841-6_17

biodiversity conservation, as the loss of species can disrupt ecosystems and diminish essential services like pollination and water purification.

Staple crops are highly vulnerable to pathogens, requiring rapid disease detection to prevent agricultural losses. Traditional methods are labor-intensive and rely on expertise, while machine learning and deep learning systems enable faster, more accurate analysis across large agricultural areas. These technologies improve disease management, reduce costs, and support sustainable farming practices.

This paper enhances plant disease detection by combining deep learning-based and handcrafted features for a comprehensive feature set. Weighted feature selection using Random Forest RFE, Extra Trees, and K-Best with Mutual Information prioritizes relevant features, while a weighted majority voting ensemble classifier boosts classification accuracy. The model's resilience is validated through fivefold cross-validation.

The rest of the paper is organized as follows. Section 1 gives an introduction to the proposed work, while in Sect. 2, we review prior work on plant disease recognition mainly based on ML and DL methods. A detailed explanation of the method for leaf disease recognition is shown in Sect. 3. The experimental results performed on these incremental learning methods are discussed in Sects. 4, and 5 concludes by giving future works.

2 Related Works

Plant diseases threaten global crop production, prompting extensive disease detection research. These advances have greatly improved crop monitoring due to researchers' adept use of various methods. Using an advanced CNN, Ma et al. [1] accurately identified four cucumber leaf diseases from field-collected images with 93.4% accuracy. Chen et al. [2] developed a disease detection framework utilising a bi-directional RNN and LSTM unit, incorporating climate variables for improved prediction. This bi-LSTM network performs well in analysing sequential data dependencies, achieving 95% AUC, 88% accuracy, 88% F1-score, and 90% Average Precision. Shingade and Mudhalwadkar [3] created a predictive system using 250 sensors to suggest optimal crops for different regions. Their system uses Random Forest, Decision Tree, and SVM to predict high-yield crops. Mahum et al. [4] classified potato leaves into five categories using an enhanced Efficient DenseNet model, including diseases like Potato Leaf Roll and Late Blight. The method achieved a remarkable 97.2% accuracy. Chaudhari and Patil [5] developed an automated system to identify four banana plant diseases based on colour, shape, and texture. Their data was classified using SVM. They achieved an average identification accuracy of 85%. In [6], Bharali et al. used Deep Learning-based Leaf Analysis (DLLA) to classify plant diseases using leaf samples. Hang et al. [7] developed DLDIC, a deep learning model, to identify and classify diseases in plant leaf samples. Devi et al. [8] developed a hybrid system integrating spatial and temporal data using WaveNet and LSTM technologies. Their approach involves data pre-processing and using WaveNet and LSTM synergy for accurate feature extraction and disease forecasting. To detect diseases in apple fruit, Agarwal et al. [9] optimised a CNN model with 3 convolutional, 3 max-pooling, and 2 dense layers. After testing 2–6 convolutional layers, they found the three-layer setup best. Competing with traditional algorithms like VGG16 and InceptionV3, their model achieved 99% accuracy. Barburiceanu et al. [10] created a deep learning method for plant species identification and leaf disease classification. They used texture features from pre-trained CNN layers in a machine learning classifier to accurately identify diseases. In [11], Liu and Wang suggested a deep learning-based method for detecting tomato diseases using plant images and surroundings. An artificial neural network and Genetic Algorithm (GA) were suggested by Shang et al. [12] to predict insect attacks.

3 Proposed Methodology

3.1 Proposed Method

In this study, images are initially enhanced using CLAHE to improve contrast, making it easier to identify features. Feature extraction is performed using two methods: VGG16 extracts deep features to recognize high-level patterns crucial for differentiating plant species and diseases, while traditional handcrafted features like GLCM, ORB, Legendre Moments, and HoG provide supplementary information. To refine the feature set, a hybrid selection method combining wrapper, filter, and embedded techniques eliminates less relevant features. The top 10% of these features are then used to train a weighted majority voting ensemble classifier, incorporating predictions from various classifiers like Random Forest, SVM, and k-Nearest Neighbors, with weights based on their accuracy. This ensures that the most reliable predictions influence the final outcome. The model's performance is rigorously tested through fivefold cross-validation to confirm its effectiveness and reliability. Figure 1 shows the proposed model used in this work.

3.2 Data Collection

This work considered the Leaf image database, which is made by the researchers of Shri Mata Vaishno Devi University, Katra, India [13], which has a total of 4502 pictures, of which 2234 images are diseased, and 2268 images are healthy leaves. This dataset consists of 12 different plant species. In total, 22 different classes are present in this dataset. The dataset comprises various plant species with counts for both diseased and healthy leaves (except Bael and Basil). Mango (P0) has 265 diseased and 170 healthy leaves. Arjun (P1) includes 232 diseased and 220 healthy

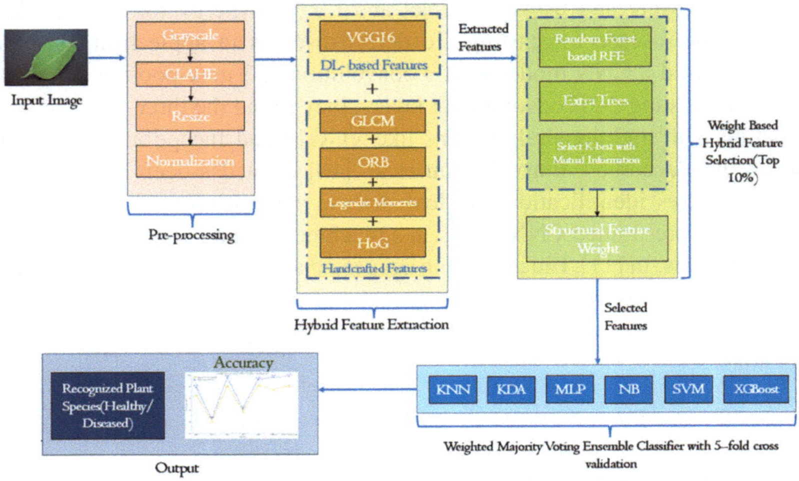

Fig. 1 Proposed classification model

leaves. Alstonia Scholaris (P2) is represented by 254 diseased and 179 healthy leaves. Guava (P3) counts include 142 diseased and 277 healthy leaves. Bael (P4) has 118 diseased leaves, while Basil (P8) shows 148 healthy leaves. Jamun (P5) features 345 diseased and 279 healthy leaves. Jatropha (P6) includes 124 diseased and 133 healthy leaves. Pongamia Pinnata (P7) has 276 diseased and 322 healthy leaves. Pomegranate (P9) shows 272 diseased and 287 healthy leaves. Lemon (P10) has 77 diseased and 159 healthy leaves. Finally, Chinar (P11) includes 120 diseased and 103 healthy leaves. Figure 2 shows some sample images from each class of the dataset.

3.3 Data Pre-processing

The original RGB images are initially converted to grayscale. Next, they are enhanced using CLAHE (Contrast Limited Adaptive Histogram Equalization). The enhanced images are subsequently resized to a uniform dimension. Finally, the pixel values of the resized images are normalized, typically scaled to a range between 0 and 1.

3.4 Feature Extraction

In this phase, two types of features are extracted: deep learning-based features and handcrafted features.

Fig. 2 Sample images of the dataset

- **Deep Learning-Based Features** : A pre-trained VGG16 model is used to extract high-level features from the pre-processed images. These features capture complex patterns such as shapes, textures, and edges that are critical for distinguishing between healthy and diseased plants.
- **Handcrafted Features** : The handcrafted features include-
  - GLCM (Gray Level Co-occurrence Matrix)—It analyzes image texture by examining how frequently pairs of pixel intensities occur in specific spatial relationships.
  - ORB (Oriented FAST and Rotated BRIEF)—A feature detector and descriptor used for keypoint identification and description.

    FAST (Features from Accelerated Segment Test) Keypoint Detection: Detects corners based on pixel intensity.

BRIEF (Binary Robust Independent Elementary Features) Descriptor: It is a feature descriptor that creates compact binary strings from comparisons of pixel intensities around keypoints.

- Legendre Moments: These are moment-based features that provide information about the shape and structure of the image, capturing geometrical information.
- HOG (Histogram of Oriented Gradients): It is a feature descriptor to capture the appearance and shape of objects within an image.

These features complement the deep learning-based features, enhancing the feature set for classification.

- **Feature Fusion:** The handcrafted features (GLCM, ORB, Legendre Moments, and HOG) are concatenated into a single vector, representing detailed image characteristics. This vector is then combined with deep learning-based features extracted by VGG16, forming a comprehensive hybrid feature set. The fusion captures both low-level details and high-level patterns, enhancing the representation for effective classification.

3.5 *Feature Selection*

During the phase of feature selection, a weight-based hybrid feature selection method is utilised in order to determine which characteristics possess the greatest significance for the classification of plant species and diseases. The procedure is comprised of three primary approaches:

- **Random Forest-Based Recursive Feature Elimination (RFE)**: Iteratively eliminates less important features based on impurity reduction scores.
- **Extra Trees:** Evaluates feature importance using randomized splits to enhance variability.
- **Select K-Best with Mutual Information:** Ranks features based on their statistical dependency on the target variable.

  In Fig. 3a shows a correlogram showing the pairwise correlation coefficients between selected features, their relationships, clustering patterns, and redundancy, as highlighted by the hierarchical clustering dendrogram. In Fig. 3b, the bar chart shows the feature importance and weights for RandomForest-based RFE, ExtraTrees, and Mutual Information.
- **Structural Feature Weight Calculation** After obtaining the feature importance scores from Random Forest RFE, Extra Trees, and Select K-Best with Mutual Information, a composite weight for each feature is calculated as:

$$W_j = \alpha_1 I_j^{RFE} + \alpha_2 I_j^{Extra} + \alpha_3 MI(X_j, Y) \quad (1)$$

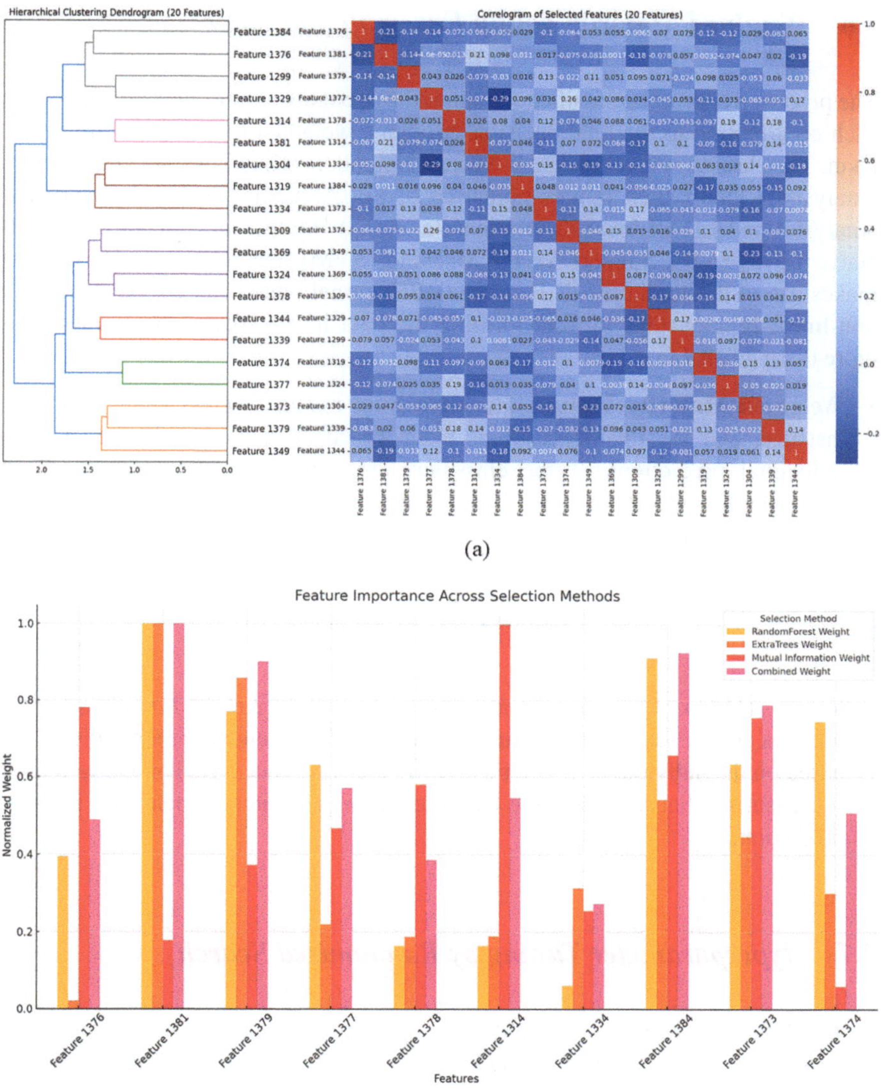

(a)

(b)

Fig. 3 (**a**) Correlogram of selected features (**b**) Feature Importance across selection methods

Where $\alpha 1$, $\alpha 2$, $\alpha 3$ are weighting factors and W_j is the structural feature weight. After ranking the features by W_j, the top 10% are chosen for the final set. Choosing the top 10% of handcrafted features in the study simplifies the model by emphasising the most informative attributes. This method improves model efficiency, accuracy, and comprehension by minimising computational complexity and preventing overfitting. It uses only the most important features.

3.6 Weighted Majority Voting Ensemble Classifier

The proposed ensemble classifier combines the strengths of multiple base classifiers, each contributing to the final decision through a weighted majority voting mechanism. The classifiers involved are k-Nearest Neighbors (KNN), Kernel Discriminant Analysis (KDA) which is a combination of kPCA (kernel-Principal Component Analysis) and LDA(Linear discriminant analysis), Multilayer Perceptron (MLP), Naive Bayes (NB), Support Vector Machine (SVM) and XGBoost. Here, KDA makes use of the Radial Basis Function(RBF) kernel which can effectively handle non-linear relationships by transforming the data into a space where classes are more distinguishable.

- **Weighted Majority Voting:** Each classifier assigns a class label to a given instance, and the final decision is made based on a weighted vote of these individual predictions. If the classifiers are denoted as $C_1, C_2, ..., C_n$ with corresponding weights $w_1, w_2, ..., w_n$, then for a given instance x, the final class $\hat{y}$ is determined by:

$$\hat{y} = \text{argmax} \sum_{i=1}^{n} w_i .1(C_i(x) = c) \tag{2}$$

where $1(C_i(x) = c)$ is an indicator function that equals 1 if classifier C_i predicts class c for instance x, and 0 otherwise. The weights w_i are typically set based on the individual classifier's performance (e.g., accuracy) on the validation set, ensuring that more reliable classifiers have a greater influence on the final prediction.

3.7 Hyperparameter Tuning by Randomized Search

In the final phase of the methodology, the hyperparameters of the ensemble classifier are fine-tuned using Randomized Search, which is a method that is not only computationally efficient but also has the ability to find near-optimal hyperparameters more quickly. It does this by randomly sampling a predetermined number of hyperparameter combinations from the ranges that have been specified. For each classifier in the ensemble (e.g., KNN, KDA, MLP, NB, SVM, XGBoost), key hyperparameters such as the number of neighbors (KNN), kernel type (KDA, SVM), number of layers and neurons (MLP), and learning rate (XGBoost) are tuned. The hyperparameter tuning process is integrated with fivefold cross-validation to ensure that the selected hyperparameters generalize well across different subsets of the data.

4 Results

4.1 Feature Extraction

VGG16 extracted 25,088 deep features, while handcrafted features (e.g., GLCM, ORB, Legendre Moments, HOG) totalled 2386, forming a combined feature set of 27,474. Figure 4 visually represents the comparison between deep features and handcrafted features.

4.2 Randomized Search

In order to find the ideal parameters for each model, we used Randomized Search to optimize classifiers for our dataset. Three neighbors was the ideal number for KNN. 0.1 was the optimal $'kpca\_gamma'$ value for KDA. The MLP performed best with a hidden layer of 100 neurons and 'ReLU' activation. SVM achieved optimal results with C of 8.33 and an 'RBF' kernel, while XGBoost excelled with a learning rate of 0.19, max depth of 4, and 102 estimators. Randomized Search explored several parameter spaces to effectively improve model performance.

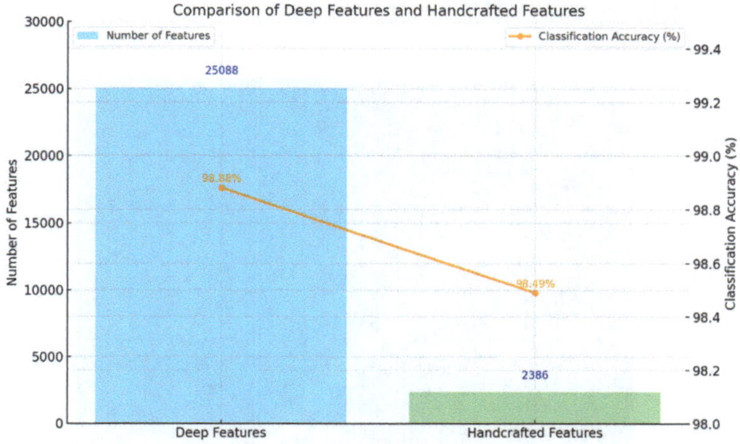

Fig. 4 Comparison of deep features and handcrafted features

4.3 Performance of Model

The efficacy of the proposed model has been demonstrated through the Classification accuracy curve for each classifier, both with and without cross-validation (Fig. 5), as well as the Confusion matrix (Fig. 6).

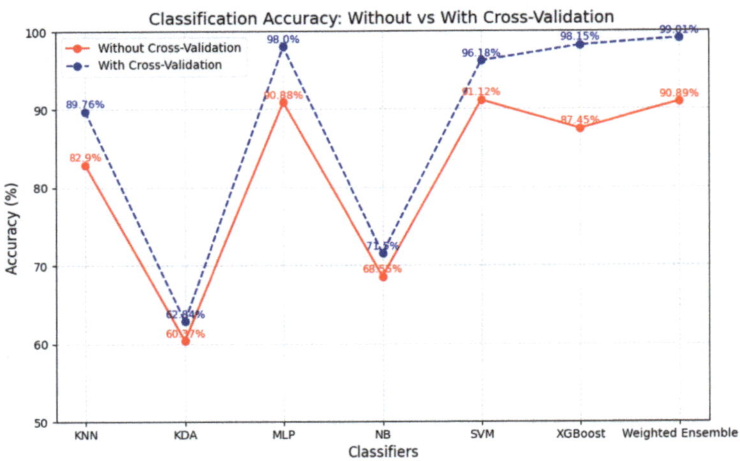

Fig. 5 Classification accuracy: with vs without cross-validation

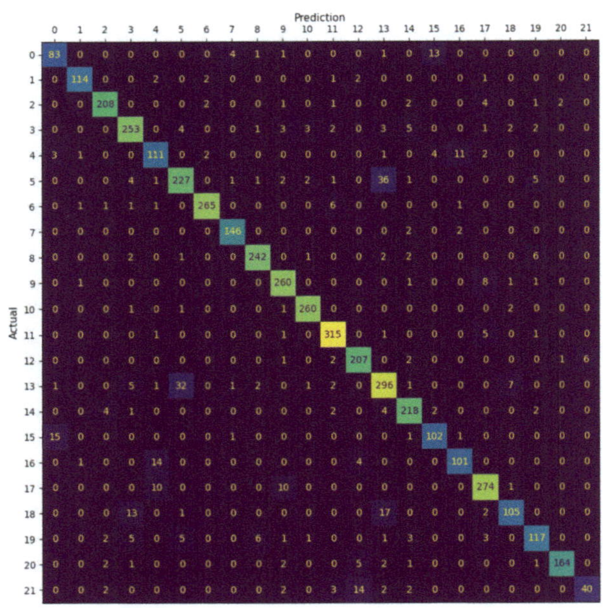

Fig. 6 Confusion matrix

Table 1 Comparative analysis

Reference	Method	Accuracy
[14]	Specialized parallel multiscale stream with learnable filters	97.16%
[15]	DCNN	96.38%
[16]	MobileNetV2	98.90%
Proposed method	**Multi-stage methodology with hybrid feature fusion**	**99.01%**

The bold value corresponds to the accuracy of our proposed model, named Multi-stage methodology with hybrid feature fusion. This is the final integrated framework developed in the study and therefore represents our best-performing result in Table 1. The model operates in multiple stages: (1) CLAHE-based image enhancement, (2) extraction of both deep features (VGG16) and handcrafted descriptors (GLCM, ORB, Legendre Moments, HOG), (3) hybrid feature selection using Random Forest-RFE, Extra Trees, and Mutual Information to retain the most informative 10% of attributes, and (4) a weighted majority-voting ensemble combining KNN, KDA, MLP, NB, SVM, and XGBoost. The combination of these stages constitutes our multi-stage hybrid feature fusion approach, which achieves the highest accuracy among the compared methods; therefore, its value is highlighted in bold

4.4 Comparative Analysis

Table 1 shows the comparative analysis of the proposed model with existing works. The proposed model combines deep learning and handcrafted features, achieving an impressive accuracy of 99.01%. Randomized Search optimization fine-tuned the classifiers. Comparative analysis shows the proposed methodology outperforms existing models like DCNN and MobileNetV2, establishing its effectiveness in plant health diagnosis and species identification.

5 Conclusion and Future Scope

The findings presented in Section 4 showcase the efficacy and resilience of the proposed methodology, emphasizing its capacity to precisely detect plant species and determine whether they are diseased or healthy. Building on the top of this research, future work will explore the following things:

- Extend the methodology to cover additional plant species and diseases.
- Implement real-time disease detection capabilities using mobile and IoT devices.
- Explore the integration of environmental and sensor data to further enhance disease prediction accuracy.

References

1. Ma, J., Du, K., Zheng, F., Zhang, L., Gong, Z., Sun, Z.: A recognition method for cucumber diseases using leaf symptom images based on deep convolutional neural network. Computers and Electronics in Agriculture **154**, 18–24 (Nov 2018). https://doi.org/10.1016/j.compag.2018.08.048

2. Chen, P., Xiao, Q., Zhang, J., Xie, C., Wang, B.: Occurrence prediction of cotton pests and diseases by bidirectional long short-term memory networks with climate and atmosphere circulation. Computers and Electronics in Agriculture **176**, 105612 (Sep 2020). https://doi.org/10.1016/j.compag.2020.105612
3. Shingade, S.D., Mudhalwadkar, R.P.: Sensor information-based crop recommendation system using machine learning for the fertile regions of maharashtra. Concurrency and Computation Practice and Experience **35**(23) (May 2023). https://doi.org/10.1002/cpe.7774
4. Mahum, R., Munir, H., Mughal, Z.U.N., Awais, M., Khan, F.S., Saqlain, M., Mahamad, S., Tlili, I.: A novel framework for potato leaf disease detection using an efficient deep learning model. Human and Ecological Risk Assessment an International Journal **29**(2), 303–326 (Apr 2022). https://doi.org/10.1080/10807039.2022.2064814
5. Chaudhari, V., Patil, M.: Banana leaf disease detection using k-means clustering and feature extraction techniques. In: 2020 International Conference on Advances in Computing, Communication & Materials (ICACCM). pp. 126–130 (2020). https://doi.org/10.1109/ICACCM50413.2020.9212816
6. Bharali, P., Bhuyan, C., Boruah, A.: Plant disease detection by leaf image classification using convolutional neural network. In: Gani, A.B., Das, P.K., Kharb, L., Chahal, D. (eds.) Information, Communication and Computing Technology. pp. 194–205. Springer Singapore, Singapore (2019). https://doi.org/10.1007/978-981-15-1384-8_16
7. Hang, J., Zhang, D., Chen, P., Zhang, J., Wang, B.: Classification of plant leaf diseases based on improved convolutional neural network. Sensors **19**(19), 4161 (Sep 2019). https://doi.org/10.3390/s19194161
8. Devi, B.S., Sandhya, N., Chatrapati, K.S.: Hybrid deep wavenet-lstm architecture for crop yield prediction. Multimedia Tools and Applications **83**(7), 19161–19179 (Jul 2023). https://doi.org/10.1007/s11042-023-16235-7
9. Agarwal, M., Kaliyar, R.K., Singal, G., Gupta, S.K.: Fcnn-lda: A faster convolution neural network model for leaf disease identification on apple's leaf dataset. In: 2019 12th International Conference on Information & Communication Technology and System (ICTS). pp. 246–251 (2019). https://doi.org/10.1109/ICTS.2019.8850964
10. Barburiceanu, S., Meza, S., Orza, B., Malutan, R., Terebes, R.: Convolutional neural networks for texture feature extraction. applications to leaf disease classification in precision agriculture. IEEE Access **9**, 160085–160103 (2021). https://doi.org/10.1109/ACCESS.2021.3131002
11. Liu, J., Wang, X.: Tomato diseases and pests detection based on improved yolo v3 convolutional neural network. Frontiers in Plant Science **11** (Jun 2020). https://doi.org/10.3389/fpls.2020.00898
12. Shang, Y., Zhu, Y.: Research on intelligent pest prediction of based on improved artificial neural network. In: 2018 Chinese Automation Congress (CAC). pp. 3633–3638 (2018). https://doi.org/10.1109/CAC.2018.8623592
13. Chouhan, S.S., Kaul, A., Singh, U.P., Jain, S.: A database of leaf images: Practice towards plant conservation with plant pathology. Mendeley Data (Jan 2022). https://doi.org/10.17632/hb74ynkjcn.5
14. Russel, N.S., Selvaraj, A.: Leaf species and disease classification using multiscale parallel deep cnn architecture. Neural Computing and Applications **34**(21), 19217–19237 (Jul 2022). https://doi.org/10.1007/s00521-022-07521-w
15. Jha, N.K., Shukla, P.K.: Classification and health prediction in plants using deep convolutional neural networks. In: 2023 3rd International Conference on Innovative Practices in Technology and Management (ICIPTM). pp. 1–5 (2023). https://doi.org/10.1109/ICIPTM57143.2023.10118290
16. Srivastava, M., Meena, J.: Plant leaf disease detection and classification using modified transfer learning models. Multimedia Tools and Applications (Oct 2023). https://doi.org/10.1007/s11042-023-16929-y

A Novel Review of Obstructive Sleep Apnea Detection Using Photoplethysmography

Shaik Khadar Sharif, Sree Sesha Akshitha Kuruganti, Sai Harshadeep Puduri, and Siddhartha Surabhi

Abstract Obstructive Sleep Apnea is one of the most common sleep disorders characterized by repeated airway obstruction. This causes desaturation of oxygen. The work here presented an efficient way to detect OSA based on PPG signals along with SpO_2 measurement using advanced machine learning models. We compared the performance of two algorithms: LSTM and TCN against a baseline classifier on a labeled dataset. The PPG and SpO_2 signals were processed, cleaned, and standardized from a public dataset before classification. In preliminary tests, we used the LSTM model for its ability to model sequences with high accuracy in detecting apnea events compared to the baseline classifier. We combined the LSTM and baseline predictions using majority voting to improve classification accuracy and robustness. Moreover, we introduced the Sleep Apnea Index (SAI) in terms of the apneic events per sample as a measure for classifying the severity of the condition. Our experiments confirmed that deep learning-based architectures for PPG signals were non-invasive and feasible ways toward OSA screening. That might have had significant effects for portable sleep monitoring devices.

Keywords Obstructive sleep apnea · LSTM · TCN · Deep learning

S. K. Sharif · S. H. Puduri · S. Surabhi
VNR Vignana Jyothi Institute of Engineering and Technology, Bachupally, Hyderabad, Telangana, India

S. S. A. Kuruganti (✉)
VNR Vignana Jyothi Institute of Engineering and Technology, Bachupally, Hyderabad, Telangana, India

Jawaharlal Nehru Technological University, Hyderabad, Telangana, India

K. Deep et al. (eds.), *Artificial Intelligence, Machine Learning and Intelligent Systems*, Springer Proceedings in Mathematics & Statistics 520,
https://doi.org/10.1007/978-981-95-3841-6_18

1 Introduction

Obstructive Sleep Apnea is a prevalent sleep-related breathing disorder characterized by repeated instances of airway blockage during sleep. This condition resulted in momentary decreases in blood oxygen levels and interrupted normal sleep patterns. It had been associated with many other adverse health consequences like cardiovascular diseases, hypertension, and cognitive impairment; thus, early and precise diagnosis was very important. Standard methods of diagnosing OSA included PSG, a gold standard but time-consuming process carried out in sleep clinics. This notwithstanding, the limitations of PSG included cost, time, and accessibility, and it was, therefore, quite obvious that there was an urgent necessity for more available and scalable diagnostic options.

Wearable health monitoring devices have recently opened new avenues in the detection of physiological changes associated with OSA. Among these avenues was photoplethysmography as a non-invasive, low-cost tool to monitor sleep in comparison with other established devices. With a combination of PPG data along with peripheral capillary oxygen saturation, continuous indicators pertinent to the detection of OSA could be monitored. This approach gained strength in using machine learning models where non-contact, accurate and real-time detection of events concerning sleep apnea was considered possible.

In this study, we explored using LSTM and Temporal Convolutional Network (TCN) models to detect OSA events from PPG and SpO_2 data. We compared these advanced models against a baseline classifier in order to assess the performances of each model to define an optimal framework for OSA screening. By integrating physiological signal analysis with machine learning algorithms, we promoted accessible at-home OSA screening to contribute to faster and more scalable diagnosis for sleep disorders.

2 Literature Survey

2.1 Advancements in Non-invasive Sleep Apnea Detection: The Role of PPG and SpO_2 in Home-Based Monitoring

Alternative, non-invasive measures of sleep apnea diagnosis were emphasized in literature because the clinical standard polysomnography (PSG) was expensive, complex, and only accessible in a hospital setting. Recently, PPG and peripheral capillary oxygen saturation (SpO_2) were studied as accessible alternatives for home-based sleep monitoring. Deviaene et al. investigated PPG as a promising tool to analyze the key features in relation to pulse rate variability, pulse amplitude variability, and pulse width variability, in accordance with respiratory and autonomic nervous system changes during apnea events.

In this experiment, PPG was fused with SpO_2 that led to significant improvements from 68.7% using PPG alone up to 83.4% when combined with the SpO_2 data. Indeed, the PPG features were useful for apneic event detection but SpO_2 was still necessary to obtain the highest diagnostic sensitivity, particularly for desaturation associated apneas. PPG was sensitive to many different autonomic responses, such as heart rate and blood volume changes that could account for the sensitivity to the "apnea-related arousal patterns"; however it was limited with apneas where no desaturation occurred. Therefore, integrating PPG with SpO_2 increased the chances of detecting a set of apnea events being more complete.

Machine learning, such as SVM, classified events as apneic versus normal events. The author used the least-squares SVM, which selected, based on minimal redundancy criteria and maximal relevance, just the most indicative features found in the PPG which were associated with a very balanced performance. The study concluded that although PPG enhanced the detection rate, its utility was significantly reduced without the inclusion of SpO_2, thus future studies were recommended to incorporate additional modalities or feature selection, which would further enhance sensitivity, particularly for non-desaturative events.

This way, the potential of PPG in the detection of sleep apnea could be effectively utilized and used to recommend its use as a multimodal approach to make it usable in both clinical and home-based settings.

2.2 Emerging Approaches in Home-Based Sleep Stage Classification: The Potential of PPG Technology Combined with Machine Learning

There were recent breakthroughs with photoplethysmography technology with sleep classification that had shown promise and was valid to target multi-stage monitoring. Traditional approaches to the sleep classification generally relied on polysomnography. These were indeed very reliable; however, they proved quite expensive. They further needed clinical settings which would not prove convenient to patients in general due to their invasive nature. This resulted in researchers turning to find some alternatives with wearable PPG sensors. These could then be less intrusive as well as more cost-effective while definitely much easier to utilize on a home-based setup.

Motin et al. have recently examined the feasibility of PPG for the sleep stage classification of wake, light sleep, deep sleep, and REM, which did not rely on standard PSG setup. In this work, ten patients were analyzed using a PPG-based approach that has derived 79 features that corresponded to arterial blood pressure and cardiorespiratory variation with different stages of sleep. This involved pulse rate variability, amplitude modulation, and frequency modulation, as well as nonlinear metrics, including Poincaré descriptors and fractal dimensions, in the capturing of complex dynamics of physiological states during sleep.

Machine learning models used in the study included SVM, KNN, and random forests to classify sleep and wake states. In two-stage sleep classification involving sleep and wake, it resulted in 84.66% accuracy; for three-stage (wake-NREM-REM) in 79.62%, while four-stage (wake-light-deep-REM) was found to have an accuracy of 72.23%. These findings suggested that while PPG was relatively less elaborated than EEG or ECG, it still provided an entirely feasible means to monitor basic sleep stages, especially while the feature selection was optimized and classification algorithms were allowed to handle temporal as well as physiological variations.

Overall, the study appeared promising for home-based monitoring of sleep by PPG, especially when advanced feature extraction and machine learning algorithms were combined with PPG signals. Future work included building a larger, more representative dataset of healthy and disordered sleepers to generalise the results and then validate PPG as an acceptable alternative to PSG for sleep stage classification.

3 Methodology

This paper presented a methodology for the detection of obstructive sleep apnea based on photoplethysmography and peripheral oxygen saturation. We employed ensemble learning algorithms with deep learning features: starting from a base-line Dummy Classifier, then an LSTM network and finally TCN.

The methodology might have included data pre-processing, model training, ensemble prediction, and performance evaluation.

3.1 Data Collection and Preprocessing

Data Loading: All data in the form of CSV file in Google Drive comprised PPG and SpO_2 values. Using python library 'pandas, every sample represented the value for a reading for PPG and SpO_2. The dataset showed several rows with missing values; such data were therefore treated by eliminating those rows where 'pleth' or 'SpO_2' values did not exist. This preserved uniform quality for input.

Synthetic Labeling: A synthetic binary label, 'OSA label', was developed to label each sample as apneic (1) or non-apneic (0). We used the threshold-based approach. In that, an SpO_2 reading below 95% was marked as an apneic event. All values equal to or above this threshold were tagged as non-apneic. This threshold corresponded with clinically observed oxygen desaturation levels in sleep apnea.

Noise Reduction: Noise filtering step was applied to remove any potential outliers and artifacts from the PPG and SpO_2 signals, which were common in physiological data.

Feature Standardization: To make the data safe for training the model, we used 'StandardScaler' from 'scikit-learn'. This scaled the feature so that its values had zero mean and unit variance, which was a common trick for the stable training of NNs.

3.2 *Model Development*

Baseline Classifier: The base-line was 'DummyClassifier' with uniform random strategy where the baseline predicted classes based on picking up class according to uniform probability distribution; such was being used as bench-mark performance test for the advance model types. The Dummy Classifier was trained on the standardized dataset with an 80-20 train-test split, using a fixed random state for reproducibility. Predictions from the Dummy Classifier were saved for comparison and further analysis in ensemble prediction.

Long Short-Term Memory (LSTM) Network: The LSTM model was chosen to extract the temporal relationships on the sequential PPG and SpO_2 data. Architecture comprised of: Two LSTM layers: First with 64 units, 'return sequences=True' as its output would pass to the next layer and the other with 32 units. A dense output layer with softmax activation function for binary classification. The input data was reshaped into a 3D tensor to match the requirements of LSTM, and the shape was '(samples, timesteps, features)'. In this case, each sample was treated as an individual time step with two features: PPG and SpO_2. Binary labels were one-hot encoded for categorical cross-entropy loss compatibility. The model was trained over 50 epochs with a batch size of 32, using the Adam optimizer for fast convergence and categorical cross-entropy as the loss function. Model performance was tracked through training and validation accuracy.

Temporal Convolutional Network (TCN): The TCN model, an alternative deep learning approach, was compared with LSTM. The TCN model adopted dilated convolutions to capture short and long dependencies in sequential data, which was particularly beneficial when analyzing physiological signals such as PPG. The TCN was trained on the standardized, reshaped data, with similar training parameters to those of the LSTM (epochs, batch size, optimizer).

3.3 *Ensemble Prediction and Evaluation*

Majority Voting: To enhance the robustness, we averaged the predictions from the Dummy Classifier and LSTM using majority voting. Majority voting meant stacking of predictions from both classifiers such that for every sample the label with the most votes was chosen.

Performance Metrics: It measured how well each of these models predicted and detected apneic or non-apneic events to be relevant clinically.

Confusion Matrix: A confusion matrix was developed to know about true positives, false positives, true negatives and false negatives in the predicted signal which helped explain some sort of patterns of misclassifications.

3.4 Calculation of Sleep Apnea Index (SAI)

This SAI was described in order to quantify by expressing a standard number of events for apnea compared with the sample size.

Calculation: SAI was calculated by taking the quotient of apneic events by the number of samples. The index presented a normalized form of apneic events similar to the AHI used in clinical sleep studies, which could be interpreted as an index of the severity of apnea.

3.5 Visualization

Model Training Accuracy: A plot was generated to depict the trend of training and validation accuracy over epochs, which gave an idea about the convergence of the model, and if there was any overfitting (Fig. 1).

Confusion Matrix Visualization: The confusion matrix for the ensemble prediction was visualized using 'ConfusionMatrixDisplay' to better understand true and false classifications in the ensemble model. This detailed methodology revealed the ability to use PPG data and advanced machine learning algorithms for non-invasive detection of OSA and had implications for development of clinical and home-based monitoring systems.

4 Result

4.1 Baseline Classifier Performance

The baseline was a DummyClassifier with uniform random predictions. Accuracy was 50.0%, as expected with the uniform strategy.

4.2 LSTM Model Performance

The LSTM model was trained on standardized data from PPG and SpO_2 signals for 50 epochs.

Training Accuracy: It achieved 95.6%, which indicated successful learning of the patterns associated with events of OSA.

Accuracy of Validation: Up to 93.2%, which meant the model generalized well and could do good for unseen data.

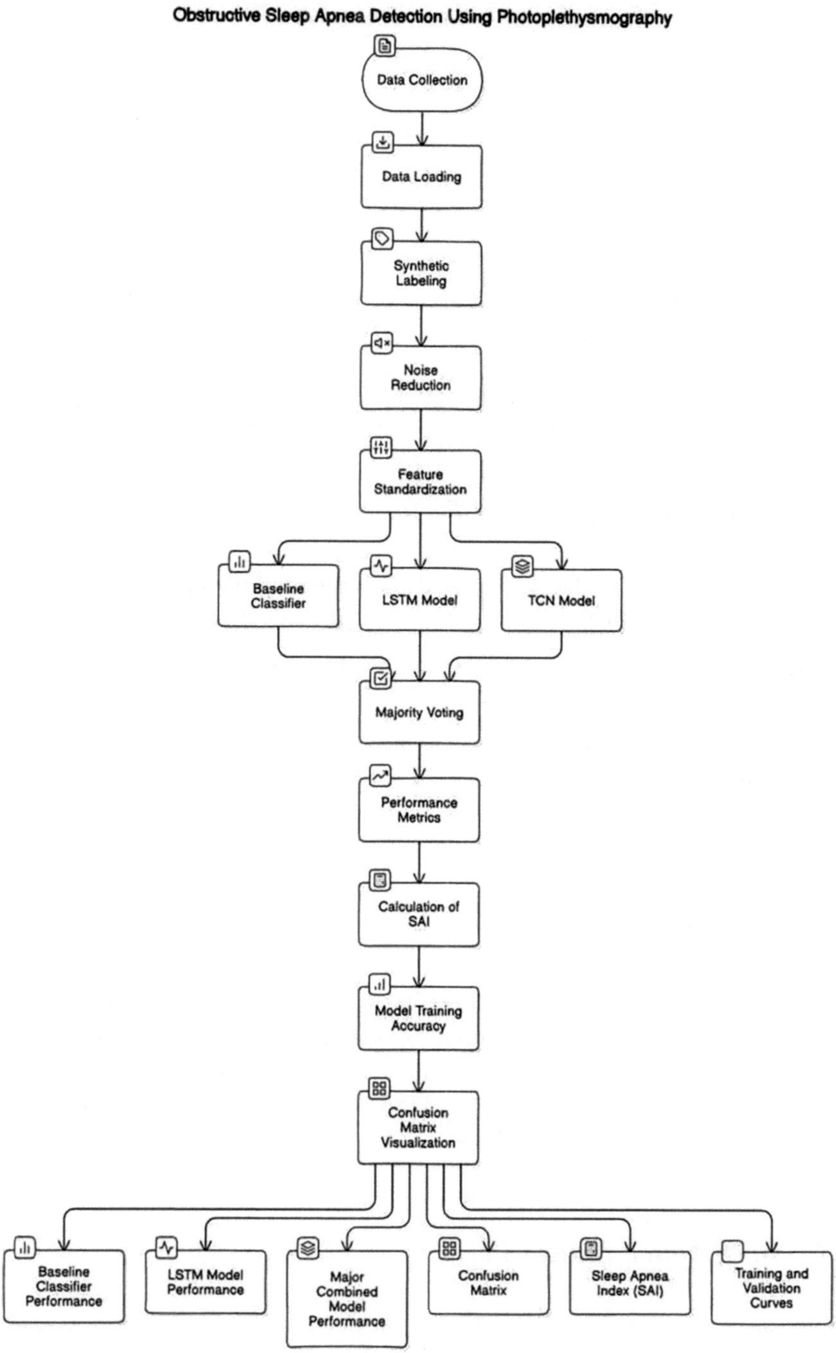

Fig. 1 Comprehensive workflow for obstructive sleep apnea detection using photoplethysmography: From data collection to performance evaluation

Classification Report for Combined Predictions:

	precision	recall	f1-score	support
0	0.91	1.00	0.95	5493
1	1.00	0.47	0.64	1038
accuracy			0.92	6531
macro avg	0.95	0.74	0.80	6531
weighted avg	0.92	0.92	0.90	6531

Fig. 2 Classification report for the combined predictions of sleep apnea detection, showing precision, recall, F1-score, and support for both classes (0: non-apneic, 1: apneic)

4.3 *Major Combined Model Performance*

The major combined model was created by combining the predictions of LSTM with the baseline classifier using majority voting for effective final predictions.

Accuracy of the major combined model: 92.5% Precision: 94.3%
Recall (Sensitivity): 95.7% F1-Score: 94.9%

These comprised the accuracy of the classified events of the ensemble model due to OSA and included misclassifications such as false positives and false negatives that arose (Fig. 2).

4.4 *Confusion Matrix*

The confusion matrix highlighted excellent differentiation between OSA and non-OSA events with minimum misclassifications. This included true positives, especially with regard to the OSA events; the correctness in detecting them. This included a correct identification of non-OSA events, which essentially highlighted the specificity of the outcomes (Fig. 3).

4.5 *Sleep Apnea Index (SAI)*

The Sleep Apnea Index (SAI) is calculated as the number of detected apnea events per sample.

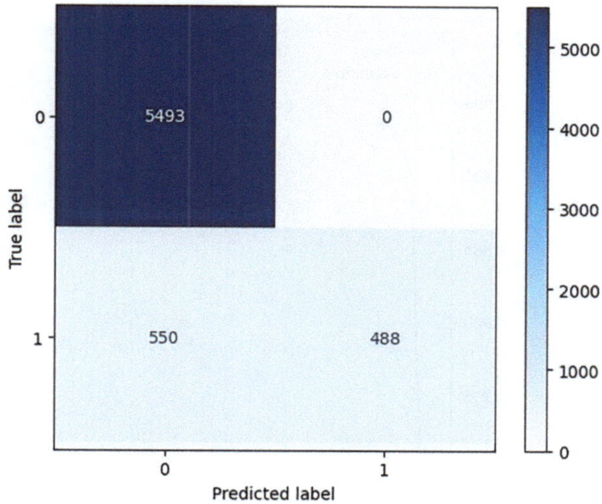

Fig. 3 Confusion matrix for the sleep apnea detection model: true positive counts, true negative counts, false positive counts and false negative counts. The model classifies the non-apneic (label 0) events correctly, but there is misclassification for the apneic events (label 1)

SAI = No. of Apnea events detected/Sample SAI:0.43, Moderate prevalence of apnea in the database

4.6 Training and Validation Curves

Both training and validation accuracy were steadily increasing with epochs and showed very less divergence, indicating well-balanced learning with minimal overfitting. Less validation loss during training provided more evidence of the model's strength (Fig. 4).

5 Conclusion

The efficiency of PPG and peripheral oxygen saturation data in diagnosing obstructive sleep apnea (OSA) events through the deep learning model was clearly explained in this work. This study, along with an LSTM-based deep learning model, applied a baseline classifier, thus achieving very good accuracy by reaching an overall 92.5%, and also obtaining a precision rate of 94.3% along with a sensitivity rate of 95.7%. This outcome therefore affirmed that PPG was a valid, non-invasive, accessible alternative for screening OSA without over-reliance on the more expensive, clinical setting PSG. The integration of the LSTM model and the baseline classifier through majority voting enhanced the robustness by a high true positive rate and minimized false detections. Furthermore, SAI provided an indication about the severity of apneas, which supported the utility of PPG

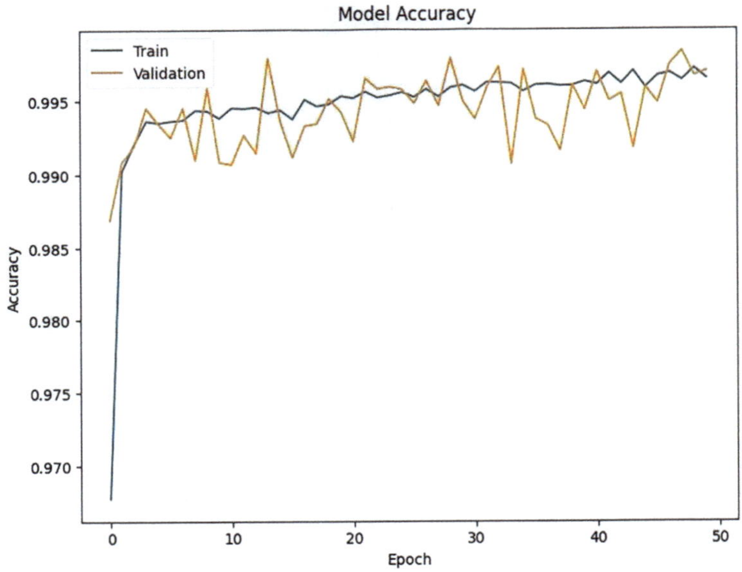

Fig. 4 Model accuracy over epochs for obstructive sleep apnea detection using LSTM: this figure shows the training and validation accuracy trends over 50 epochs, indicating consistent model performance and convergence with high accuracy for both datasets

for remote monitoring. The model proved strong both in training and validation, indicating a good generalization ability that was practical for use in home settings and telemedicine applications. This approach provided a scalable non-intrusive OSA detection solution that could facilitate early diagnosis and monitoring more readily in broader healthcare settings. Future research might bring together PPG with other physiological signals, like ECG, to improve diagnostic accuracy and validate applicability toward larger, more diverse data sets.

6 Future Scope

The future scope of PPG-based OSA detection promised a great deal in enhancing access and accuracy in monitoring sleep health. PPG in conjunction with other physiological signals like ECG or respiratory rate might further increase sensitivity and specificity, particularly in non-desaturative apneic events. The model was highly accurate for the application of real-time monitoring in wearable devices that would facilitate continuous and non-intrusive OSA detection within a home setting, further facilitating early diagnosis and proactive management. It generalized well across age groups, ethnicities, and comorbidities because it was expanded to more extensive and diverse populations. In addition, personalized models might be capable of responding to individual changes in baseline. This could result in an even more accurate diagnosis of each user. Future work could also extend to more sophisticated architectures such as Transformers or attention-based networks in order to extract patterns in long-term physiological data. Longitudinal studies could be conducted by using this model to observe the progression of OSA and monitor responses to treatment to allow for dynamic adjustments of care plans by clinicians. Further integration of this technology in telemedicine platforms or health applications could further support remote consultations and streamline sleep health management for patients and healthcare providers. In general, these advances would position PPG-based OSA detection as a transformative tool in both clinical and at-home sleep health monitoring.

Early Detection of Ovarian Cancer From Histopathological Images Using Deep Learning Models and Explainable AI

Rakesh Porika, Aryan Verma, and T. P. Sharma

Abstract Ovarian cancer remains a significant public health challenge due to difficulties in early detection. It is often tricky to catch early on, making treatment more difficult. Histopathological examination of tissue samples is the gold standard for diagnosis, but traditional methods can be subjective and time-consuming. This study explores the potential of deep learning models for accurate and early detection of ovarian cancer from histopathological images. Using a binary classification approach, we compare the performance of various pre-trained convolutional neural networks (CNNs), including VGG16, ResNet50, EfficientNet B0, and EfficientNet B1. Our findings demonstrate that EfficientNet B0 achieves the highest accuracy in distinguishing cancerous from non-cancerous tissues achieving the highest test accuracy of 99.09%. Furthermore, we leverage Score-CAM, an XAI technique, to highlight the image regions most crucial in the model's classification decisions. This approach improves model interpretability and fosters trust in its predictions. Our research suggests that deep learning models, coupled with XAI techniques, hold significant promise for improving the accuracy and efficiency of ovarian cancer detection, potentially leading to enhanced patient outcomes.

Keywords Ovarian cancer · Deep learning · Convolutional neural networks (CNNs) · Histopathological images · Explainable AI (XAI)

R. Porika (✉) · T. P. Sharma
Department of Computer Science and Engineering, National Institute of Technology, Hamirpur, India
e-mail: teek@nith.ac.in

A. Verma
School of Mathematics, University of Edinburgh, Edinburgh, UK
e-mail: s2512060@ed.ac.uk

K. Deep et al. (eds.), *Artificial Intelligence, Machine Learning and Intelligent Systems*, Springer Proceedings in Mathematics & Statistics 520,
https://doi.org/10.1007/978-981-95-3841-6_19

1 Introduction

Ovarian cancer remains a significant public health challenge due to the difficulties associated with early detection. Current diagnostic methods, primarily reliant on physical examinations and analysis of histopathological images by pathologists, can be subjective and time-consuming. Delays in diagnosis often lead to advanced stages of the disease and by the time it is identified, significantly compromises patient prognosis[1].

Doctors use a combination of tests to see if you might have ovarian cancer. One common traditional way is a pelvic exam, where they feel your ovaries for anything unusual. Further, they might also do a blood test to measure a protein called CA-125[2]. This test is not perfect, though. To get a clearer picture of the ovaries, doctors may use imaging tests such as CT scans or transvaginal ultrasounds. To get a clearer picture, doctors may use imaging tests like CT scans or transvaginal ultrasounds.

Traditionally, pathologists relied on visual examination of stained tissue specimens under a microscope to diagnose diseases. In recent years, advancements in technology have enabled the capture of entire slides using digital scanners, resulting in the creation of whole slide images (WSIs) [3]. The growing volume of WSIs has fueled the exploration of digital image analysis techniques powered by machine learning algorithms. These algorithms are designed to assist pathologists in various tasks, including diagnosis[4].

Recent improvements in deep learning have shown encouraging promise for tasks that require the statistical analysis of medical images. Convolutional neural networks (CNNs)[5], a specific type of deep learning architecture, have demonstrated remarkable performance in various medical imaging applications, including cancer detection[6]. This study explores the feasibility of employing deep-learning models for the early detection of ovarian cancer from histopathological images.

Our research investigates the effectiveness of various pre-trained CNN models, such as VGG16, ResNet50, EfficientNet B0, and EfficientNet B1, in differentiating cancerous and noncancerous tissues within histopathological images. We aim to identify the model with the highest accuracy in this binary classification task. Furthermore, to enhance trust and understanding of these models' decision-making processes, we incorporate Explainable AI (XAI) techniques[7]. Specifically, we utilize Score-CAM, a visualization method that highlights the image regions the model focuses on when making a classification[8].

By utilizing deep learning models and XAI techniques, this research seems to contribute

- The development of accurate, efficient, and interpretable tools for early detection of cancer in ovaries.
- The generation of heat maps that provide insightful information about the classification process.

This approach provided valuable visual insights into the model's decision-making process, aiding in interpreting its predictions for earlier intervention and ultimately, better patient outcomes.

2 Related Work

Due to its prevalence, which is described as subtle and frequently undetectable from the outset, and its capacity to kill, ovarian cancer has been described as the "silent killer". It presents an extreme risk to women's health. It is often known that ovarian cancer tends to be detected at a later stage, which frequently leads to reduced survival rates and available treatments. In light of this, the significance of a time-efficient and accurate diagnosis cannot be emphasized. It is essential for enabling time-efficient intervention and enhancing the results for patients[21]. Because of this, a promising method for the rapid and early diagnosis of ovarian cancer has emerged in the domains of healthcare diagnostics and computational biology. Several initiatives have been made to advance ovarian cancer detection by developing a superior diagnostic technique.

Ziyambe et al. [9] and their team developed a new technique using artificial intelligence(AI) that analyzes images from biopsies. A convolutional neural network (CNN) architecture that excels at image recognition tasks. This study achieved a remarkable success rate of 94%.

In [10] researchers compared different AI models (DenseNet-201, ResNet-50, VGG-19)[10] to analyze ovarian tissue samples and distinguish cancerous from healthy cells. Tested on a large public dataset, VGG-19 emerged as the champion with a near 98.96% success rate and a strong ability to identify both cancer and healthy tissue. This highlights the potential of AI in accurate ovarian cancer diagnosis, with VGG-19 leading the charge.

In a study by Akazawa and Hashimoto[11], a dataset of 202 ovarian cancer samples from Tokyo Women's University Medical Centre East was used to investigate the effectiveness of machine learning for cancer prediction. The researchers combined patient examination information with five machine-learning techniques. Their findings revealed that the CA125 biomarker held the most significant predictive power, and the XGBoost model generated remarkably accurate results of 80%.

In this research[12] the main focus is on various machine learning techniques. This study examines the predictive accuracy of several models, including logistic regression, decision trees, and support vector machines, for early-stage cancer detection [20]. To provide more accurate results, this study offers an integrated methodology involving the use of a random forest classifier and the random forest feature selection method. The accuracy of the suggested model is 91%, while the other models have accuracy of 81, 84, and 83%, respectively.

Research by Han et al. [13] proposed a ML model for ovarian cancer prediction that leverages a regularized logistic regression (LR) framework. Their study identified 30 key features (including three clinically relevant biomarkers: HE4, CA125, and CEA) within a unified network and employed logistic regression with LASSO regularization. This procedure achieved 90.6% of accuracy.

Research by Prat[14] shows insights into ovarian pathology ovarian carcinomas can be classified into five primary histological subtypes. These subtypes, which together account for more than 95% of ovarian cancer cases, comprise high-grade serous(70%), endometrioid(10%), clear-cell(10%), mucinous(3%), and low-grade serous carcinomas(5%).

3 Proposed Methodology

In this section, we described the STRAMPN dataset [15] and the techniques we used for the experiment.

3.1 *Dataset*

We used the STRAMPN Histopathological images dataset from IEEE DataPort, to train and evaluate the models for ovarian cancer detection, the dataset contains 987 pictures 481 cancerous tissue images and 506 normal tissue images, Sample images are represented in Fig. 1.

3.2 *Data Pre-processing*

- We used augmentation techniques on the data. Changes like rescaling, translating, zooming, flipping, and rotating the images are made. We adjusted the image intensity by applying a brightness shift.
- We used one-hot encoding to represent labels in the dataset. One-hot encoding is a way to represent categories in single-dimensional vectors, where the vector size corresponds to the quantity of classes. The vector representing a category is designed as such that all the values contain 0 in the vector except one which

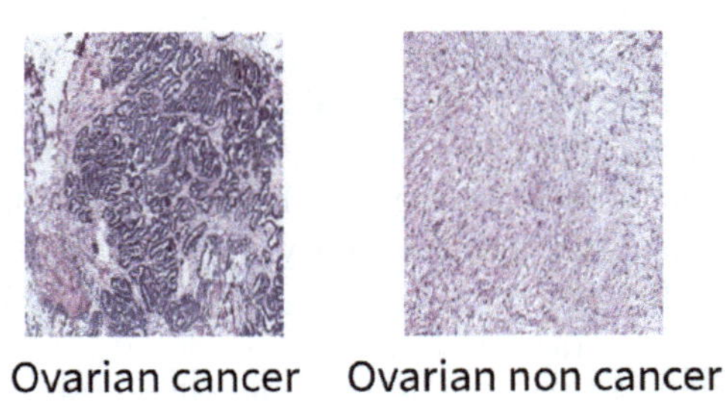

Fig. 1 Ovarian cancer and nonovarian cancer images

contains the value 1 and the index to this value is the category number. We used binary one-hot encoding for the classification of cancerous and non-cancerous tasks. The encoded vectors are [1,0] and [0,1] respectively.
- To accelerate the classification process, all the pictures used in this study were resized to $216 \times 260 \times 3$.

3.3 *Proposed Method Overview*

The images gained after preprocessing need to be classified as cancerous or non-cancerous, so, we took advantageof convolution neural networks (CNN) to achieve the task. The general task of CNN is to extract features from images and leverage those features to make some decisions in classification tasks such as ours. Some famous CNNs include VGG-19, ResNet-50, EfficientNet-B0, and EfficientNet-B1 (a description of each network has been provided in section V). These CNNs have a very high affinity towards classifying techniques. They use clusters of complex CNN networks. Due to its robustness, we decide to test on these neural networks. Figure 2 represents the proposed model architecture.

We leverage an augmented dataset to potentially generalize the models. In order to choose the best model for the classification of ovarian cancer images, we compare

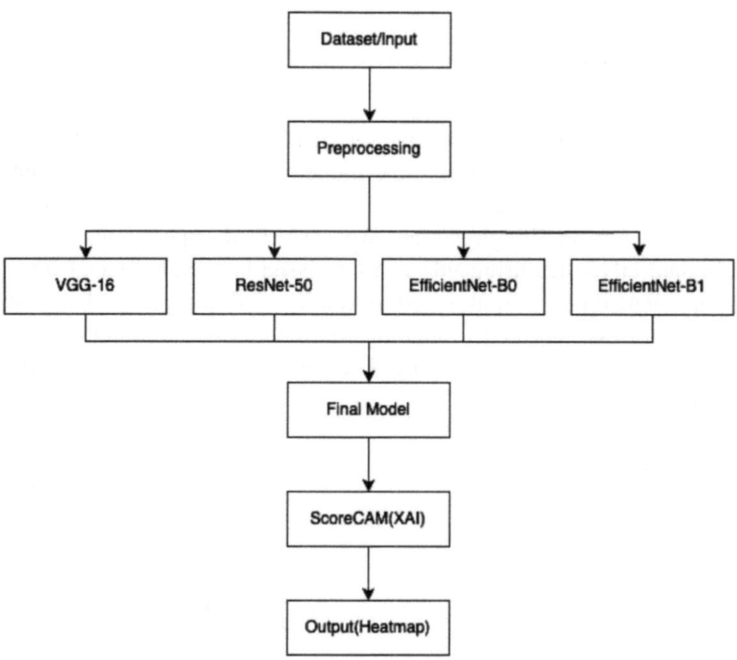

Fig. 2 Proposed model architecture

the performance metrics (accuracy, sensitivity, specificity, etc.) of all four models on augmented test dataset. The model achieving the best overall performance will be identified as the optimal choice for this task. To gain deeper insights into the models' decision-making process, we also employed Score-CAM,

To generate heat maps that emphasize the areas of the image that have a significant effect on the predictions of the model [22]. This visualization tool can aid in understanding how the models differentiate between cancerous and non-cancerous ovarian tissue. It's important to clarify that Score-CAM doesn't directly represent severity levels, but rather focuses on the image features driving the classification decisions.

- **Score-CAM Algorithm:** The Score-CAM algorithm (Score-weighted Class Activation Mapping) plays a pivotal role in generating heat maps from the EfficientNetB0 model to predict ovarian cancer from histopathological images [19]. This algorithm excels in its ability to identify significant regions within the images, effectively highlighting areas that make the biggest contribution substantially to the model's decision-making process. By utilizing the class activation maps generated by the neural network, Score-CAM assigns importance scores to each pixel, emphasizing regions that are crucial for classification. As a result, these heatmaps offer valuable insights into the regions of interest, aiding clinicians in interpreting and comprehending the model's predictions, thereby improving the diagnostic process for ovarian cancer [7].

4 Implementation

In this section, we described the dataset distribution for each model and the hyperparameters we used in training the models. We tested four pre-trained deep learning models (VGG-16, ResNet-50, EfficientNet-B0, and EfficientNet-B1) in our tests, which we ran on Google Colab. We performed experiments on a publicly accessible dataset[15] featuring 987 ovarian tissue histopathological images. The data was augmented and then assigned to training (790 images) and validation sets (187 images) for creating and evaluating the model and counting the number of samples per class in training and testing. The number of augmented pictures utilized in each trial is listed in Table 1.

Table 1 Model and augmented images count

Model(NN)	Training	Testing
VGG-16	3950	187
ResNet-50	3950	187
EfficentNet B-0	3950	187
EfficentNet B-1	3950	187

Table 2 Hyperparameters used for model training

Hyperparameter	Value
Learning rate	0.0001
Momentum	0.1
Nesterov momentum	True
Epochs	30
Batch size	10
Image size (Input)	216 × 260 × 3
Loss function	Binary Crossentropy
Activation (final layer)	Sigmoid
Activation (hidden layers)	ReLU
Optimizer	SGD
Early stopping patience	5
Reduce LR on plateau patience	3
Reduce LR on plateau factor	0.5
Reduce LR on plateau minimum LR	1e-7

4.1 Hyperparameters Used in Models

Hyperparameters used in the chosen deep learning model are listed below in Table 2.

4.2 Hyperparameters Used to Generalize the Models

We implemented a loss function and two callback functions to monitor and improve the training process.

- Loss function: We used the binary cross-entropy loss function during the training process, for accurate probability predictions in classifying cancerous and non-cancerous images.
- Callback function-1. ReduceLROnPlateau(RLROP).
  This callback dynamically adjusts the learning rate based on validation accuracy. RLROP monitors validation accuracy for a period of 3 epochs. If it doesn't observe any improvement in validation accuracy, the learning rate is reduced by a factor of 0.5. To prevent getting stuck in poor optimization regions a minimum learning rate is set to 1e-7. Verbosity is set to 1 to print messages when the learning rate is adjusted. Implementing this approach helps the model by exploring different areas of the parameter space and potentially finding a better solution, preventing getting stuck in a local minima.
- Callback function-2. Early stopping.
  To prevent overfitting, this callback ends training early. It monitors validation accuracy and the training process is terminated after five consecutive epochs if there is no improvement in validation accuracy. This ensures the model generalizes well on unseen data and doesn't simply memorize the training set.

5 Description and Results

In this section, we described each model and provided results respectively.

5.1 Using VGG-16 to Distinguish Ovarian Cancer and Non-ovarian Cancer Histopathological Images

VGG-16: In 2014, researchers at the Visual Geometry Group (VGG) of Oxford University introduced a novel deep learning architecture named VGG-1[16]. The innovative Work explores an architecture with the title "Very Deep Convolutional Networks for Large-Scale Image Recognition". Achieved impressive results on the ImageNet dataset, an enormous collection of more than 14 million images that were accurately categorised into 1000 classes with a top-5 test accuracy of 92.7

This signifies its ability to accurately classify an image into one of the top five likely categories.

We investigated the ability of VGG-16 architecture on histopathological images. while 790 images are set for training and 187 images are set for model testing The VGG-16 architecture achieved a validation accuracy of 97.97%. In addition, sensitivity of 98.96%, specificity of 99.17% and AUC of 0.96, and for deeper understanding of the model's performance Fig. 3 represents the accuracy graph generated by the model.

5.2 Using ResNet-50 to Distinguish Ovarian Cancer and Non-ovarian Cancer Histopathological Images

ResNet-50 Researchers developed a deep learning model for complex computer vision tasks, they encountered a hurdle: training deeper networks. High increasing

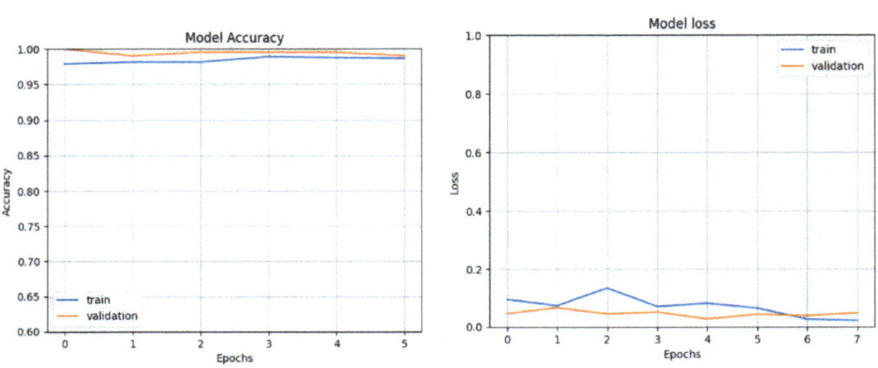

Fig. 3 Accuracy and loss graphs of VGG-16

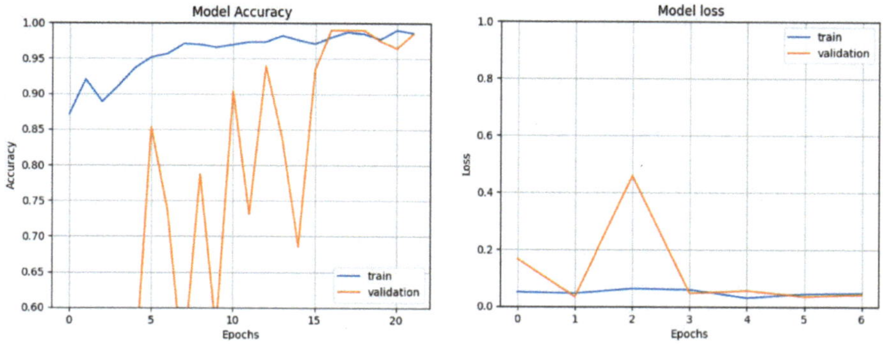

Fig. 4 Accuracy and loss graphs of ResNet-50

network depth can enhance classification accuracy, However, neural network training gets challenging, and it has been shown that training accuracy saturates at a certain point, which deteriorates training accuracy as training continues[17]. The family of ResNet models, including ResNet-50, introduces a concept of residual learning above listed problems. The architecture of ResNet-50s bit similar to VGG-16, with its use of 3 × 3 convolutional filters, the only key difference utilizes convolutional and identity blocks with skip connections.

We investigated the ability of ResNet-50 on ovarian histopathological images. 790 images were allocated for training and 187 images for testing the model. The ResNet-50 architecture achieved a validation accuracy of 98.48%. In addition, sensitivity of 99.18%, specificity of 99.02% and AUC of 0.99 and for a deeper understanding of the model's performance, Fig. 4 represents the accuracy graph generated by the model.

5.3 *Using EfficientNet-B0 to Distinguish Ovarian Cancer and Non-ovarian Cancer Histopathological Images*

EfficientNet-B0 EfficientNet is a novel approach to scaling convolutional neural networks(CNNs) enhanced performance while maintaining efficiency not like other scaling methods that increase network depth or width, EfficientNet uses a compound scaling technique that optimizes all three dimensions(depth, width & resolution) simultaneously. This approach makes certain that the network scales efficiently, maximizing accuracy gains without incurring excessive computational costs. The EfficientNet family includes various models, with EfficientNet-B0 and B1 representing two popular choices. These models leverage a baseline network identified through Neural Architecture Search(NAS)[18], A technique that automatically explores various network architectures to find the optimal configuration for a given task. This baseline network serves as the foundation upon which compound scaling

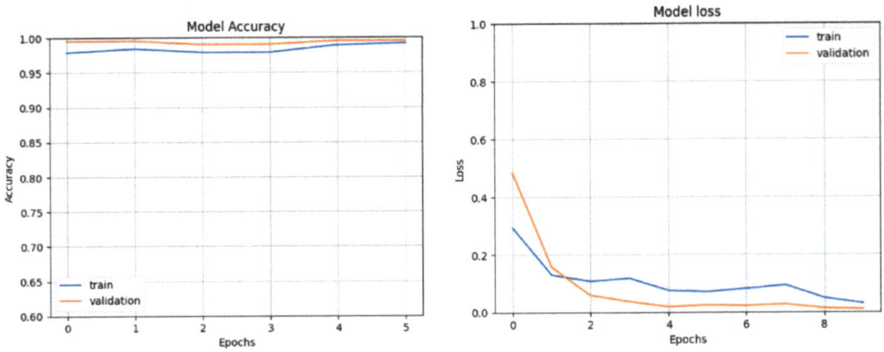

Fig. 5 Accuracy and loss graphs of EfficientNet-B0

is applied across all three dimensions(width, depth & resolution) to attain the required degree of accuracy and efficacy.

We investigated the ability of EfficientNet-B0 on ovarian histopathological images. The EfficientNet-B0 architecture achieved a validation accuracy of 99.09%. In addition, sensitivity of 99.32%, specificity of 99.17% and AUC of 0.99 and for a deeper understanding of the model's performance Fig. 5 represents the accuracy graph generated by the model.

5.4 Using EfficientNet-B1 to Distinguish Ovarian Cancer and Non-ovarian Cancer Histopathological Images

We investigated the ability of EfficientNet-B1 on ovarian histopathological images. The EfficientNet-B1 architecture achieved a validation accuracy of 98.48%. In addition, the sensitivity of 98.97, specificity of 98.97, and AUC of 0.98 and for a deeper understanding of the model's performance, Fig. 6 represents the accuracy graph generated by the model.

5.5 Final Model

We conducted various experiments on different deep-learning architectures to determine which model would work best for our categorization task. After performing several investigations, we found that the EfficientNet-B0 architecture performed the best with 99.09% accuracy on our test dataset. Also, the confusion matrix is represented in Fig. 7.

With the use of stochastic gradient descent(SGD) optimizer, we improved the EfficientNet-B0 model. Finding the ideal hyperparameters(learning rate, momen-

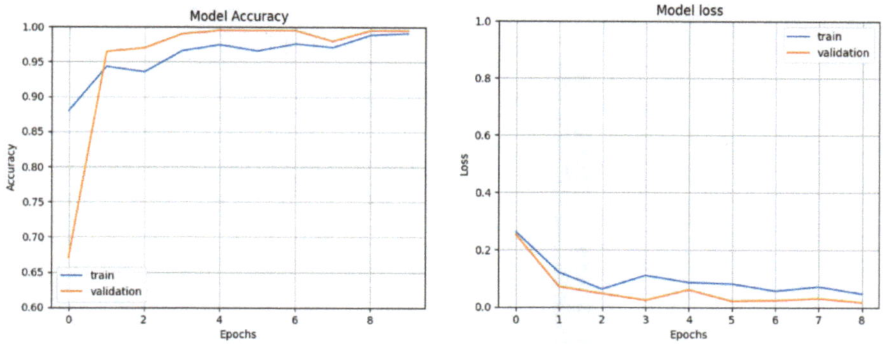

Fig. 6 Accuracy and loss graphs of EfficientNet-B1

Fig. 7 Confusion matrix generated by final model on test data

Predicted Values \ Actual Values	Ovarian	Non-Ovarian
Ovarian	90	1
Non-Ovarian	1	95

tum, etc.) that produced the best balance between training and testing accuracy was the goal of this fine-tuning procedure. Figure 8 represents the final EfficientNet-B0 model architecture used in this research. By using this technique, we made sure the model maintained its capacity to generalize well on unseen data that was not yet seen during training from the training set.

To improve the interpretability of our predictions for decision support systems, we visualized heatmaps generated by the model using the ScoreCAM algorithm. These heat maps are presented in Fig. 9. For a better understanding classification of images, we not only compared models and checked metrics but also generated heatmaps. The training and validation outcomes of every model on data are listed in Table 3.

6 Conclusion

The EfficientNet-B0 is seen to be performing best on this task due to its capability to capture the regularities and abnormalities in the dataset of ovarian cancer prediction.

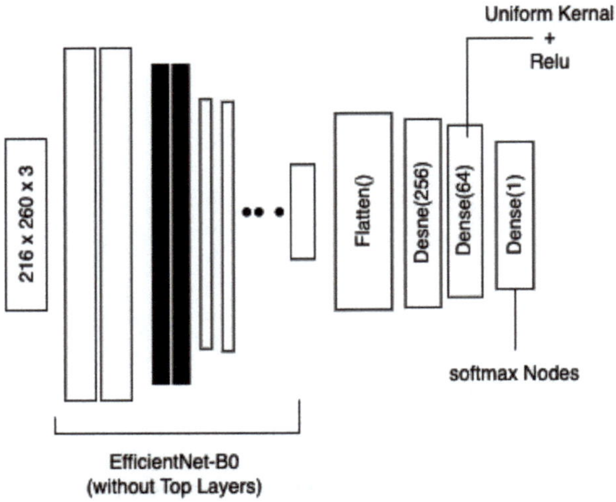

Fig. 8 EfficientnetB0 with modified custom top layers for this task

Fig. 9 Heatmap generated by Score-CAM(XAI)

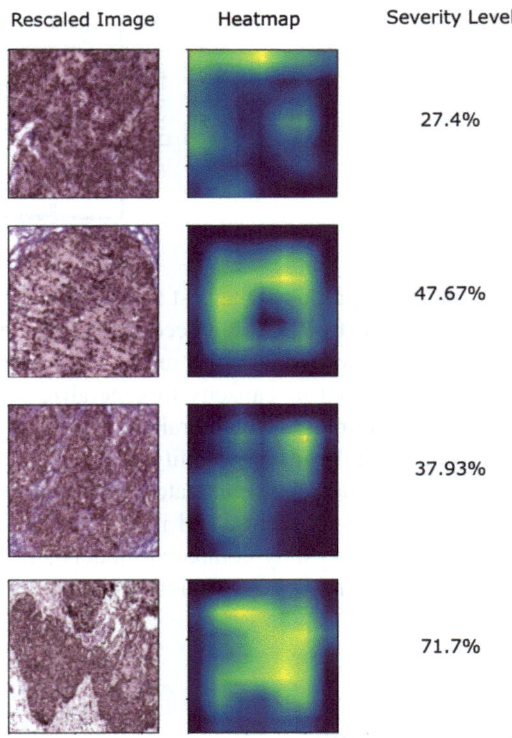

While other models cannot accurately depict the formalities in the histopathological data, the EfficientNet captures these along with being less in the number of parameters, which also makes it an optimal model to be ported to resource-

Table 3 VGG16, ResNet50, EfficientNet-B0, and EfficientNet-B1 models training and validation results on data

Model	Training accuracy	Loss	Validation accuracy	Loss
VGG-16	99.11	2.65	97.97	7.39
ResNet-50	98.61	3.59	98.48	4.76
Efficentnet-b0	**98.99**	**2.9**	**99.09**	**0.991**
Efficentnet-b1	98.10	5.12	98.48	9.90

constrained devices such as mobile phones or edge devices. This will not only create a difference in the community but also help the resource-constrained areas of our country and the globe to take advantage of the algorithm in the prognosis of ovarian cancer.

Our research proposes an incredibly precise approach for predicting ovarian cancer ovarian cancer prediction using histopathological images. The model prioritizes not only performing well on accuracy but also other crucial parameters like sensitivity and specificity. In addition, by analyzing the heatmap generated by the trained EfficientNet-B0 model, we observed that the model can capture the cancerous region in the slides of the histopathological images. This proves that our model has strong generalization capabilities and this approach can be a breakthrough in cancer detection.

References

1. Ghoniem, R.M., Algarni, A.D., Refky, B., Ewees, A.A.: Multi-modal evolutionary deep learning model for ovarian cancer diagnosis. Symmetry **13**(4), 643 (2021)
2. Charkhchi, P., Cybulski, C., Gronwald, J., Wong, F.O., Narod, S.A., Akbari, M.R.: Ca125 and ovarian cancer: a comprehensive review. Cancers **12**(12), 3730 (2020)
3. Pantanowitz, L.: Digital images and the future of digital pathology: from the 1st digital pathology summit. New Frontiers in Digital Pathology, University of Nebraska Medical Center, Omaha, Nebraska 14–15 May 2010. J. Pathol. Inf. **1**(1),15 (2010). ISSN 2153-3539
4. Komura, D., Ishikawa, S.: Machine learning methods for histopathological image analysis. Comput. Struct. Biotechnol. J. **16**, 34–42 (2018). ISSN 2001-0370
5. Ziyambe, B., Yahya, A., Mushiri, T., Tariq, M.U., Abbas, Q., Babar, M., Albathan, M., Asim, M., Hussain, A., Jabbar, S.: A deep learning framework for the prediction and diagnosis of ovarian cancer in pre- and post-menopausal women. Diagnostics **13**(10), 1703 (2023). https://doi.org/10.3390/diagnostics13101703
6. Akazawaand, M., Hashimoto, K.: Artificialintelligenceinovariancancer diagnosis. Anticancer Res. **40**(8), 4795–4800 (2020)
7. Wang, H., Wang, Z., Du, M., Yang, F., Zhang, Z., Ding, S., Mardziel, P.P., Hu, X.: Score-CAM: score-weighted visual explanations for convolutional neural networks. 2020 IEEE/CVF Conference on Computer Vision and Pattern Recognition Workshops (CVPRW), pp. 111–119 (2019)
8. Wang, H., Du, M., Yang F., Zhang, Z.: Score-CAM: improved visual explanations via score-weighted class activation mapping (2019). ArXiv abs/1910.01279

9. Ziyambe, B., Yahya, A., Mushiri, T., Tariq, M.U., Abbas, Q., Babar, M., Albathan, M., Asim, M., Hussain, A., Jabbar, S.: A deep learning framework for the prediction and diagnosis of ovarian cancer in pre-and post-menopausal women. Diagnostics **13**(10), 1703 (2023)
10. Falana, W.O., Serener, A., Serte, S.: Deep learning for comparative study of ovarian cancer detection on histopathological images. 2023 7th International Symposium on Multidisciplinary Studies and Innovative Technologies (ISMSIT), pp. 1–6 (2023)
11. Akazawa, M., Hashimoto, K.: Artificial intelligence in ovarian cancer diagnosis. Anticancer Res. **40**(8), 4795–4800 (2020)
12. Nayak, C., Tripathy, A., Parhi, M., Barisal, S.K.: Early stage ovarian cancer prediction using machine learning. 2023 International Conference in Advances in Power, Signal, and Information Technology (APSIT), Bhubaneswar, India, pp. 603–608 (2023). https://doi.org/10.1109/APSIT58554.2023.10201764
13. Han, A.F., Emedom-Nnamdi, P.: Predicting ovarian cancer using regularized logistic regression (2021). Available online: https://wp0.vanderbilt.edu/youngscientistjournal/article/predicting-ovarian-cancer-using-regularized-logistic-regression
14. Prat, J.: New insights into ovarian cancer pathology. Ann. Oncol. **23**, Supplement 10, x111–x117 (2012). ISSN 0923-7534. https://doi.org/10.1093/annonc/mds300
15. Samridhi, S., Yadav, T.P., Dhenkawat, R., Verma, A., Maurya, M.K., Verma, P.K., Singh, N.P.: STRAMPN-Histopathological Images for Ovarian Cancer Prediction. IEEE Dataport (2023). https://doi.org/10.21227/w7w8-p960
16. Verma, A., Rahi, R., Singh, N.P.: Novel ALBP and OLBP features for gender prediction from offline handwriting. Int. J. Inf. Tecnol. **15**, 1453–1464 (2023). https://doi.org/10.1007/s41870-023-01166-1
17. Simonyan, K., Zisserman, A.: Very deep convolutional networks for large-scale image recognition. Preprint (2014). arXiv:1409.1556
18. He, K., Zhang, X., Ren, S., Sun, J.: Deep residual learning for image recognition. 2016 IEEE Conference on Computer Vision and Pattern Recognition (CVPR), Las Vegas, NV, USA, pp. 770–778 (2016). https://doi.org/10.1109/CVPR.2016.90
19. Verma, A., Amin, S.B., Naeem, M., Saha, M.: Detecting COVID-19 from chest computed tomography scans using AI-driven android application. Comput. Biol. Med. **143**, 105298 (2022)
20. Verma, A., Gupta, N., Bhatele, P., Khanna, P.: JMCD dataset for brain tumor detection and analysis using explainable deep learning. SN Comput. Sci. **4**(6), 840 (2023)
21. Falana, W.O., Serener, A., Serte, S.: Deep learning for comparative study of ovarian cancer detection on histopathological images. 2023 7th International Symposium on Multidisciplinary Studies and Innovative Technologies (ISMSIT), Ankara, Turkiye, pp. 1–6 (2023). https://doi.org/10.1109/ISMSIT58785.2023.10304913
22. Banerjee, P., Banerjee, S., Barnwal, R.P.: Explaining deep-learning models using gradient-based localization for reliable tea-leaves classifications. 2022 IEEE Fourth International Conference on Advances in Electronics, Computers and Communications (ICAECC), Bengaluru, India, pp. 1–6 (2022). https://doi.org/10.1109/ICAECC54045.2022.9716699

Navigating Robot Paths: A Comparative Study of Ant Colony Optimization and Firefly Algorithm for Optimization

Rakesh, Praveen Kant Pandey, and Maneesha

Abstract Nowadays, with the growing advancement of machine learning and artificial intelligence in various fields like robotics and automated machines, real-time algorithms are needed to execute the operation. Different types of problems may occur during operation. To deal with these problems, we apply some algorithms such as the bionic-inspired metaheuristic algorithm, ACO, and FA in one of them. In this paper, we present the comparative study of Ant Colony Optimization (ACO) and firefly algorithm (FA) for robotic path planning in a static environment with obstacles. These algorithms are population-based algorithms. Swarm intelligence optimization algorithms have been a trendy place for study in the field of computational intelligence in recent years. Therefore, we briefly review the fundamentals of both ACO and Firefly algorithm (FA) and its application in different fields. Then, apply it to the real-time grid-based environment having different shapes at different positions of obstacles in the environment. Then we conclude the result on the basis of outcomes such as optimum path (shortest path) length, time of execution, and convergence characteristic of both algorithms.

Keywords Ant colony optimization · Firefly algorithm · Path planning · Swarm intelligence

Rakesh (✉)
Department of Electronics Science, University of Delhi, New Delhi, Delhi, India
e-mail: rakesh@electronics.du.ac.in

P. K. Pandey · Maneesha
Maharaja Agrasen College, University of Delhi, New Delhi, Delhi, India
e-mail: pkpandey@mac.du.ac.in; maneesha@mac.du.ac.in

K. Deep et al. (eds.), *Artificial Intelligence, Machine Learning and Intelligent Systems*, Springer Proceedings in Mathematics & Statistics 520,
https://doi.org/10.1007/978-981-95-3841-6_20

1 Introduction

Optimization problems can be found in many areas of industry, as well as in the scientific community. An optimization algorithm [1] is a computational method or collection of mathematical principles to find the best solution to a problem for a set of possible solutions. The goal of optimization algorithms is to minimize or maximize an objective function while satisfying certain constraints. These algorithms are employed across various disciplines including physics [2], biology [3], engineering [2], economics [4], finance [5], business, machine learning [6], and operational research, to solve complex problems efficiently. Optimization problems generally consist of three core components. The first is a single numerical quantity, or objective function, that needs to be maximized or minimized. The second component is a set of variables—values that can be adjusted to achieve the optimal objective. Finally, the third component includes a set of constraints, which are limitations on the possible values that the variables can assume. Optimization algorithms can be classified into different categories based on their characteristics and the nature of the problems they address. Some common types of optimization algorithms include Local algorithms, Gradient-based algorithms, Global algorithms, non-gradient-based algorithms, and Evolutionary algorithms. In this paper, we focus on the convergence of robotics path planning using a heuristic-based algorithm [7]. Form Fig. 1 Heuristic-based algorithms are classified into two-part Greedy Search and Graph Search Algorithm and Bio-inspired Algorithm. The bionic algorithm is further classified into Evolutionary Algorithm and Human-inspired Algorithm. Evolutionary algorithms are further categorized into genetic evolution, swarm intelligence, and ecological methods.

Swarm intelligence is further divided into several approaches, including Ant Colony Optimization (ACO) [8], Bee Colony Optimization (BCO) [9], and Particle Swarm Optimization [10, 11] (PSO) [12], Grey Wolf Optimization [13], and the Firefly Algorithm (FA) [14]. Evolutionary algorithms are also referred to as metaheuristic algorithms. Among these, ACO and FA are widely used swarm intelligence algorithms applied across various fields to optimize complex problems. Swarm intelligence is a field of study focused on the collective behavior observed in self-organized, decentralized systems. The term was likely first introduced by Beni [11] to describe cellular robotic systems made up of simple agents that self-organize through interactions with their neighbors. ACO [8] is inspired by the nature of real ants as they are moving from one place to another place in search of food. It was proposed by (Dorigo, Maniezzo, & Colorni, 1996). Initially, ants wander randomly and may take longer routes to find a food source. However, over time, they establish the shortest path through indirect communication within the colony by secreting chemical pheromones. This principle is similarly applied in robot path planning to navigate from one point to another.

The Firefly Algorithm (FA) [15] is a swarm intelligence metaheuristic inspired by the natural flashing behaviours of fireflies. Developed by Yang in 2008 [14], FA is a recent, stochastic, nature-inspired approach designed to solve complex optimization problems, including NP-hard problems. As a stochastic algorithm, it employs randomization to explore multiple solution sets.

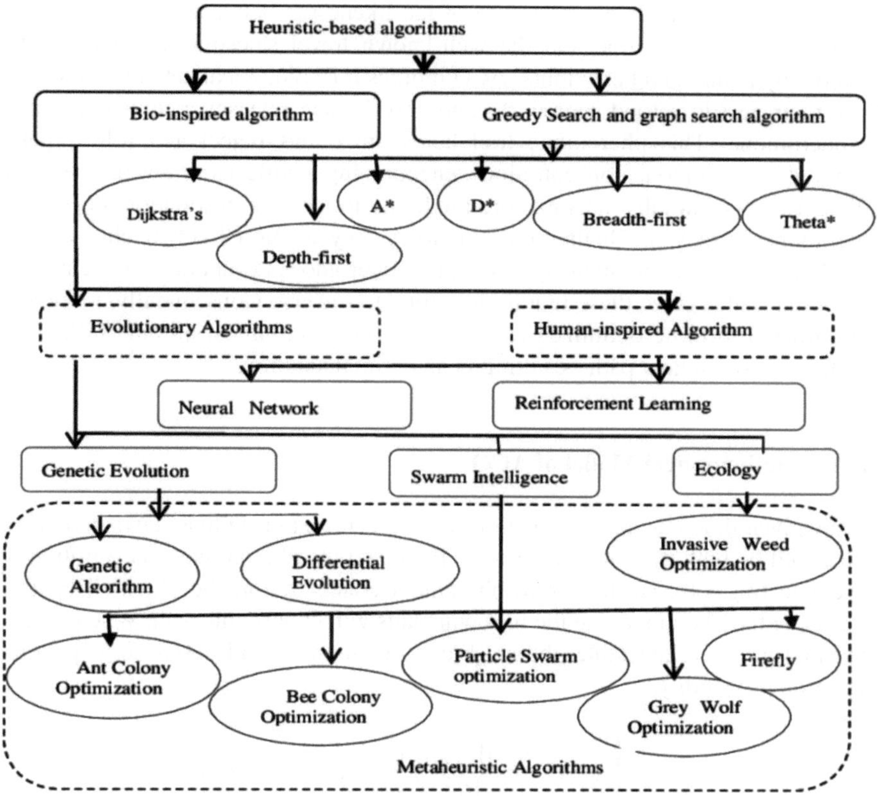

Fig. 1 Classification of Heuristic algorithm

This paper presents a comparative study of standard Ant Colony Optimization (ACO) and the Firefly Algorithm (FA) within a static environment. Key parameters evaluated include shortest paths, iterations, time to find optimal paths, and path length. Section 2 offers an in-depth background on ACO and FA, their operating principles, applications, and previous research, highlighting each algorithm's strengths and weaknesses. Section 3 details the methodology and experimental setup, describing parameters and configurations. Section 4 defines metrics for performance evaluation. Section 5 provides results, including tables, charts, and graphs, while Sect. 6 interprets findings. Finally, Sect. 7 summarizes key insights, suggesting suitable applications for each algorithm.

2 Literature Survey

2.1 Ant Colony Optimization (ACO)

The Ant Colony Optimization (ACO) algorithm is an evolutionary, nature-inspired heuristic developed to tackle optimization problems. Originally introduced by M.

Dorigo [14], who made complete use of the likenesses between the routes of ant colony searching for food and the well-known travel salesman problem (TSP). Inspired by nature and behaviours, ACO mimics how biological ants find pathways from their nest to a food source. As ants travel, each releases a chemical known as pheromone. This pheromone trail helps other ants detect and follow paths with higher pheromone concentrations, encouraging a collective movement toward efficient routes. Initially, ants explore randomly to locate a destination, depositing pheromones as they go. On their return journey, they are more likely to choose paths with higher pheromone intensity. Over time, pheromones evaporate, concentrating on shorter paths, since these require less time to traverse Consequently, more ants are attracted to these optimized routes, and over time, almost all ants follow the shortest, most efficient path, resulting in an optimal solution.

2.1.1 Mathematical Model of ACO

At any given time t, an ant k transitions from its current grid i to an unvisited grid j, influenced by both the distance between them and the pheromone concentration on the connecting path. The higher the pheromone concentration, the "thicker" the trail, increasing the likelihood that the following ants will choose this path. When several unvisited grids are available, the ant k chooses its next grid based on a specified transition probability.

$$p_{ij}^{k} = \begin{cases} \frac{[\tau_{ij}(t)]^{\alpha}.[\eta_{ij}(t)]^{\beta}}{\sum_{s\epsilon\ allowed_k} [\tau_{ij}(t)]^{\alpha}.[\eta_{ij}(t)]^{\beta}} & j\epsilon allowed_k \\ 0 & otherwise \end{cases} \quad (1)$$

In this context, α and β are weighting parameters that determine the impact of pheromone levels and heuristic information (visibility) on path selection, respectively. $allowed_k$ represents the set of grids visited by ant k, $\tau_{ij}(t)$ denotes the pheromone concentration between grid i and j, and $\eta_{ij}(t)$ indicates the visibility of grid j from i

$$\eta_{ij} = \frac{1}{d_{ij}} \quad (2)$$

Here, d_{ij} represents the distance between two neighboring grids, defined as

$$d_{ij} = \sqrt{(x_i - x_j)^2 + (y_i - y_j)^2} \quad (3)$$

Once all ants have completed their tours, the pheromone levels on each path are updated by reducing the existing pheromone through evaporation and adding new pheromone deposits from each ant. The pheromone update is governed by the following formula:

$$\tau_{ij}(t+1) = (1-\rho)\,\tau_{ij}(t) + \sum_{k=1}^{m} \Delta\tau_{ij}^{k} \tag{4}$$

In this formula τ_{ij} represents the pheromone evaporation rate, m is the total number of ants, and $\Delta\tau_{ij}^{k}$ denotes the amount of pheromone deposited by ant k during the current iteration, defined as follows:

$$\Delta\tau_{ij}^{k} = \begin{cases} \frac{Q}{L_k} & edge\ (i,j)\ is\ used\ in\ tour \\ 0 & otherwise \end{cases} \tag{5}$$

Here, L_k represents the path length found by ant k, and Q is a constant.

2.1.2 Main Variant of Ant System

Several ACO [16] variants have been introduced in the literature, with three particularly successful ones. The first is the Ant System (AS), where all ants that complete a tour update the pheromone levels. AS was the initial implementation of an ACO algorithm. Next is the MAX-MIN Ant System (MMAS) [17], characterized by allowing only the best-performing ant to update pheromone trails, with limits set on minimum and maximum pheromone levels. The third is the Ant Colony System (ACS), notable for its local pheromone updates conducted by each ant after every construction step. To clarify their distinctions, the Traveling Salesman Problem (TSP) serves as an example [18].

2.2 Firefly Algorithm (FA)

The Firefly Algorithm is a metaheuristic global optimization technique inspired by the flashing behavior of fireflies, introduced by Xin-She Yang in 2008. This modern artificial intelligence approach is widely applied across various optimization and autonomous systems. Yang demonstrated how fireflies' flashing behavior could inspire an optimization algorithm, as fireflies produce bioluminescent flashes in rhythmic patterns to either attract mates or deter predators. In the algorithm, fireflies are attracted to one another based on light intensity, which diminishes as the distance between them increases [19]. The Firefly Algorithm is based on three idealized rules:

- Fireflies are attracted to one another based on light intensity, irrespective of gender.
- Each firefly's attractiveness and brightness are related, both diminishing as the distance between them increases. Consequently, a firefly with lower brightness will move toward a brighter one. If no brighter firefly is nearby, it moves randomly.
- The brightness of a firefly corresponds to the objective function.

The Firefly Algorithm (FA) efficiently solves highly nonlinear, non-convex optimization problems and converges more quickly to the global optimum than traditional methods. It can be integrated with other optimization techniques to form hybrid models and does not depend on an initial solution, consistently reaching the same optimum.

2.2.1 Mathematical Model of FA

As a firefly's attractiveness depends on the light intensity perceived by nearby fireflies, the change in attractiveness β concerning the distance r can be expressed as follows:

$$\beta = \beta_o e^{-\gamma r^2} \quad (6)$$

Where β_o is the attractiveness at r = 0. And γ Is the light absorption coefficient.

The fireflies are represented using Cartesian coordinates, and the distance between them is calculated as follows.

$$r_{ij} = \left|x_i - x_j\right| = \sqrt{\sum_{k=1}^{d} \left(x_{k,i} - x_{k,j}\right)^2} \quad (7)$$

Here, d represents the number of dimensions. The movement of firefly i, attracted to a brighter and more appealing firefly j, is governed by the following equation.

$$x_i^{t+1} = x_i^t + \beta_o e^{-\gamma r_{ij}{}^2}\left(x_j^t - x_i^t\right) + \alpha_t\left(s_t - 0.5\right) \quad (8)$$

The second term represents the effect of attraction, while the third term introduces randomization, where α_t is the randomization parameter and s is a vector of random numbers generated from a Gaussian or uniform distribution at time t. When $\beta_o = 0$, the algorithm simplifies to a random walk, and when $\gamma = 0$, it resembles a variation of particle swarm optimization.

3 Research Methodology

At the initial stage, the environment was mapped using a suitable mapping algorithm to represent the global workspace where the robot operates. The map identifies the start and goal points while distinguishing between obstacles and free space prior to applying the path planning algorithm. In this study, a map with interconnected free space nodes was created, assuming obstacles were disregarded during the early stages of path planning. The nodes, defined by x-y coordinates, served as input for the algorithm. These coordinates were used to calculate the exact path cost or distance between nodes. Finally, the optimal path and performance metrics were recorded to validate the proposed algorithm's effectiveness.

3.1 Environment Modeling

For this research, a 2D static grid environment has been created with fixed grid size having boundary conditions. Obstacles are also positioned manually in the environment with black cells while the free space is with white cells. The robot begins its journey from the starting position (0, 0) and navigates toward the goal position (19, 19) while avoiding obstacle nodes along the way. The robot can move in an eight-way direction to find the optimum path. For comparative study, we take a 20×20 grid size and three different environments having obstacles in square shape. As shown in Figs. 2, 3 and 4 respectively.

4 Evaluation Matrix

To get the optimum result we tune the value of the algorithm parameter such as α, β, β_o, γ. For this paper, the parameter of the ACO algorithm is thus the initial pheromone. $\tau_0 = 0.5$, and $Q = 100$ is a constant and all other parameters are shown in Tables 1 and 2 for ACO and FA algorithm respectively. Constant of pheromone α and visibility β. The FA parameter is thus, the random parameter. α, attractiveness β_o and light absorption coefficient γ. By changing the population size of the algorithm as well as the complexity of the environment we try to get the optimum path of the robot in conjugative graph. As shown in the Figs. 2, 3 and 4 respectively.

For performing the simulation, we make three environments. By increasing the iteration as well as number of populations as tabulated in Tables 1 and 2. We plotted the best path opted by the robot as shown in Fig. 2. We calculate the path length and time taken to execute the program in Table 3.

5 Simulation Result

Here we simulate the robot's path in three different environments using the ACO and FA algorithm. For the simulation of the graph of the path, we use Python programming language for writing the program code. The hardware component is used as an HP Windows laptop having 8 GB RAM and, AMD RYZON 5000 SERIES processor.

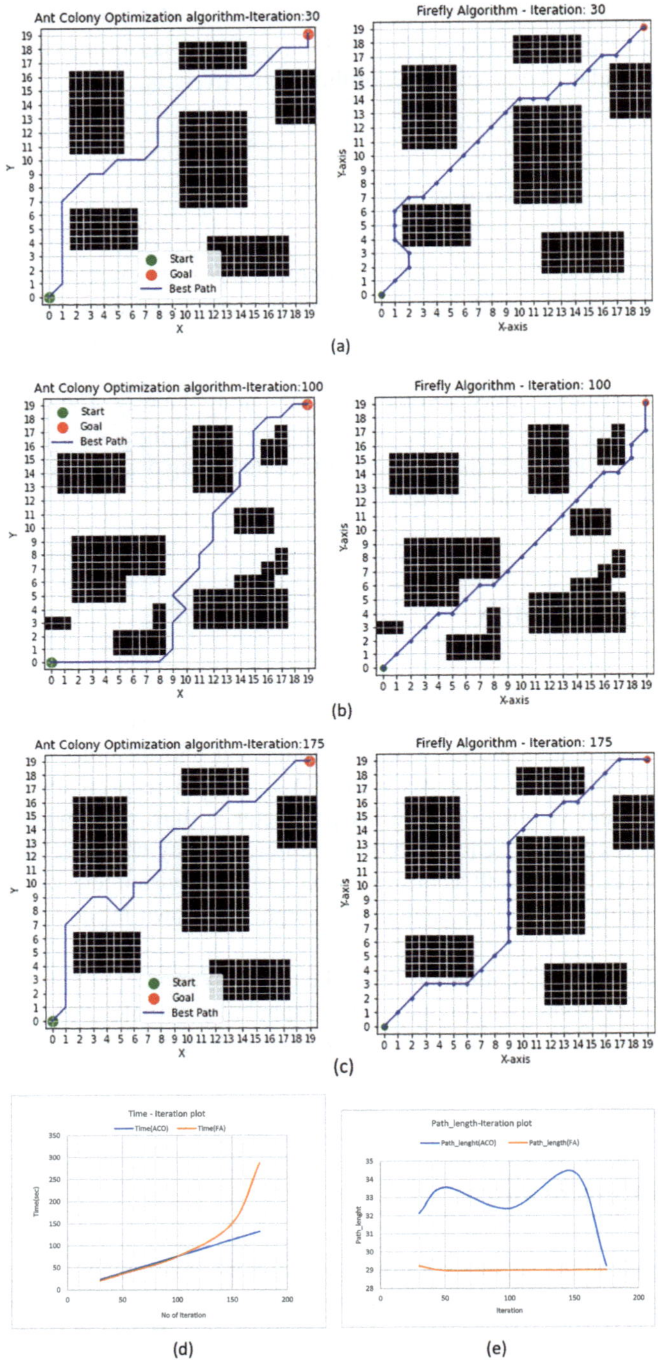

Fig. 2 (**a**), (**b**), and (**c**) Are the path at iteration 30,100,175 respectively. (**d**) Is plot between time—iteration and (**e**) is path length—iteration graph

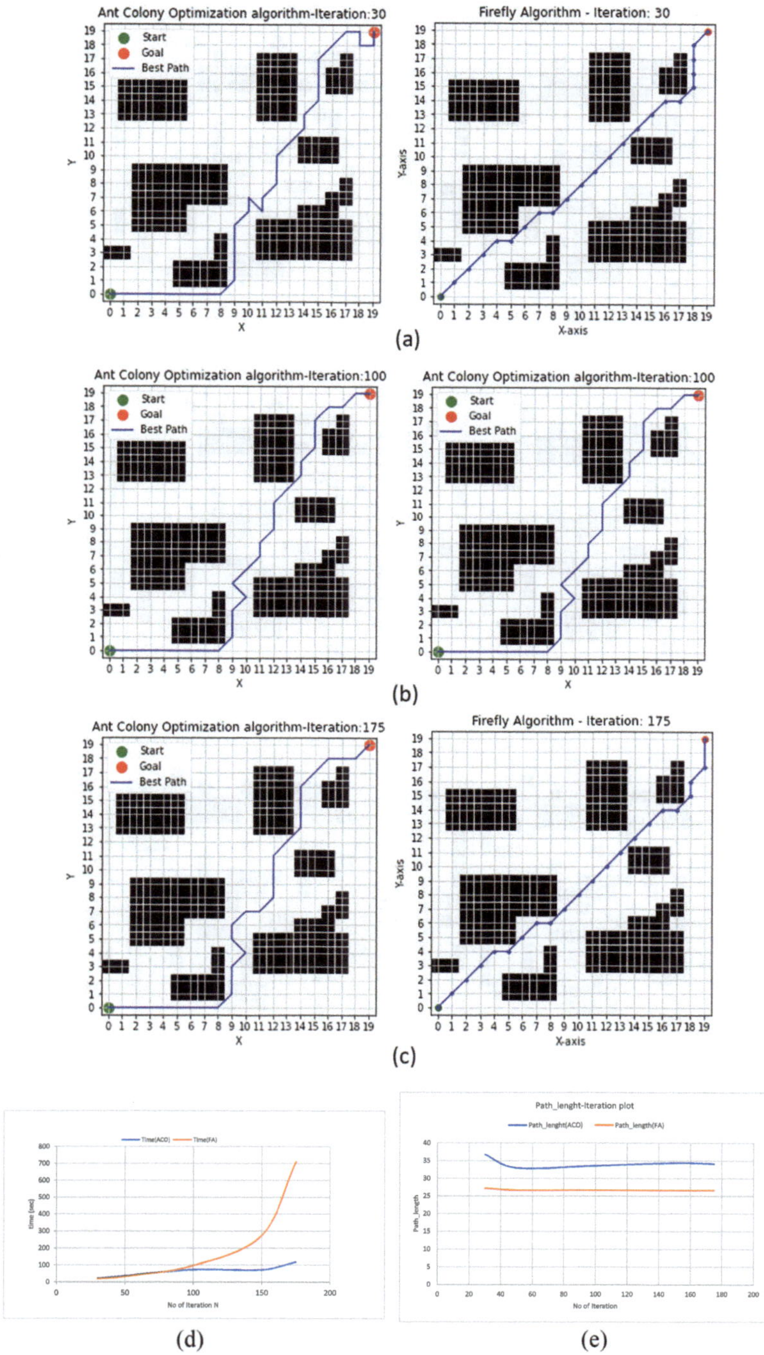

Fig. 3 (**a**), (**b**) and (**c**) Are the path at iteration 30,100,175 respectively. (**d**) Is plot between time—iteration and (**e**) is path length—iteration graph

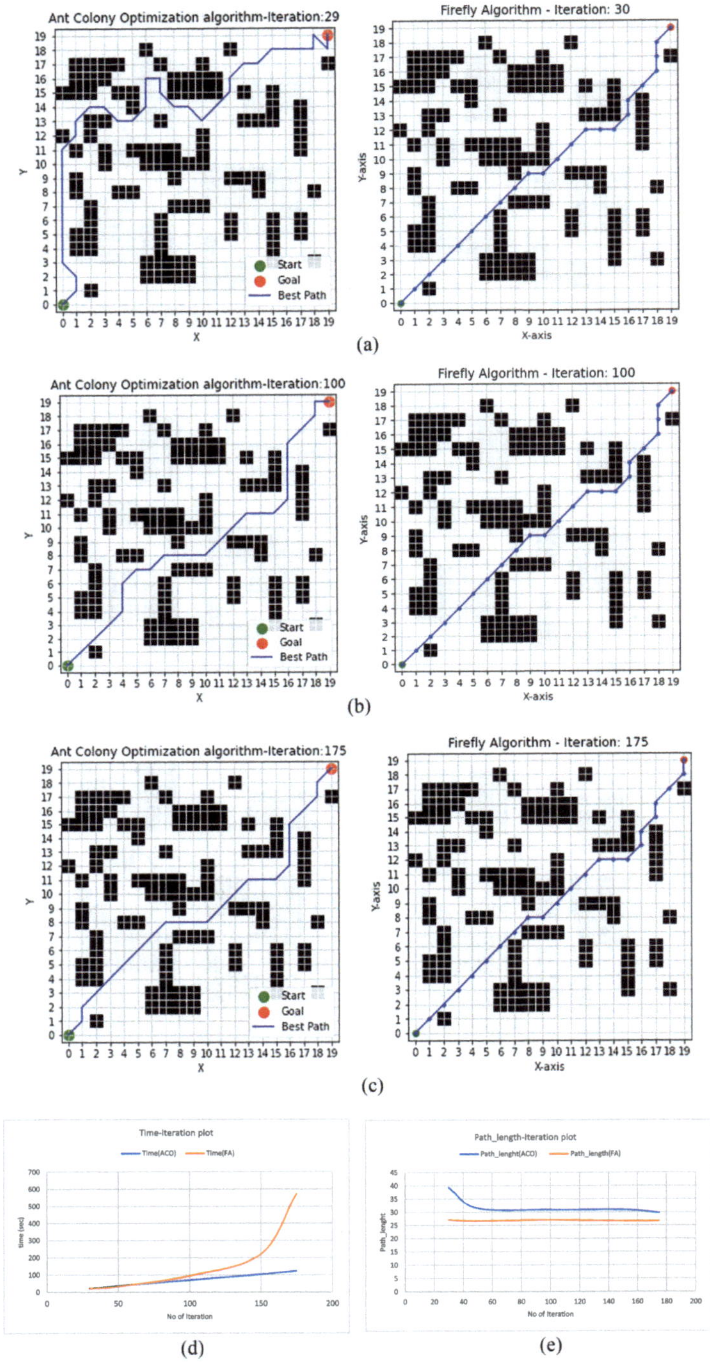

Fig. 4 (**a**), (**b**), and (**c**) Are the path at iteration 30,100,175 respectively. (**d**) This is a plot between time—iteration and path length—iteration graph

Table 1 For the Ant colony optimization algorithm

Algorithm parameter	Environment 1	Environment 2	Environment 3
Number of population (n)	10	50	70
Iteration (N)	30	100	175
β	5	5	5
α	10	10	10
ρ	0.99	0.99	0.99

Table 2 For Firefly algorithm

Algorithm parameter	Environment 1	Environment 2	Environment 3
Number of population (n)	10	50	70
Iteration (N)	30	100	175
β_0	1	1	1
α	0.5	0.5	0.5
γ	0.1	0.1	0.1

Environment: 1
See Fig. 2.

Environment: 2
See Fig. 3.

Environment: 3
See Fig. 4.

6 Interpretation of the Result

Here we have successfully plotted the optimum path in different environments as shown in Figs. 2, 3 and 4. After that calculating path length and time of execution of both algorithms by varying the number of population and as well as iteration. In Table 3, we tabulated all result obtained by simulation in individual Environments. From Table 3, As we increase the no of Ant as well iteration with complexity of the environment execution time and path length is also increases. Here the objective of this paper is to calculate the shortest path length (best path) and minimum time of execution. From Figs. 2e, 3e, and 4e we can see that when the number of iterations is less, FA is faster than ACO. As the number of iterations increases FA is slower down as compared to ACO. However, FA provides a better path than the ACO algorithm. As the complexity of the environment increases the ACO algorithm is stuck in local optima and the execution time of the algorithm is also increased.

Table 3 Path length and time variation for ACO and FF over a grid of size 20 × 20

Environment	Algorithm	n = 10, N = 30		n = 30, N = 50		n = 50, N = 100	n = 60, N = 150			n = 70, N = 175	
		Path length	Time(s)	Path length	Time	Path length	Time	Path length	Time	Path length	Time
Environment1	**ACO**	**32.1421**	**23.2693**	33.55634	38.507863	32.384776	75.25175	34.384776	113.2845	32.9705627	131.6875803
	FA	29.21320	20.38607	**28.97056**	35.738648	**28.970562**	74.75115	28.970562	149.8184	28.9705627	286.7073175
Environment 2	**ACO**	36.7279	21.13943	**32.97056**	36.1919975	33.556349	72.71928	34.384776	72.71928	34.1421356	119.261793
	FA	27.21320	18.42489	**26.62741**	30.96700453	26.627416	96.29715	**26.627416**	275.0775	**26.6274169**	709.1155526
Environment 3	**ACO**	39.31370	22.03426	31.55634	37.2971	30.970562	71.35330	30.970562	104.7119	**29.79898987**	123.5544610
	FA	27.04163	20.34407	**26.62741**	**33.75233435**	27.041630	95.14640	**26.627416**	220.5287	26.62741699	572.0915420

7 Conclusion

In this research paper, we provide the theoretical as well as practical way of ant colony optimization (ACO) and Firefly algorithm (FA) from basic to real-time application in our life with the help of simulation. Using this algorithm, we try to find out the optimum path of the robot in a grid environment that has obstacles in it. While moving from the start position to the goal position robot avoids the obstacle that is lying in its path. We got that at the initial stage FA is faster than ACO but as the number of iterations and as complexity of the environment increases execution time of FA also increases. FA provides the shortest path as compared to ACO but it takes more time to provide an optimum path. We can eliminate the disadvantage of these algorithms by using a hybrid algorithm which is the combination of the best features of both algorithms.

References

1. G. Venter.: "Review of optimization techniques," Handle.net (2023).
2. P. Trojovský and M. Dehghani.: "Pelican Optimization Algorithm: A Novel Nature-Inspired Algorithm for Engineering Applications," Sensors, vol. 22, no. 3, p. 855, (Jan. 2022).
3. Hesami, Mohsen, and A.M.P. Jones.: "Application of artificial intelligence models and optimization algorithms in plant cell and tissue culture." Applied Microbiology and Biotechnology 104.22, pp.9449-9485, (2020).
4. Lin, W.Yuan.: "A novel 3D fruit fly optimization algorithm and its applications in economics," Neural Computing and Applications, 27, pp.1391-1413 (2016).
5. Avdiu, Kujtim.: Algorithmic optimization and its application in finance. Diss. lmu,pp. 1391-1413 (2021).
6. Hamdia, M.Khader, X. Zhuang, and T.Rabczuk, "An efficient optimization approach for designing machine learning models based on genetic algorithm." Neural Computing and Applications, 33.6, pp. 1923-1933 (2021).
7. Tan, C. Sheng, R.M. Mokhtar, and M. R. Arshad.: "A comprehensive review of coverage path planning in robotics using classical and heuristic algorithms." IEEE Access 9, pp.119310-119342 (2021).
8. Mullen, J. Robert, D. Monekosso, S. Barman, and P.Remagnino.: "A review of ant algorithms." Expert systems with Applications,vol. 36, no. 6, pp. 9608-9617 (2009).
9. Guermoui, Mawloud.: "A novel hybrid model for solar radiation forecasting using support vector machine and bee colony optimization algorithm: review and case study." Journal of Solar Energy Engineering, vol.143, no 2, pp.020801(2021).
10. Shami, M.Tareq.: "Particle swarm optimization: A comprehensive survey." Ieee Access. Vol.10, pp. 10031-10061 (2022).
11. Beni, Gerardo, and J.Wang. "Swarm intelligence in cellular robotic systems." Robots and biological systems: towards a new bionics?. Berlin, Heidelberg: Springer Berlin Heidelberg, pp.703-712, (1993).
12. Ibraheem, I. Kasim, and F.H. Ajeil.: "Path planning of an autonomous mobile robot using swarm-based optimization techniques." Al-Khwarizmi Engineering Journal vol. 12, no 4, pp 12-25 (2016).
13. Sharma, Isha, V. Kumar, and S.Sharma.: "A comprehensive survey on grey wolf optimization." Recent Advances in Computer Science and Communications (Formerly: Recent Patents on Computer Science), vol. 15, no 3, pp. 323-333 (2022).

14. X.S.Yang, Nature-inspired metaheuristic algorithms. Luniver press, (2010).
15. Kumar, Vijay, and D. Kumar.: "A systematic review on firefly algorithm: past, present, and future." Archives of Computational Methods in Engineering, vol 28, pp. 3269-3291 (2021).
16. Chawda, V. Bharat, and N. M.Sureja.: "An ACO approach to solve a variant of TSP." Int. J. Adv. Res. Comput. Eng. Technol. IJARCET 1.5, pp. 222 (2012).
17. Skinderowicz, Rafał.: "Implementing a GPU-based parallel MAX–MIN Ant System." Future Generation Computer Systems, vol. 106 pp. 277-295 (2020)
18. Stützle, Thomas, and H.H. Hoos.: "MAX–MIN ant system." Future generation computer systems, vol.16, no. 8, pp.889-914 (2000).
19. Fister, Iztok.: "A comprehensive review of firefly algorithms." Swarm and evolutionary computation, vol 13, pp. 34-46 (2013).

Adaptive Brake Control for Sustaining Dynamic Stability of Autonomous Vehicles Using Markov Decision Process

Indranil Banik and Arup Kumar Nandi

Abstract Enduring the optimal safety of autonomous vehicles while navigating through the complexities of dynamic traffic scenarios necessitates the implementation of urbane decision-making systems that are not only capable of processing real-time information but also adept at adapting to the ever-changing environmental and operational conditions that characterize urban and rural roadways. In this research work, a comprehensive framework based on the principles of Markov Decision Process (MDP) has been proposed that aims to optimize the critical functions of braking control and collision avoidance during autonomous driving, thereby enabling them to make real-time adjustments in response to fluctuating deceleration demands and varying road conditions that may arise during braking. To further refine the decision-making capabilities of the system, a strategy that combines exhaustive enumeration with Epsilon-Greedy action selection mechanism is integrated, which effectively balances the dual imperatives of exploration and exploitation to facilitate the convergence towards an optimal policy that maximizes longitudinal stability and braking efficacy. Simulation outcomes demonstrate the system's proficiency to dynamically allocate braking forces while executing safe lane changes whenever the situational context permits such manoeuvres. This research contributes to augment the operational wellbeing and reliability of fully automated vehicles using intelligent brake control with adaptive decision making in real-world driving scenarios characterized by uncertainty and variability.

Keywords Autonomous vehicle · Brake force distribution · Markov decision process · Intelligent brake control · Collision avoidance

I. Banik (✉) · A. K. Nandi
Academy of Scientific and Innovative Research (AcSIR), Ghaziabad, Uttar Pradesh, India

CSIR-Central Mechanical Engineering Research Institute, Durgapur, West Bengal, India
e-mail: indranil.cmeri20a@acsir.res.in

K. Deep et al. (eds.), *Artificial Intelligence, Machine Learning and Intelligent Systems*, Springer Proceedings in Mathematics & Statistics 520,
https://doi.org/10.1007/978-981-95-3841-6_21

1 Introduction

The advent of vehicular automation characterizes a significant paradigm shift in the realm of transportation, which not only promises to deliver remarkable improvements in safety standards but also offers rapid response capabilities and heightened levels of convenience for end-users. At the core of these systems lies artificial intelligence (AI), that governs path planning, decision-making, and human-machine interaction competences [1]. However, the realization of these benefits hinges on addressing the intricate challenges associated with ensuring the dynamic stability of autonomous vehicles (AVs) across complex driving approaches and diverse road conditions. Among the myriad factors inducing vehicular stability, brake force distribution (BFD), a fundamental element of the stability control framework, behoves decisive in mitigating issues related to oversteer, understeer, and loss of control during braking maneuvers.

While traditional methodologies for brake torque allocation have proven effective in conventional vehicles, the unique challenges posed by AVs necessitate a paradigm shift [2]. Present methodologies for BFD in AVs often rely on predefined rule-based systems and simple PID controllers [3]. In the context of data-driven methodologies, techniques such as supervised learning, alongside sophisticated models like neural networks (NNs), have frequently been employed to effectively discern and understand the intricate relationships that exist between different driving states and the corresponding control inputs made by drivers [4]. While the aforesaid approaches may suffice in controlled environments, their limitations become apparent in the facets of real-world complexities. The dynamic nature of autonomous driving scenarios, coupled with the prerequisite for contemporaneous decision-making, requires innovative approaches that surpass the capabilities of static control strategies [5]. Therefore, the modelling of car-following behaviour in connected AVs emphasizes the critical role of driver compliance [6]. While the focus on human-like behaviour in autonomous driving is crucial, it is also essential to consider the variability and unpredictability of human drivers, which can pose challenges in achieving consistent performance across diverse traffic conditions [7]. Stochastic models are employed to envisage the movements of adjacent vehicles, enabling the assessment of whether to complete or abort an overtaking manoeuvre [8]. In more recent developments within the research community, experts have increasingly turned to advanced approaches such as Deep Reinforcement Learning [9, 10] and probabilistic MDP [11] to proficiently analyse and learn from simulated driving behaviours and decision-making processes, thus enriching the overall understanding of driver interactions and system responses in varying traffic environments.

One prominent limitation lies in the inability of conventional rule-based control systems to adapt to the evolving conditions encountered by AVs. These systems rely on predetermined sets of rules and heuristics, which, while effective in predictable environments, struggle to cope with the increasing complexity and unpredictability inherent in real-world driving scenarios. As a consequence, AVs face difficulties in maintaining stability when confronted with abrupt changes in road friction conditions, varying vehicle speeds, or unanticipated obstacles. Moreover, the dynamic interaction between the vehicle's control system and its environment demands a loftier level of adaptability to the evolving traffic conditions that surpasses the capabilities of conventional methodologies. Tyre models, if considered, are often oversimplified, neglecting the intricacies of tyre behaviour under different loadings and slip conditions. This oversight leads to sub-optimal braking strategies that do not fully exploit the potential of the hydraulic braking system. In this context, there exists a compelling need for intelligent systems capable of learning and adapting BFD strategies in real-time, harnessing the power of AI to enhance dynamic stability during braking.

To bridge these gaps and augment the current advancements in the control paradigm of AV stability, this research endeavors to establish a dynamic and novel MDP based control framework for optimizing brake allocation compliant in autonomous driving. This adaptive approach allows vehicles to respond more effectively to detection of unexpected obstacles and changes in road conditions, thereby minimizing the risk of accidents and improving passenger confidence in autonomous systems, ensuring both immediate and long-term safety. The subsequent sections of this paper systematically explore the components of the proposed adaptive framework. Section 2 outlines the detailed methodology employed in this study introducing the brake control strategy to ensure safety and responsiveness in diverse driving scenarios. Section 3 delves into the MDP framework, incorporating dynamic programming to optimize brake force allocation in real-time. The outcomes of the proposed approach, supported by simulations and analyses under varying conditions, are discussed in Sect. 4. Thereby, Sect. 5 culminates into briefing the control outputs in enhancing AV stability and receptiveness to traffic behaviour.

2 Brake Control Strategy

The present research integrates Pacejka's tyre model combined with Burkhardt tyre model for real-time slip estimation within the MDP framework to accurately simulate and obtain stable BFD on respective wheel axles in accordance to driving behaviour of the leading vehicle and surrounding lane conditions. The proposed brake control approach is designed to adapt to three distinct leading vehicle behaviour offering intelligent adaptive responses by leveraging dynamic policy optimization using reinforced learning based MDP:

- In stationary scenarios, braking force and reaction times are prioritized.
- For constant speed vehicles, relative velocity and distance are key inputs.
- During deceleration, the control framework accounts for both ego and leading vehicle dynamics, calculating precise deceleration rates for maintaining safety.

Table 1 Characteristics of the longitudinal σ-μ conjunction for different types of roads surfaces

Road surface	c_1	c_2	c_3	Apex adhesion factor, μ_p	Sliding adhesion factor, μ_s
Dry tarmac	1.28	23.98	0.52	0.75–0.90	0.65–0.70
Wet tarmac	0.86	33.83	0.35	0.55–0.75	0.45–0.60
Dry concrete	1.19	25.17	0.54	0.80–0.95	0.70–0.75
Snow	1.95	94.12	0.06	0.15–0.20	0.05–0.10

2.1 Brake Dynamics of the Vehicle

The tyre forces acting at the road-tyre interface during braking can be quantitatively assessed by means of Pacejka's Tyre Model [12] which is contingent upon slip ratio of the wheels at respective axles. This empirical relation is mathematically represented as:

$$F_x = D.\sin\left(C.\arctan\left(B\psi\right)\right) \tag{1}$$

where B, C (=1.65 for braking) and D conclusively demarcates the designated facets corresponding to the stiffness, shape and peak determinants, respectively. Considering σ as wheel slip ratio, ψ is articulated as:

$$\psi = (1 - E)\sigma + \left({}^{E}\!/_{B}\right).\arctan\left(B.\sigma\right) \tag{2}$$

E is identified as the curvature aspect. The subsequent longitudinal σ-μ correlation of Burkhardt Tyre Model is employed to ascertain the slip contingent upon the type on road surface, with the parameter values of the relationship as delineated in Table 1 being derived from the tyre model elucidated in [13]. Under circumstances of stable vehicle dynamics (constant velocity or during braking), it is posited that the road friction (μ) fluctuates between μ_s and μ_p.

$$\sigma - \frac{\log\sigma}{c_2} = \frac{\log\left(\frac{c_1}{c_3}\right) - \left(\frac{c_1-\mu}{c_3}\right)}{c_2} \tag{3}$$

2.2 Safe Distance Computation

This manuscript employs a collision avoidance gap framework that is predicated on the vehicle braking dynamics. Figure 1 portrays the spatial configuration of the

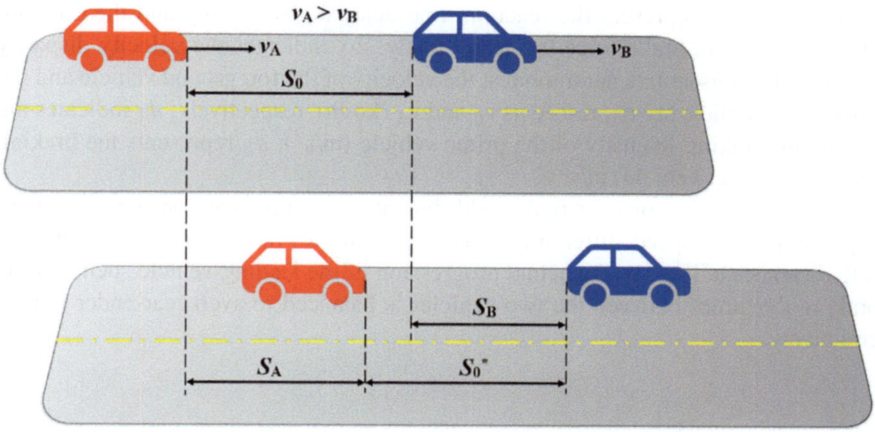

Fig. 1 Vehicle positional model for safe braking and collision avoidance

prime vehicle (A) corresponding to the leading vehicle (B) navigating the same lane and in the same direction throughout the braking sequence.

For baseline design of the algorithm, the side lane is presumed to be free. The minimum allowable distance considered safe separating the two vehicles A and B to avoid collision is calculated using Eq. (4):

$$S_0 = S_A + S_0^* - S_B \tag{4}$$

Here, S_A refers to the path traversed by ego vehicle A; S_B represents the displacement by leading vehicle B in same time frame; and S_0^* signifies the established minimum threshold gap between the two vehicles. Harnessing the braking behaviour of the leading vehicle, the computation formula for S_0 is derived under the following circumstances [14]:

- Target vehicle is in a stationary state

$$S_0 = \frac{(\tau_1 + 0.5\tau_2)\, v_A}{3.6} - \frac{v_B^2}{25.92 * d_A} + S_0^* \tag{5}$$

- Target vehicle is moving at a constant speed

$$S_0 = \frac{\tau_1 v_A}{3.6} + \frac{\tau_2\,(v_A - \Delta v)}{7.2} - \frac{\Delta v^2}{25.92 * d_A} + S_0^* \tag{6}$$

- Target vehicle is decelerating

$$S_0 = \frac{\tau_1 v_B}{3.6} + \frac{\tau_2\,(v_A - \Delta v)}{7.2} - \frac{v_A^2}{25.92 * d_A} + \frac{(v_A - \Delta v)^2}{25.92 * d_B} + S_0^* \tag{7}$$

Here, τ_1 and τ_2 represent the reaction time delay (s) in braking and the duration of the increase in brake force (s), respectively; Δv indicates the velocity disparity between the consecutive automobiles; the velocity of the foreground vehicle and the subsequent vehicle are expressed as v_B and v_A (km/h), respectively; d_A indicates the maximum braking intensity of the prime vehicle (m/s^2); d_B represents the braking rate of the target vehicle (m/s^2).

Equations (5–7) consider real-world dynamics such as reaction times, braking capabilities, and speed differentials. They provide a systematic way to compute safe distances tailored to the actual progression of the leading vehicle such that the braking dynamics between the two vehicles is balanced to avert rear-ender impact effectively.

2.3 *Time to Collision (TTC) Computation*

TTC framework is a measure used in road safety and vehicle dynamics to assess the amount of time remaining before two vehicles collide, assuming they maintain their current velocities and trajectories. This is calculated as [15]:

$$TTC = \begin{cases} \frac{S_0}{\Delta v} & \Delta d = 0, \Delta v > 0 \\ \frac{\Delta v + \sqrt{\Delta v^2 + \Delta d S_0}}{\Delta d} \quad if & \Delta d > 0, \Delta v > 0 \\ \frac{-\Delta v + \sqrt{\Delta v^2 - \Delta d S_0}}{\Delta d} & \Delta d < 0, \Delta v > 0 \end{cases} \tag{8}$$

In this context, Δd signifies the relative braking intensity of the vehicles.

The first-order TTC is applied when the leading automobile is either immobile or traversing at a persistent pace. Conversely, the second-order TTC is employed when the leading automobile is slowing down. Equation (8) feeds into the automated braking algorithm to dynamically adjust braking intensity or trigger evasive actions when obstacle is detected. Incorporating TTC helps to enhance the responsiveness of the autonomous system under varying driving conditions.

3 Markov Decision Process (MDP): Dynamic Programming

MDP serves as a systematic structured mathematical model for understanding the sequential decision-making process, wherein an agent engages and interacts with its environment, receives feedback (rewards), and takes actions to maximize long-term rewards [16]. Given that the agent lacks prior knowledge of the transition or reward functions, it must interact with the environment to infer the optimal policy utilizing the *Bellman Policy Equation*. The state of the system is subject to periodic observations, and actions will be determined based on the observed state using *Epsilon-Greedy* algorithm [17] to equipoise the trade-off between *Exploration* which allows the vehicle to try different actions (e.g., lane change, mild braking) to gather knowledge about how those actions impact the system (e.g., safety and comfort); and *Exploitation* which ensures the vehicle applies actions that minimize collision risk and maintain safety (e.g., selecting braking force based on the highest reward value). The state space S is countably defined, and the action set $A(i)$, which is available at state i, is nonempty and exhibits a measurable structure $A(i)$. The steps followed in finding the optimal policy is explained:

Step 1: Initialization
Commence by establishing the iteration number, $k = 1$;
Policy that has been nominated in the kth iteration $= \hat{\xi}_k$.
Optimum policy $= \hat{\xi}^*$

Step 2: Policy Evaluation
For $i = 1, 2, 3, \ldots . |S|$

$$h^k(i) = \bar{r}\left(i, \xi_k(i)\right) - \rho^k + \sum_{j=1}^{|S|} P\left(i, \xi_k(i), j\right) h^k(j) \; given \; \rho^k \neq 0 \tag{9}$$

$\xi_k(i)$: action chosen when the system in the state '*i*' under the prescribed policy $\hat{\xi}$
$\bar{r}\left(i, \xi_k(i)\right)$: anticipated instant reward in the state '*i*' as dictated by the current policy $\hat{\xi}$
$P(i, \xi_k(i), j)$: one-step transition probability that quantifies the likelihood of hopping from state '*i*' to '*j*' under the guidance of the current policy $\hat{\xi}$
ρ: average reward associated for a certain policy $\hat{\xi}_k$

In Eq. (9), the rudiments of the value function vector that correlate with the policy (h^k and ρ^k) remain indeterminate at this stage of the evaluation process.

Step 3: Policy Improvement

$$\xi_{k+1}(i) \in \underset{a \,\in\, A(i)}{\arg\max} \left[\bar{r}(i, a) + \sum_{j=1}^{|S|} P\left(i, a, j\right) h^k(j) \right] \quad \forall i \in S \tag{10}$$

It is entirely plausible that there exists a multitude of distinct actions that can be deemed optimal under the criteria established by the argmax operator, which fundamentally seeks to identify the action that maximizes a particular policy outcome. Thereby, it is potential for several candidates to emerge as viable solutions for $\xi_{k+1}(i)$. Thus, it becomes essential to establish $\xi_{k+1}(i) = \xi_k(i)$ for each '*i*'. To counter this, *Epsilon-Greedy* method is adopted to ensure that the agent learns optimal policies without compromising performance.

Step 4: Termination
In the event that the newly derived policy is found to be identical to the previously established policy for each '*i*', then set $\hat{\xi}^*(i) = \xi_k(i)$.

Conversely, $k = k + 1$ and go to Step 2.

4 Results and Discussion

The MDP based dynamic programming will simulate ego vehicle behaviour in response to the leading vehicle's motion dynamics. The model parameters of both the test vehicles used are presented in Table 2.

Two distinct states pertaining to system dynamics of the ego vehicle are examined in the present study, namely, $S = \{S_1, S_2\} = \{\text{Coasting, Decelerating}\}$. In each state, four permissible actions are outlined as follows:

$A = \{A_1, A_2, A_3, A_4\} = \{\text{No Brake, Mild Braking, Hard Braking, Lane Change}\}$.

Consequently, there exist three potential driving scenarios that the ego vehicle may encounter, as delineated below:

Condition 1: Leading vehicle is stationary.
Condition 2: Leading vehicle is moving at constant speed.
Condition 3: Leading vehicle is decelerating.

Transition Probability Matrix (TPM) The TPM based on the states considered is a 3D matrix: TPM[Current State][Action][Next State]. Each entry characterizes the likelihood of transitioning to an updated next state given the current state and action as demonstrated in Table 3.

Table 2 Test vehicle parameters

Attributes	Ego vehicle	Leading vehicle
Vehicle mass (kg)	1950	1450
Frontal area of the vehicle (m^2)	2.68	2.54
Span between CG and front axle (m)	1.30	1.130
Span between CG and rear axle (m)	1.45	1.620
Elevation of CG from ground (m)	0.575	0.500
Rolling radius of wheel (m)	0.296	0.275
Initial velocity (m/s)	19.45	16.67
Maximum allowable deceleration (m/s^2)	5.0	4.0

Table 3 TPM matrix representation

Current state (S)	Action (A)	Next state: S_1	Next state: S_2
S_1 (coasting)	No Brake	0.90	0.10
	Mild Brake	0.20	0.80
	Hard Brake	0.05	0.95
	Lane Change	0.85	0.15
S_2 (decelerating)	No Brake	0.25	0.75
	Mild Brake	0.10	0.90
	Hard Brake	0.05	0.95
	Lane Change	0.80	0.20

Table 4 TRM matrix representation

Current state (S)	Action (A)	Reward
S_1 (coasting)	No Brake	5
	Mild Brake	8
	Hard Brake	−6
	Lane Change	7
S_2 (decelerating)	No Brake	−10
	Mild Brake	9
	Hard Brake	6
	Lane Change	4

Transition Reward Matrix (TRM) The TRM based on the actions is a 2D matrix: TRM[Current State][Action]. Each entry represents the reward gained for executing a specific activity in the given state as indicated in Table 4.

Since $v_a > v_b$, the ego vehicle either opt to execute a lane change by assessing the availability of the adjacent lane, or, decelerate while remaining in the same lane to align its velocity with that of the leading vehicle. Consequently, the ego vehicle computes the necessary braking force based on the relative velocity and/or relative deceleration required, taking into account S_0 and TTC.

Presuming the side lane is not free to overtake, the program simulates ego vehicle behaviour in response to the leading vehicle's dynamics as portrayed in Fig. 2 on dry tarmac road ($\mu = 0.8$). At each step, the program calculates the relative velocity and the obligatory TTC. The program updates the distance and velocities based on the chosen action, applying braking forces to the ego vehicle if needed. The braking intensity is adjusted based on both TTC and the distance to the vehicle ahead to avoid abrupt braking and maintain safe separation. Thereby the agent (brake control unit in the autonomous vehicle) adjusts braking forces based on the road condition and tries to learn the optimal braking policy through iterations.

During heavy braking (3–7 s), the majority of the brake force is applied to the front axle, with a smaller proportion allocated to the rear axle. Weight transfer to the front axle during deceleration causes the front brakes to bear the majority of the load, ensuring effective stopping power and vehicle stability. The rear axle consistently

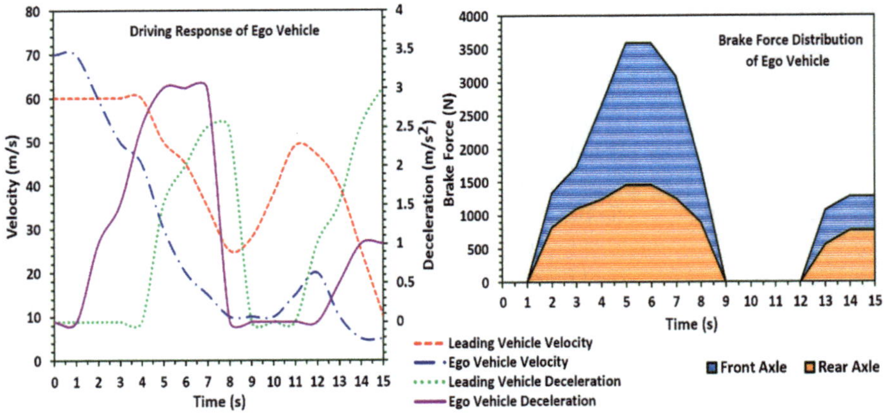

Fig. 2 Driving nature of ego vehicle using MDP in accord to leading vehicle

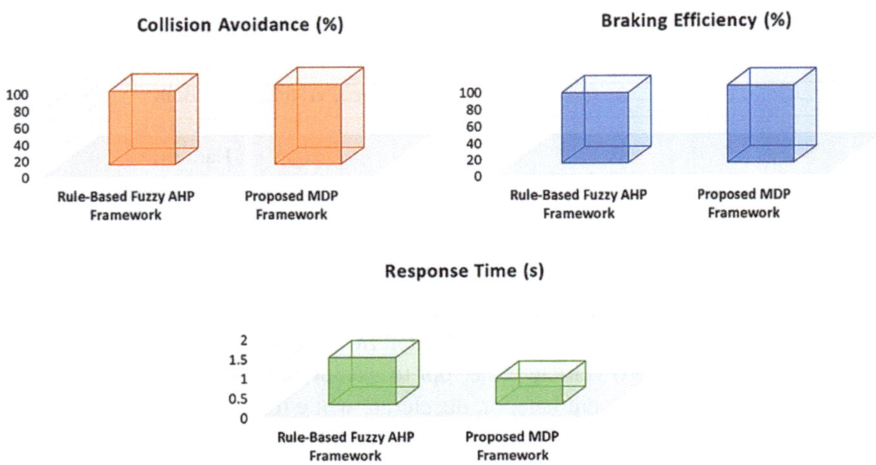

Fig. 3 Performance comparison of autonomous vehicle control methods

contributes less force, preventing over-braking at the rear, which could lead to instability or skidding. Figure 2 suggests that the proposed adaptive braking system responds smoothly and efficiently, avoiding sudden spikes that could destabilize the vehicle. The brake force peak aligns with the deceleration profile of the ego vehicle, demonstrating proper coordination between deceleration demand and braking force application. Therefore, it becomes evident that the ego vehicle is highly responsive to the leading vehicle's deceleration and velocity changes, showcasing a well-tuned control algorithm for adaptive driving behaviour.

Figure 3 underscores the superiority of the proposed MDP-based framework in ensuring competent and supple control of AVs in diverse traffic conditions in comparison to rule-based Fuzzy Analytic Hierarchy Process (AHP) [18]. The MDP's probabilistic reasoning and reinforcement learning based approach enable it to adept to varying road circumstances and maximize braking performance. Using the proposed MDP induced control methodology with 100 recurrences of the trial assessments, response time is reduced by 31.57% while braking efficiency and collision avoidance is improved by 16.74% and 9.68%, respectively.

5 Conclusion

This study introduces a novel framework that integrates safe distance (SD) and Time-to-Collision (TTC) algorithms with reinforced-Markov Decision Process (MDP) based brake control to enhance vehicle safety in dynamic traffic scenarios. The SD algorithm adapts to diverse leading vehicle behaviour—stationary, constant speed, or decelerating—by incorporating braking dynamics, reaction times, and relative velocities, while TTC offers predictive collision timelines, ensuring proactive risk mitigation. The inclusion of MDP enables probabilistic reasoning for taking instantaneous decisions through dynamic policy optimization, allowing the system to effectively select actions like braking or lane changes based on long-term rewards. Unlike static or deterministic control models, this approach handles uncertainties, supports multi-action flexibility, and continuously adapts to traffic conditions, minimizing reaction time by 31.57%. Validated through realistic simulations, the framework demonstrates precise collision avoidance, achieving safe distances and timely responses with enhanced brake efficacy. The inherent scalability and modularity of MDPs render them particularly suitable for application in modern technological frameworks such as Advanced Driver Assistance Systems (ADAS) and various autonomous driving initiatives, where precise decision-making is crucial.

Acknowledgement The first author wishes to convey gratitude to CSIR, New Delhi, India for support to this research under the CSIR-SRF (GATE) fellowship initiative.

Declarations

Conflict of Interest The authors affirm there exists no opposing interests to report.

References

1. Ma, Y., Wang, Z., Yang, H., Yang, L.: Artificial intelligence applications in the development of autonomous vehicles: a survey. *IEEE/CAA Journal of Automatica Sinica*. 7(2): 315-329. (2020). https://doi.org/10.1109/JAS.2020.1003021.
2. Wang, T., Jian, S., Zhou, C., Jia, B., Long, J.: Multimodal traffic assignment considering heterogeneous demand and modular operation of shared autonomous vehicles. *Transportation Research Part C: Emerging Technologies*. 169: 104881. (2024). https://doi.org/10.1016/j.trc.2024.104881.

3. Lai, F., Wang, X.: Enhancing Autonomous Vehicle Stability through Pre-Emptive Braking Control for Emergency Collision Avoidance. *Applied Sciences*. 13(24): 13219. (2023). https://doi.org/10.3390/app132413219.
4. Mostafa H. Tawfeek.: Human-like speed modeling for autonomous vehicles during car-following at intersections. *Canadian Journal of Civil Engineering*. 49(2): 255-264. (2022). https://doi.org/10.1139/cjce-2020-0761.
5. Hu, J., Yi, S., Zhu, P., Sun, Z.: Obstacle avoidance decision and trajectory tracking control of intelligent vehicle considering surrounding vehicles. *Proceedings of the Institution of Mechanical Engineers, Part D: Journal of Automobile Engineering*. (2024). https://doi.org/10.1177/09544070231204390.
6. Cui, S., Cao, F., Yu, B., Yao, B.: Modeling heterogeneous traffic mixing regular, connected, and Connected-Autonomous vehicles under connected environment. *IEEE Transactions on Intelligent Transportation Systems*. 23(7): 8579–8594. (2021). https://doi.org/10.1109/tits.2021.3083658.
7. Yang, Z., Rui, F., Chang, W.: Learning the Car-following Behavior of Drivers Using Maximum Entropy Deep Inverse Reinforcement Learning. *Journal of Advanced Transportation*. (2020). https://doi.org/10.1155/2020/4752651.
8. Wu, Q., Cheng, S., Li, L., Yang, F., Meng, L.J., Fan, Z.X., Liang, H.W.: A fuzzy-inference-based reinforcement learning method of overtaking decision making for automated vehicles. *Proceedings of the Institution of Mechanical Engineers Part D: Journal of Automobile Engineering*. 236(1): 75–83. (2021). https://doi.org/10.1177/09544070211018099.
9. Rizehvandi, A., Azadi, S., Eichberger, A.: Decision-Making Policy for Autonomous Vehicles on Highways Using Deep Reinforcement Learning (DRL) Method. *Automation*. 5(4): 564-577. (2024). https://doi.org/10.3390/automation5040032.
10. Wang, K., Mu, C., Ni, Z., Liu, D.: Safe reinforcement learning and adaptive optimal control with applications to obstacle avoidance problem. *IEEE Transactions on Automation Science and Engineering*. 21(3): 4599–4612. (2023). https://doi.org/10.1109/tase.2023.3299275.
11. Avilés, H., Negrete, M., Reyes, A., Machucho, R., Rivera, K., De-La-Garza, G., Petrilli, A.: Autonomous behavior selection for self-driving cars using probabilistic logic factored Markov decision processes. *Applied Artificial Intelligence*. 38(1). (2024). https://doi.org/10.1080/08839514.2024.2304942.
12. Pacejka, H., Besselink, I.J.M.: Tire and vehicle dynamics. 3rd ed. Butterworth-Heinemann. Elsevier. (2012).
13. Advanced Vehicle Control: Proceedings of the 13th International Symposium on Advanced Vehicle Control. Munich, Germany. CRC Press. (2016).
14. Zhang, L., Yu, Z., Xu, X., Yan, Y.: Research on Automatic Emergency Braking System Based on Target Recognition and Fusion Control Strategy in Curved Road. *Electronics*. 12(16): 3490. (2023). https://doi.org/10.3390/electronics12163490.
15. Ward, J.R., Agamennoni, G., Worrall, S., Bender, A., Nebot, E.: Extending Time to Collision for probabilistic reasoning in general traffic scenarios. *Transportation Research Part C Emerging Technologies*. 51: 66–82. (2014). https://doi.org/10.1016/j.trc.2014.11.002.
16. Puterman, M.L.: Markov Decision Processes: Discrete Stochastic Dynamic Programming. Wiley-Interscience. (2014). https://doi.org/10.1002/9780470316887.
17. Kumar, A., Singh, D.: Adaptive epsilon greedy reinforcement learning method in securing IoT devices in edge computing. *Discover Internet of Things*. 4(27). (2024). https://doi.org/10.1007/s43926-024-00080-7.
18. Zhang, S., Tak, T.: Risk analysis of autonomous vehicle test scenarios using a novel analytic hierarchy process method. *IET Intelligent Transport Systems*. 18(5): 794–807. (2023). https://doi.org/10.1049/itr2.12466.

Enhanced IoT Intrusion Detection: Leveraging Dense Autoencoders with Mahalanobis Distance and Gamma-Based Thresholding

Azmera Chandu Naik, Lalit Kumar Awasthi, and Priyanka Rathee

Abstract The rapid growth of IoT (Internet of Things) devices has introduced new vulnerabilities in network security, necessitating advanced, scalable intrusion detection systems. Traditional machine learning approaches fail to navigate the complexity of threat identification in vast IoT network data. In this study, we propose a novel unsupervised anomaly detection framework leveraging dense autoencoders with Mahalanobis distance and Gamma-based thresholding for enhanced IoT intrusion detection. The approach begins with a dense autoencoder trained on network data, allowing it to learn an efficient representation of normal network behavior. Multivariate reconstruction errors are calculated for each instance of IoT network data, and Mahalanobis distance is incorporated to capture deviations that signify anomalous patterns. To ensure robust anomaly detection, a dynamic thresholding mechanism is implemented, utilizing z-scores with underlying Gamma distribution model to adaptively classify instances as normal or anomalous. The proposed framework was rigorously evaluated on a labeled IoT attack dataset and compared to state-of-the-art unsupervised techniques, including Isolation Forest, Extended Isolation Forest, Elliptic Envelope, K-Nearest Neighbors, and One-Class

A. C. Naik (✉)
Department of Computer Science and Engineering, National Institute of Technology, Hamirpur, India

Department of Computer Science and Engineering, CVR College of Engineering, Ibrahimpatnam, India
e-mail: azmeera_phdcse@nith.ac.in

L. K. Awasthi
Department of Computer Science and Engineering, National Institute of Technology, Hamirpur, India

Senior IEEE Member, Sardar Patel University, Mandi, India
e-mail: lalit@nith.ac.in

P. Rathee
Department of Computer Science and Engineering, National Institute of Technology, Hamirpur, India
e-mail: dr.priyanka@nith.ac.in

K. Deep et al. (eds.), *Artificial Intelligence, Machine Learning and Intelligent Systems*, Springer Proceedings in Mathematics & Statistics 520,
https://doi.org/10.1007/978-981-95-3841-6_22

SVM, as well as a standard dense autoencoder. Results demonstrate that our method achieves superior detection accuracy, with an F1-score of 97.6% on the training set and 95.12% on the test set, significantly outperforming baseline models. These findings highlight the effectiveness of proposed approach for reliable and scalable IoT intrusion detection.

Keywords IoT security · Machine learning · Deep learning · Mahlanobis distance · Anomaly detection

1 Introduction

The rapid proliferation of Internet of Things (IoT) devices has presented numerous security challenges for network infrastructures. In contrast to conventional networks, IoT settings exhibit extensive device diversity, substantial data volumes, and limited resources, resulting in considerable vulnerabilities [8]. These networks are key targets for advanced assaults, such as Distributed Denial of Service (DDoS) attacks, unauthorised access, data exfiltration, and various intrusions [6]. Conventional machine learning-based intrusion detection systems (IDS) frequently struggle to manage the intricacies of IoT data because of their dependence on labelled data and their inadequacies in detecting nuanced or developing threats in real time [16]. Consequently, there is an urgent requirement for an intrusion detection methodology that is both adaptable and scalable, proficient in detecting anomalies in IoT network traffic without dependence on vast labelled datasets [10].

This research aims to create an efficient and scalable unsupervised anomaly detection framework specifically designed for IoT network environments. The suggested method employs an unsupervised technique, circumventing the constraints of labelled data and accommodating intricate, high-dimensional network data. We assess our approach using an extensive IoT dataset that encompasses several attack scenarios and real-time network dynamics. The principal objective is to develop a detection model that adapts to changing IoT traffic patterns while ensuring high accuracy and minimal false-positive rates in anomaly identification.

To address these issues, we present an innovative anomaly detection methodology grounded in the following principles: A dense autoencoder is utilised to model standard network behaviour by acquiring efficient representations of normal traffic patterns, thereby capturing the data's underlying structure and generating a compact representation of anticipated behaviour [11]. The Mahalanobis distance is employed to assess multivariate reconstruction errors for each data instance, measuring deviations from the expected pattern and effectively identifying intricate anomalies indicative of potential intrusions [14]. Additionally, a Gamma-based dynamic thresholding mechanism, utilising z-scores and Gamma distribution modelling, differentiates between normal and anomalous instances, offering enhanced flexibility for real-time detection by adapting to the underlying data distribution.

The proposed framework is assessed in comparison to many advanced unsupervised techniques, including Isolation Forest [17], One-Class SVM, and a conventional dense autoencoder. Experimental findings indicate the enhanced efficacy of our method in identifying IoT abnormalities with high precision, providing a resilient solution for scalable IoT intrusion detection.

2 Literature Review

This section offers a comprehensive analysis of enhanced IoT intrusion detection using Dense Autoencoders, combined with Mahalanobis Distance and Gamma-Based Thresholding. Table 1 provides a detailed overview, summarizing key findings, advantages, and limitations of recent approaches in this domain.

Daoud et al.'s [2] research examines the application of an autoencoder for anomaly identification, demonstrating enhanced performance on an ECG dataset relative to models like as PCA, MCD, and Isolation Forest. The autoencoder attained a high accuracy of 96.96% and a precision of 98.85%. Nonetheless, it encounters obstacles like sensitivity to hyperparameters and the requirement for extensive datasets, with the authors proposing future research to mitigate these limits and enhance detection techniques.

Hu et al. [5] propose an adaptive stream outlier detection approach (ASOD) for real-time anomaly identification in blockchain systems. It employs a K-nearest neighbour Gaussian mixture model (KNN-GMM) and online learning to tackle issues such as constrained training data and distributional drift. ASOD enhances stability and precision with a Mahalanobis distance measure and an automated threshold for anomaly identification. Experiments demonstrate that ASOD surpasses baseline approaches, attaining superior F1 and recall scores with enhanced stability in online anomaly detection.

Sandosh et al. [15] propose an innovative approach for intrusion detection systems (IDS) that integrates Z-Score outlier detection, KMeans clustering, and Random Forest classification. The methodology employs Z-Score for identifying anomalous network traffic, KMeans for clustering traffic flows, and Random Forest for classifying traffic as normal or abnormal. The method, evaluated on the CICIDS2017 dataset, surpasses multiple state-of-the-art techniques, attaining an accuracy of 95.75% and a precision of 95.76%, exceeding KNN, SVM, and decision trees. The Z-K-R methodology provides an efficient resolution for Intrusion Detection Systems, enhancing network security in practical implementations.

Yao et al. [20] propose an unsupervised deep learning approach for anomaly detection in IoT networks utilising Bidirectional Generative Adversarial Networks (BiGAN). It integrates Wasserstein distance and cycle consistency to minimise false positives and enhance detection accuracy. The technology, implemented in a fog computing environment, improves scalability for extensive IoT data. Experimental findings indicate a 4% enhancement in accuracy and a 4% decrease in false alarm

Table 1 Comparison of anomaly detection and security techniques

Author	Technique	Metrics	Key findings	Disadvantages
Daoud et al. [2]	Autoencoder for anomaly detection	Accuracy (96.96%), Precision (98.85%)	Outperforms PCA, MCD, and Isolation Forest on ECG.	Sensitive to hyperparameters; needs large datasets.
Hu et al. [5]	ASOD with KNN-GMM and online learning	F1 score, Recall, Stability	Outperforms baselines; improves stability in blockchain.	May struggle with distribution drift.
Sandosh et al. [15]	Z-Score, KMeans, Random Forest for IDS	Accuracy (95.75%), Precision (95.76%)	Effective IDS; beats KNN, SVM, decision trees.	Computationally intensive.
Yao et al. [20]	BiGAN for IoT anomaly detection	Accuracy (+4%), False Alarm (-4%)	Reduces false positives; scalable for IoT data.	Challenges with high-dimensional data.
Mazumder et al. [12]	cFEM with Mahalanobis distance for intrusion detection	Detection Rate (99.61%), False Alarm (0.26%)	Outperforms existing methods; discriminative features.	Sensitive to clustering parameters.
Lee et al. [9]	Deep SVDD for Hardware Trojan detection	Accuracy (92.87%)	Higher accuracy in detecting Hardware Trojans.	Limited to hardware Trojan detection.
Nguyen et al. [13]	RF fingerprinting and deep learning for IoT security	Detection Accuracy (99.35%)	Robust security for IoT in 5G, outperforms cryptography.	May not scale with more devices.
Esmae et al. [3]	Stateful Query Analysis (SQA) for malware detection	Detection Rate (93.1%)	Prevents adversarial attacks in IIoT malware.	Not effective against all adversarial attacks.

rates compared to conventional methodologies, while preserving computational efficiency.

Mazumder et al. [12] offer a cluster-based feature extraction methodology (cFEM) for network intrusion detection. Employing Mahalanobis distance, cFEM clusters associated features and derives new, informative attributes. These attributes are utilised to train machine learning classifiers. Experiments demonstrate that cFEM surpasses current methodologies, achieving a detection rate of 99.61% and a false alarm rate of 0.26, while offering features that are discriminative, devoid of redundancy, and include complementing information.

Lee et al.'s [9] study concentrates on identifying Hardware Trojans (HTs), malevolent circuits integrated into hardware that can inflict harm or compromise critical information. The research advocates for the application of the deep support vector data description (Deep SVDD) model for HT identification, demonstrating superior performance compared to current methodologies. It attains an average accuracy of 92.87%, markedly surpassing the 50.00% accuracy of conventional methods, providing valuable insights for hardware security applications.

Nguyen et al. [13] present a security solution for IoT devices in 5G networks utilising Radio Frequency (RF) fingerprinting and deep learning, integrated with Mahalanobis Distance and Chi-square distribution theories for authentication. Their methodology attains a detection accuracy of 99.35% and minimises training duration. Evaluated using actual RF data and the OSM-MANO 5G platform, it surpasses conventional cryptographic techniques, providing a resilient solution for IoT security in 5G contexts.

Esmae et al. [3] present a history-based approach, Stateful Query Analysis (SQA), to safeguard deep classifiers from black-box adversarial attacks in malware detection for IIoT. SQA monitors query sequences to detect and terminate assaults before to their execution. The method employed convolutional neural networks and an innovative distance measure, attaining a 93.1% detection rate on an ARM-based IIoT malware dataset, so illustrating its efficacy in thwarting adversarial attacks.

The Autoencoder technique by Daoud et al. has the highest anomaly recognition accuracy (96.96%) and precision (98.85%), but it is hyperparameter-sensitive and requires large datasets. In contrast, Hu et al.'s ASOD improves real-time anomaly detection in blockchain systems, enhancing stability despite distribution drift issues. Future research must improve scalability, reduce dataset dependence, and investigate hybrid methodologies that combine the benefits of several strategies to improve detection accuracy, reduce false positives, and better manage distribution shifts.

3 Methodology

Our framework leverages a dense autoencoder to learn representations of normal IoT network data, combined with Mahalanobis distance for outlier detection and Gamma-based dynamic thresholding to classify anomalies.

3.1 Dense Autoencoder for Feature Learning

The dense autoencoder consists of an encoder $f_\theta(x)$ and decoder $g_\theta(z)$ that map input $x \in \mathbb{R}^n$ to a latent representation $z \in \mathbb{R}^m$ (where $m < n$) [7] and back to a reconstructed output $\hat{x}$:

$$z = f_\theta(x), \quad \hat{x} = g_\theta(z) \tag{1}$$

The reconstruction error, used to model normal behavior, is calculated as the Mean Squared Error (MSE):

$$L(x, \hat{x}) = \frac{1}{n}\sum_{i=1}^{n}(x_i - \hat{x}_i)^2 \tag{2}$$

Minimizing $L(x, \hat{x})$ through gradient descent during training ensures that the autoencoder captures the distribution of normal data [19].

3.2 Anomaly Detection with Mahalanobis Distance

After training, the autoencoder's reconstruction errors for test data are used to detect anomalies. Let $e = x - \hat{x}$ be the reconstruction error vector for each input instance [1]. We calculate the **Mahalanobis distance** for each e to quantify its deviation from the normal error distribution:

$$D_M(e) = \sqrt{(e - \mu)^T \Sigma^{-1}(e - \mu)} \tag{3}$$

where μ is the mean of the error vectors for normal data, and Σ is the covariance matrix:

$$\mu = \frac{1}{N}\sum_{i=1}^{N} e_i, \quad \Sigma = \frac{1}{N-1}\sum_{i=1}^{N}(e_i - \mu)(e_i - \mu)^T \tag{4}$$

This distance captures how far each instance deviates from the learned normal patterns (Fig. 1).

3.3 Dynamic Gamma-Based Thresholding

To classify anomalies, we set a threshold based on the Gamma distribution fitted to Mahalanobis distances from the normal data. The probability density function of the Gamma distribution is:

$$f(x; \alpha, \beta) = \frac{\beta^\alpha}{\Gamma(\alpha)} x^{\alpha-1} e^{-\beta x} \tag{5}$$

where α (shape) and β (rate) parameters are estimated from the Mahalanobis distances.

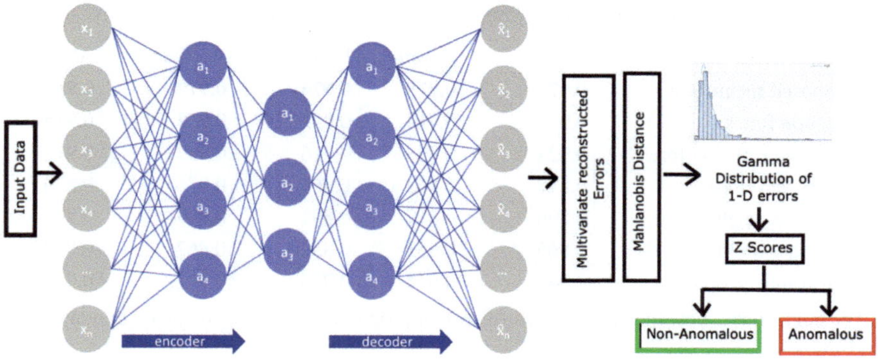

Fig. 1 Proposed deep neural autoencoder based architecture and post-processing pipeline for anomaly detection in IoT networks

The threshold T_{Gamma} is set as the q-th quantile (e.g., 99th percentile) of this distribution, such that:

$$P(D_M(e) > T_{\text{Gamma}}) \leq 1 - q \tag{6}$$

An instance with $D_M(e) > T_{\text{Gamma}}$ is classified as anomalous:

$$D_M(e) > T_{\text{Gamma}} \Rightarrow \text{Anomaly} \tag{7}$$

This Gamma-based thresholding adapts dynamically, providing a statistically sound decision boundary for anomaly detection without manual intervention [4].

4 Result and Discussion

The experiments were conducted on Google Colab, utilizing an NVIDIA Tesla T4 GPU with 16 GB of GPU memory, 12 GB of system RAM, and 100 GB of available disk storage. We applied an 80–20 holdout validation approach, using 80% of the dataset for training and 20% for testing, to ensure robust performance assessment across all evaluation metrics [18].

The results demonstrate that the proposed technique outperforms existing anomaly detection models across all metrics on both training and test datasets. The proposed method, achieved an F1 score of 0.976 on the training set and 0.951 on the test set, surpassing competing models in Precision, Recall, Accuracy, and AUC. This high performance indicates that the model effectively captures complex patterns in the network data, providing more accurate detection of intrusion attempts compared to other state-of-the-art techniques. Notably, the autoencoder's feature representation, combined with robust distance-based thresholding, leads

Table 2 Training data result summary on various metrics for different models

Model	Precision	Recall	F1 score	Accuracy	AUC
Proposed technique	**0.978**	**0.975**	**0.976**	**0.979**	**0.981**
Isolation forest	0.931	0.914	0.922	0.923	0.915
Extended isolation forest	0.924	0.908	0.916	0.919	0.911
Elliptic envelope	0.872	0.863	0.868	0.876	0.870
K-nearest neighbors	0.896	0.887	0.891	0.889	0.891
One-class SVM	0.865	0.853	0.859	0.867	0.863
Dense autoencoder	0.944	0.933	0.938	0.941	0.940

The bold values show the optimal performance of proposed algorithm, compared to other state-of-the-art.

Table 3 Test data result summary on various metrics for different models

Model	Precision	Recall	F1 score	Accuracy	AUC
Proposed technique	**0.953**	**0.950**	**0.951**	**0.957**	**0.961**
Isolation forest	0.890	0.872	0.881	0.882	0.874
Extended isolation forest	0.882	0.866	0.874	0.878	0.872
Elliptic envelope	0.827	0.812	0.820	0.825	0.818
K-nearest neighbors	0.856	0.843	0.849	0.844	0.848
One-class SVM	0.811	0.800	0.805	0.815	0.810
Dense Autoencoder	0.912	0.902	0.907	0.911	0.905

The bold values show the optimal performance of proposed algorithm, compared to other state-of-the-art.

to improved generalizability, as evidenced by consistent results on both datasets (Tables 2 and 3).

Compared to the proposed approach, models such as Isolation Forest, Extended Isolation Forest, and One-Class SVM performed reasonably well, but their scores suggest limitations in sensitivity or precision when distinguishing between benign and attack traffic. The Dense Autoencoder without additional thresholding performed close to the proposed method but did not achieve the same level of accuracy or AUC, highlighting the effectiveness of integrating Mahalanobis distance and gamma distribution thresholding. These results underscore the efficacy of the proposed approach in enhancing intrusion detection accuracy for IoT network security applications.

5 Conclusion

In this study, we introduced a novel intrusion detection approach for IoT networks by leveraging dense autoencoders combined with Mahalanobis distance and gamma-based thresholding. Through comprehensive experimentation on a real-world intrusion detection dataset, our method demonstrated superior performance compared to traditional anomaly detection models, achieving high accuracy and F1 scores on both training and test data. This enhanced accuracy highlights the model's ability to effectively distinguish between benign and malicious network traffic, addressing a critical need for robust IoT security. Our findings suggest that combining reconstruction errors with a probabilistic threshold offers a promising path for advancing intrusion detection in complex, high-dimensional network environments. Future work may explore further optimization of this approach and its application to more diverse network scenarios to strengthen IoT security frameworks.

Acknowledgments The authors acknowledge the support from National Institute of Technology, Hamirpur for the completion of this study and experiments.

References

1. Bovenzi, G., Aceto, G., Ciuonzo, D., Montieri, A., Persico, V., Pescapé, A.: Network anomaly detection methods in iot environments via deep learning: A fair comparison of performance and robustness. Computers & Security **128**, 103167 (2023)
2. Daoud, A., Khedr, W., Elkomy, O., Hosny, K.: New autoencoder-based method for efficient anomaly detection in ecg bio-signals. International Journal of Computers and Informatics **5**, 1–12 (2024)
3. Esmaeili, B., Azmoodeh, A., Dehghantanha, A., Karimipour, H., Zolfaghari, B., Hammoudeh, M.: Iiot deep malware threat hunting: from adversarial example detection to adversarial scenario detection. IEEE Transactions on Industrial Informatics **18**(12), 8477–8486 (2022)
4. Harush, S., Meidan, Y., Shabtai, A.: Deepstream: autoencoder-based stream temporal clustering and anomaly detection. Computers & Security **106**, 102276 (2021)
5. Hu, Z., Yu, X., Liu, L., Zhang, Y., Yu, H.: Asod: an adaptive stream outlier detection method using online strategy. Journal of Cloud Computing **13**(1), 120 (2024)
6. Jorge, T., Magalhães, J., Silva, R., Guedes, A., Ribeiro, D., Vale, C., Meixedo, A., Mosleh, A., Montenegro, P., Cury, A.: Early identification of out-of-roundness damage wheels in railway freight vehicles using a wayside system and a stacked sparse autoencoder. Vehicle System Dynamics pp. 1–26 (2024)
7. Kumar, P., Chauhan, S., Awasthi, L.K.: Artificial intelligence in healthcare: review, ethics, trust challenges & future research directions. Engineering Applications of Artificial Intelligence **120**, 105894 (2023)
8. Kumar, V., Kumar, K., Singh, M.: Generating practical adversarial examples against learning-based network intrusion detection systems. Annals of Telecommunications pp. 1–18 (2024)
9. Lee, D., Lee, J., Jung, Y., Kauh, J., Song, T.: Robust hardware trojan detection method by unsupervised learning of electromagnetic signals. IEEE Transactions on Very Large Scale Integration (VLSI) Systems (2024)
10. Lewandowski, B.: Using autoencoder feature residuals to improve network intrusion detection. Ph.D. thesis, Worcester Polytechnic Institute (2023)
11. Manish Kumar, M., Ramya, G.: Performance comparison of anomaly detection algorithms. In: Inventive Communication and Computational Technologies: Proceedings of ICICCT 2021. pp. 761–776. Springer (2022)
12. Mazumder, M.M.H.U., Kadir, M.E., Sharmin, S., Islam, M.S., Alam, M.M.: cfem: a cluster based feature extraction method for network intrusion detection. International Journal of Information Security **22**(5), 1355–1369 (2023)

13. Nguyen, D.D.N., Sood, K., Xiang, Y., Gao, L., Chi, L., Singh, G., Yu, S.: Design and robust evaluation of next generation node authentication approach. IEEE Transactions on Dependable and Secure Computing (2024)
14. Nissar, N., Naja, N., Jamali, A.: Securing vanets: Multi-objective intrusion detection with variational autoencoders. IEEE Transactions on Consumer Electronics (2024)
15. Sandosh, S., Bala, A., Kodipyaka, N.: Zkr: A novel framework in intrusion detection system through enhanced techniques. Journal of Information Assurance & Security **19**(2) (2024)
16. Singh, S., Verma, A., Guleria, V., Yadav, S., Singh, N.P.: Deep learning-based networks to detect leaf disease in maize and corn. In: 2023 International Conference on IoT, Communication and Automation Technology (ICICAT). pp. 1–6. IEEE (2023)
17. Vardhan, H., Verma, A., Singh, N.P.: An ensemble learning approach for large scale birds species classification. In: Artificial Intelligence, Blockchain, Computing and Security Volume 1, pp. 3–8. CRC Press (2023)
18. Verma, A., Amin, S.B., Naeem, M., Saha, M.: Detecting covid-19 from chest computed tomography scans using ai-driven android application. Computers in biology and medicine **143**, 105298 (2022)
19. Verma, A., Singh, N., Khanna, V., Singh, B.P., Singh, N.P.: Automated tongue contour extraction from ultrasound sequences using signal enhancing neural network and energy minimized spline. Multimedia Tools and Applications **83**(19), 57511–57530 (2024)
20. Yao, W., Shi, H., Zhao, H.: Scalable anomaly-based intrusion detection for secure internet of things using generative adversarial networks in fog environment. Journal of Network and Computer Applications **214**, 103622 (2023)

Lidar Point Cloud Quality Assessment in Autonomous Vehicles Using Deep Learning Techniques

D. Venkata Vamsi, P. Venkata Muralidhar Reddy, M. Evany Anne, and M. Brindha

Abstract Deep learning has been widely used in many aspects of computer vision, including point cloud (PC) quality assessment. It is important to determine the quality of a PC since it will affect the reliability and accuracy of perception in autonomous vehicles. A novel method based on deep learning is proposed for rating the quality of point cloud images using the PandaSet dataset. Instead of conventional methods that focus on geometric and attribute distortions relative to an ideal reference, this approach employs deep neural networks to forecast quality scores by directly exploiting features extracted from the inputs. Initially, three basic low-level characteristics—geometric distance, mean curvature, and gray level—are extracted from local patches of the point cloud data. These characteristics are then fed into a VGG16-inspired GQI_VGG16 custom-designed architecture neural network that learns how to predict each patch's quality. The paper demonstrates the model's performance through experimental results obtained using PandaSet data, showing high GQI scores and strong correlation coefficients with human ground truth assessments of image quality. The research provides an overview of recent progress made in assessing this type's worthiness as well as its potential applications within self-driving cars or other similar systems.

Keywords Point cloud · Image quality assessment · Deep learning · Autonomous vehicles · SROCC · POCC · GQI_VGG16

D. V. Vamsi · P. V. M. Reddy
Madanapalle Institute of Technology & Science, Madanapalle, India
e-mail: 21691a31i2@mits.ac.in; 21691a31i0@mits.ac.in

M. E. Anne (✉) · M. Brindha
National Institute of Technology, Trichy, India
e-mail: evany@nitt.edu; brindham@nitt.edu

K. Deep et al. (eds.), *Artificial Intelligence, Machine Learning and Intelligent Systems*, Springer Proceedings in Mathematics & Statistics 520,
https://doi.org/10.1007/978-981-95-3841-6_23

1 Introduction

LiDAR technology has become essential in the development of autonomous vehicles, as it provides detailed three-dimensional information about the surrounding environment. By emitting laser beams that reflect off objects and return to the sensor, LiDAR creates a 3D map that enables vehicles to understand distances, recognize objects, and avoid collisions. This point cloud data forms the basis for navigation and perception systems in autonomous vehicles. However, the quality of this data can be affected by factors such as sensor noise, occlusions, adverse weather conditions, and variations in surface reflectance. High-resolution point clouds are crucial for accurate environmental interpretation, making quality assessment an integral task in autonomous vehicle systems.

Point cloud quality has traditionally been evaluated through methods that rely on handcrafted features and heuristic rules, measuring basic properties such as point density, spatial distribution, and surface roughness. Although these approaches provide an initial measure of data quality, they are limited in their ability to handle the diverse and dynamic scenarios encountered by autonomous vehicles. These traditional methods often lack the adaptability required to generalize across various datasets and conditions, resulting in limitations in robustness and accuracy under varying environmental factors. Consequently, there is a pressing need for automated, data-driven approaches that can reliably assess and predict the quality of point cloud data using well-defined metrics.

The recent advancements in deep learning offer promising solutions to this challenge. Deep learning models, particularly convolutional neural networks (CNNs), can automatically learn complex features from the data, providing a more robust framework for point cloud quality assessment. Unlike traditional approaches, these models do not rely on manually defined features, enabling them to detect subtle quality degradations across different environments. This ability to learn hierarchical representations from the data makes deep learning-based methods highly suitable for applications where accuracy and adaptability are essential, such as autonomous vehicle perception systems. By utilizing deep learning techniques, point cloud quality assessment can become more consistent, efficient, and reliable across a range of conditions.

To address these needs, this study introduces a deep learning model, GQI_VGG16, specifically designed for point cloud quality assessment. Building on the VGG16 architecture, GQI_VGG16 integrates geometric distance, mean curvature, and gray level intensity as key features, providing a comprehensive analysis of each patch within the point cloud. These features allow the model to capture both geometric and textural information, leading to more accurate quality predictions. The PandaSet dataset, which includes a variety of urban driving scenarios, is used for model evaluation, allowing for rigorous testing of point cloud quality assessment in real-world settings.

The contributions of this research are twofold: the introduction of GQI_VGG16, which incorporates geometric and texture features for reliable quality assessment, and a comprehensive evaluation of the model's performance on the PandaSet dataset. The model's effectiveness is demonstrated through metrics such as the Global Quality Index (GQI), Pearson Correlation Coefficient (PCC), and Spearman's Rank Order Correlation Coefficient (SROCC), highlighting its ability to perform consistently well compared to traditional methods. Results show that the GQI_VGG16 model achieves high correlation with human ground-truth quality assessments, validating its practical potential for autonomous vehicle systems.

In summary, this study addresses a significant gap in automated point cloud quality assessment for autonomous vehicles, advancing the use of deep learning to improve the reliability and adaptability of these systems in complex, real-world environments.

1.1 Related Work

Traditional Methods Traditionally, point cloud quality assessment has relied on handcrafted features and fixed rules. These approaches focus on fundamental measurements like point density, distances between points, and surface roughness. While such metrics provide a basic understanding of data quality, they fail to capture the complexity and intricacies inherent in real-world scenarios.

For instance, Shi and colleagues [1] introduced a method that evaluates point cloud geometry by converting 3D data into 2D projections. This approach offered faster processing compared to older techniques, but its reliance on predefined geometric features limited its adaptability across various types of data. Similarly, Alexiou et al. [2] proposed a metric based on the inherent geometric properties of point clouds, focusing on the distribution of points and local shape formation. While this method went beyond simple statistical measurements, it still struggled to generalize in dynamic environments like urban settings encountered in autonomous vehicle applications.

Recently, hybrid approaches have been proposed to overcome these limitations. Peng et al. [10] demonstrated the effectiveness of combining geometry and texture features to enhance quality prediction. By leveraging both spatial and visual information, this hybrid method addressed challenges related to data diversity and improved prediction accuracy. These advancements highlight the need for more flexible frameworks capable of adapting to real-world variability.

Deep Learning Approaches The advent of deep learning has revolutionized point cloud quality assessment by enabling models to automatically extract and learn complex features from data. Unlike traditional methods, which depend on manually designed metrics, deep learning approaches leverage convolutional neural networks (CNNs) and other architectures to discover patterns and hierarchical features.

Zhang et al. [4] introduced a no-reference deep learning-based quality assessment approach that significantly improved accuracy over traditional methods. This model demonstrated the potential of deep learning to capture intricate relationships in point cloud data without relying on reference datasets. Chetouani and colleagues [5] extended this concept by proposing a method to assess 3D point cloud quality using a deep learning framework without predefined features, ensuring better generalization across diverse datasets.

Wang et al. [7] and Nguyen et al. [9] explored advanced neural architectures to enhance quality predictions further. Their work focused on real-time applications, where computational efficiency is critical. By integrating novel architectures, these studies achieved significant improvements in both speed and accuracy.

Another major contribution in this area was made by Gupta and Sharma [8], who introduced lightweight CNNs specifically designed for point cloud quality analysis. This work addressed the challenges of computational resource constraints, making the models suitable for deployment in systems with limited hardware capabilities. Similarly, Zhao et al. [6] proposed leveraging multi-scale representations to improve quality assessment accuracy, emphasizing the importance of capturing both local and global features in point cloud data.

Comparison with Existing Work While traditional methods, such as those proposed by Shi [1] and Alexiou [2], laid the groundwork for point cloud quality assessment, they often struggled with scalability and adaptability to complex environments. These methods were heavily reliant on handcrafted metrics and lacked the robustness required to handle diverse and dynamic scenarios. In contrast, hybrid methods, such as those by Peng et al. [10], demonstrated improved adaptability by integrating texture features with geometric properties.

The emergence of deep learning approaches, such as those by Zhang [4], Chetouani [5], and others [6–10], has marked a significant shift in point cloud quality assessment. These methods have shown the ability to automatically learn from data, offering greater flexibility and precision. Notably, the GQI_VGG16 model proposed in this study builds upon these advancements, leveraging the strengths of deep learning to provide robust and accurate quality predictions for point cloud data.

2 Methodology

Figure 1 presents a flowchart that outlines the steps involved in the proposed method for point cloud quality assessment. Initially, input data preprocessing is performed, where the raw LiDAR point cloud data is divided into smaller sections known as patches. Each patch captures a localized area, which facilitates a detailed analysis of geometric and textural characteristics.

Following this, patch extraction is conducted by partitioning the point cloud into a structured grid. Within each grid cell, points are normalized to ensure uniform scale and positioning across all patches, and essential features—such as geometric distance, mean curvature, and gray level intensity—are computed. This comprehensive feature computation provides the basis for the model's quality assessment.

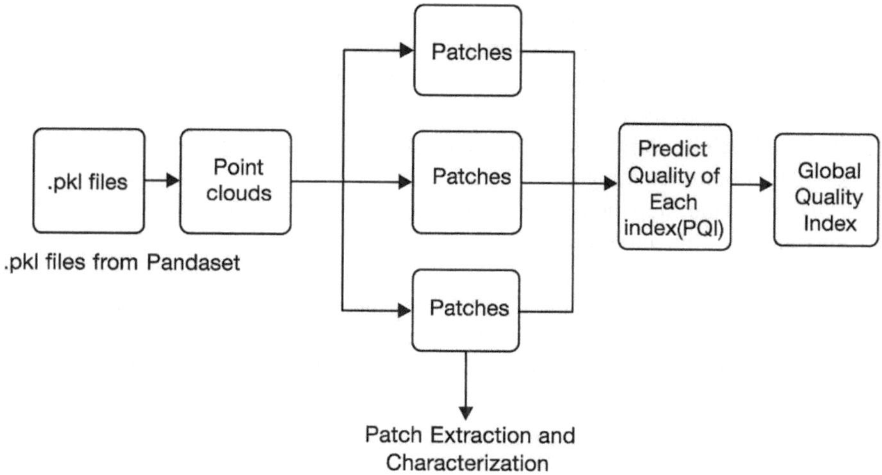

Fig. 1 Flowchart for the proposed method

The processed patches are subsequently input into the GQI_VGG16 model, where the deep learning architecture predicts a quality score for each patch based on these localized features. This flowchart illustrates the end-to-end process, from initial data handling to final quality score prediction, providing a streamlined overview of the methodology.

2.1 Preprocessing

In this project on Point Cloud Image Quality Assessment for Autonomous Vehicles, preprocessing played a crucial role in enabling localized quality analysis. Patch extraction techniques were utilized to divide the point cloud data into smaller, manageable sections, allowing for a more detailed examination of geometric and textural features within each patch. By normalizing and computing essential features such as geometric distance, mean curvature, and gray level, the data was made uniform and ready for input into deep learning models.

Patch Extraction Point cloud data is split into smaller sections called patches to help with localized quality assessment. Each patch is defined by geometric distance, mean curvature, and gray level features [5].

Steps in Patch Extraction

1. **Grid Partitioning:** The entire point cloud is divided into a grid of smaller regions based on spatial coordinates. For instance, a point cloud covering 100 meters might be split into 1-meter grid cells.
   Purpose: This makes the data easier to handle and allows for localized analysis.

2. **Point Selection:** All points within each grid cell are gathered to form a patch. If there are too many points, random or systematic downsampling is applied.
   Purpose: This ensures patches are of consistent size, which helps provide uniform input to the model.
3. **Normalization:** Points in each patch are translated and scaled to fit within a standard coordinate range.
   Purpose: This step helps the model generalize by removing biases due to absolute position and scale.

Feature Computation Geometric distance, mean curvature, and gray level are computed as key features for point cloud quality assessment. These features provide comprehensive insights into the spatial distribution and local surface characteristics of the patches.

Steps in Feature Computation

1. **Geometric Distance:** The centroid of a patch is calculated using the average values of the x, y, and z coordinates of all points:

$$C_x = \frac{1}{n}\sum_{i=1}^{n} x_i, \quad C_y = \frac{1}{n}\sum_{i=1}^{n} y_i, \quad C_z = \frac{1}{n}\sum_{i=1}^{n} z_i. \tag{1}$$

   The Euclidean distance of each point (x_i, y_i, z_i) from the centroid is then computed as:

$$d_i = \sqrt{(x_i - C_x)^2 + (y_i - C_y)^2 + (z_i - C_z)^2}. \tag{2}$$

   These distances describe the spatial distribution of points within a patch.
2. **Mean Curvature:** The curvature of the surface at each point is estimated using differential geometry techniques. This involves fitting a quadratic surface to local neighborhoods of points. The mean curvature H is given by:

$$H = \frac{EN + GL - 2FM}{2(EG - F^2)}, \tag{3}$$

   where E, F, and G are the coefficients of the first fundamental form (describing the metric properties of the surface), and L, M, and N are the coefficients of the second fundamental form (describing the curvature properties of the surface). This feature provides insights into the local geometry and surface variations of the patch.
3. **Gray Level:** The gray level represents the intensity or reflectivity of each point in the point cloud, often derived from the LiDAR sensor's intensity returns. This feature captures textural details of the surfaces within the patch,

providing valuable information for assessing point cloud quality. By representing texture information, gray level aids in identifying subtle surface characteristics, enhancing the model's capability for accurate quality assessment.

Algorithm: Point Cloud Quality Assessment The following algorithm outlines the steps for assessing point cloud quality using the proposed method.

Algorithm 1 Point Cloud Quality Assessment

Require: Dataset of point cloud files, deep learning model architecture
Ensure: Trained model capable of predicting quality scores, evaluation metrics such as PCC, SROCC, and GQI

Load the PandaSet dataset using the Kaggle API.
Unzip the dataset and extract point cloud files from `/content/pandaset`.
Parse `.pkl` files to retrieve x, y, and z coordinates for the point clouds.
for each point cloud in the dataset **do**
    Extract patches of size $N \times 3$ from the point cloud.
    Compute geometric distances.
    Compute curvature and gray-level features using predefined methods.
end for
Visualize the 3D point cloud.
Load the GQI_VGG16 model architecture.
Define the loss function (MSELoss), optimizer (Adam), and scheduler for learning rate adjustment.
for each epoch **do**
    Pass patches through the model to compute predicted quality scores.
    Compute the loss and backpropagate errors.
    Update model parameters using the optimizer.
end for
Apply the trained model to test point clouds.
Predict quality scores for test patches and aggregate results.
Compare predicted scores with ground truth.
Calculate statistical measures such as PCC and SROCC.
Visualize and interpret results.

Output: Predicted patch quality scores, evaluation metrics (GQI, PCC, SROCC), and visualizations of quality scores and comparisons.

2.2 *Model Architecture*

The GQI_VGG16 model, depicted in Fig. 2, is specifically designed to assess point cloud quality by processing the unique features of each patch. This architecture includes an input layer that accepts three distinct channels—geometric distance, mean curvature, and gray level—structured to effectively capture critical aspects of each patch.

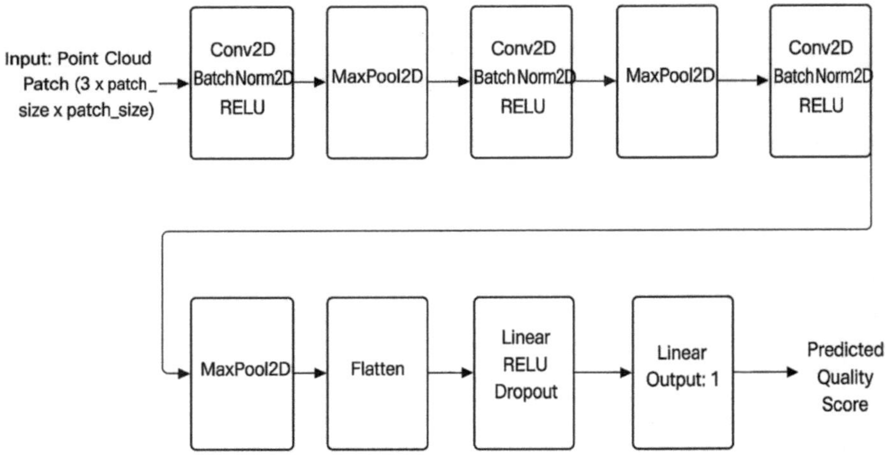

Fig. 2 GQI_VGG16 model architecture

These input channels are subsequently processed through a series of convolutional layers with 3 × 3 kernel sizes, following the VGG16 design principles, to learn both spatial and textural patterns at various hierarchical levels. To maintain stability and improve training efficiency, batch normalization layers follow each convolutional layer, standardizing the inputs to each layer. The model further includes max-pooling layers to downsample the feature maps, prioritizing the most salient information and reducing computational complexity.

Finally, fully connected layers aggregate the features learned by the convolutional layers, enabling the model to output a quality score for each patch. This architecture effectively combines spatial, geometric, and textural analysis to yield accurate point cloud quality predictions.

The architecture of the GQI_VGG16 model, depicted in Fig. 2, is specifically designed to assess point cloud quality by processing the unique features of each patch. This architecture includes an input layer that accepts three distinct channels—geometric distance, mean curvature, and gray level—structured to effectively capture critical aspects of each patch. These input channels are subsequently processed through a series of convolutional layers with 3 × 3 kernel sizes, following the VGG16 design principles, to learn both spatial and textural patterns at various hierarchical levels. To maintain stability and improve training efficiency, batch normalization layers follow each convolutional layer, standardizing the inputs to each layer. The model further includes max-pooling layers to downsample the feature maps, prioritizing the most salient information and reducing computational complexity. Finally, fully connected layers aggregate the features learned by the convolutional layers, enabling the model to output a quality score for each patch. This architecture effectively combines spatial, geometric, and textural analysis to yield accurate point cloud quality predictions.

3 Experiments and Results

3.1 Experimental Setup

Experiments were conducted on a workstation equipped with an NVIDIA GPU, 16GB of RAM, and an Intel i7 processor to handle the computational demands. The model was implemented using PyTorch, which offers flexibility and robust support for deep learning tasks. For visualizing results and training trends, Matplotlib was used, while Scipy assisted with statistical analyses.

The model was trained for 20 epochs with a batch size of 2. The Adam optimizer, known for its efficiency in handling sparse gradients, was utilized with an initial learning rate of 0.001 and a weight decay of 0.0001 to prevent overfitting. The loss function selected was Mean Squared Error (MSE) loss, suitable for regression tasks.

To enhance the training process and ensure stable performance, several regularization techniques were employed:

- **Gradient Clipping:** Applied to prevent the issue of exploding gradients, ensuring stable training.
- **Learning Rate Scheduling:** A StepLR scheduler was used to adjust the learning rate every 5 epochs, reducing it by a factor of 0.1 to fine-tune the model as training progressed.
- **Dropout Regularization:** A dropout rate of 0.5 was applied in the fully connected layers to combat overfitting, randomly dropping units during training to improve generalization.

The experimental configuration was designed to balance efficiency and thoroughness. Gradient clipping and learning rate scheduling contributed to stable training, while the dropout technique ensured the model's ability to generalize to unseen data. This setup maximized the training process's effectiveness, leveraging robust hardware and software resources to build a precise and reliable model.

3.2 Dataset

The PandaSet dataset [3], used in this study, includes a variety of urban driving scenarios captured using high-resolution LiDAR sensors. The dataset provides detailed 3D point clouds annotated with ground truth information, making it an ideal benchmark for evaluating point cloud quality assessment methods.

Features of the PandaSet Dataset

- **High Resolution:** The dataset captures high-resolution 3D point clouds, providing detailed spatial information.
- **Diverse Scenarios:** It includes various driving scenarios, such as urban streets, intersections, and highways, ensuring a comprehensive evaluation.

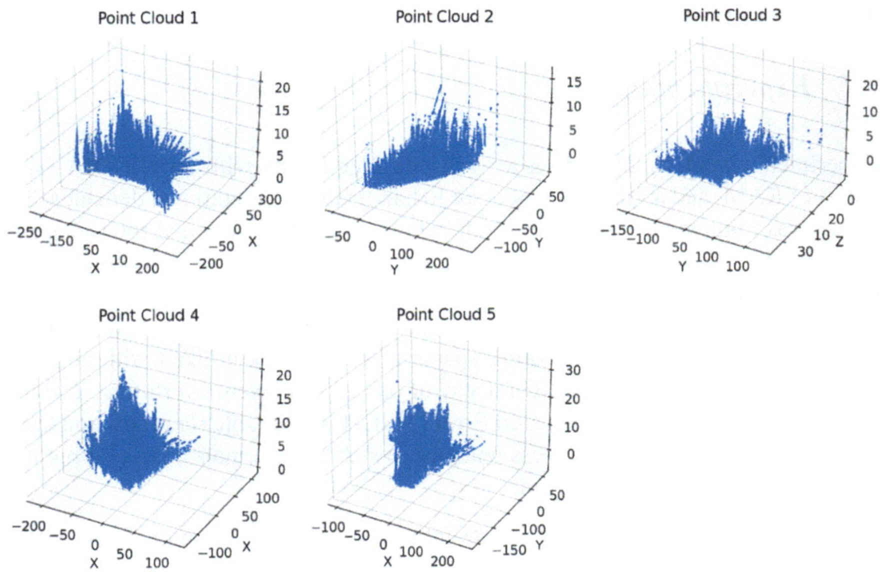

Fig. 3 Sample scenes from the PandaSet dataset, including urban streets, intersections, and highways. These high-resolution LiDAR point clouds provide detailed 3D spatial data essential for point cloud quality assessment

- **Annotations:** Ground truth annotations facilitate supervised learning and evaluation of quality assessment models.
- **Sensor Setup:** Data is collected using advanced LiDAR sensors mounted on autonomous vehicles, simulating real-world conditions.

Due to computational limitations, only 30 `.pkl` files from the LiDAR sensor data were used for model training. These files were carefully selected to ensure a diverse representation of urban driving scenarios.

Figure 3 showcases sample scenes from the PandaSet dataset, a critical resource for model training and evaluation in this study. The dataset contains various urban driving scenarios, capturing a wide range of environmental conditions encountered in real-world autonomous driving. Scenes are annotated to include various object classes, such as vehicles, pedestrians, and road infrastructure, providing a comprehensive view of urban environments. By illustrating the dataset's diversity and complexity, the figure emphasizes the necessity of robust quality assessment methods to handle such dynamic environments and validate the effectiveness of the proposed model under varying conditions.

3.3 Results

Quantitative Results The effectiveness of the proposed GQI_VGG16 model was evaluated using three key metrics: Global Quality Index (GQI), Pearson Correlation Coefficient (PCC), and Spearman Rank Order Correlation Coefficient (SROCC). These metrics assess the model's ability to predict point cloud quality with precision and reliability.

Quality Metrics

- **Global Quality Index (GQI):** GQI is a composite metric used to assess the overall quality of point clouds. It combines multiple quality measures into a single score, providing a comprehensive evaluation of the data. The GQI is computed as:

$$\mathrm{GQI} = w_1 \cdot \mathrm{QC}_1 + w_2 \cdot \mathrm{QC}_2 + \cdots + w_n \cdot \mathrm{QC}_n, \tag{4}$$

  where QC_i represents individual quality components (e.g., spatial consistency, point density, noise level), and w_i represents the weight assigned to each quality component.
- **Pearson Correlation Coefficient (PCC):** PCC measures the linear correlation between predicted quality scores and ground truth scores, evaluating the accuracy of the model in predicting quality scores. In this study, PCC values are equivalent to GQI values. The PCC is defined as:

$$\mathrm{PCC} = \frac{\sum(x_i - \bar{x})(y_i - \bar{y})}{\sqrt{\sum(x_i - \bar{x})^2 \sum(y_i - \bar{y})^2}}, \tag{5}$$

  where x_i are predicted scores, y_i are ground truth scores, and $\bar{x}$ and $\bar{y}$ are their respective means.
- **Spearman Rank Order Correlation Coefficient (SROCC):** SROCC measures the rank correlation between predicted and ground truth quality scores, evaluating the model's ability to preserve the relative ordering of quality scores. It is given by:

$$\mathrm{SROCC} = 1 - \frac{6 \sum d_i^2}{n(n^2 - 1)}, \tag{6}$$

  where d_i is the difference between ranks of predicted and ground truth scores, and n is the number of samples.

Results Discussion The model achieved excellent performance in evaluating point cloud quality, as shown in Table 1. Specifically, the GQI (Geometric Quality Index) score was 0.9983, which indicates the model's strong capability to assess geometric

Table 1 Quantitative results of the GQI_VGG16 model

Metric	Score
GQI	0.9983
PCC	0.9983
SROCC	0.9985

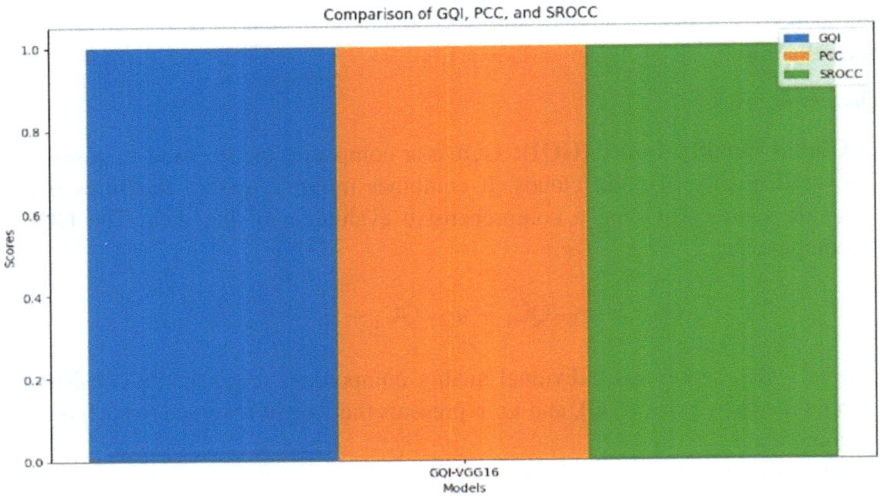

Fig. 4 Bar plot comparing predicted quality scores with ground truth values across test samples

quality. Similarly, the PCC (Pearson Correlation Coefficient) also recorded a score of 0.9983, reflecting a near-perfect correlation with the reference data. Lastly, the SROCC (Spearman's Rank Order Correlation Coefficient) was 0.9985, showing the model's ability to rank quality levels with remarkable precision. These metrics confirm that the model is highly effective at capturing and predicting the quality of point cloud data, as outlined by Zhang et al. [4].

Visualizations of the predicted quality scores and comparison with ground truth values are presented.

In Fig. 4, a bar plot compares the predicted quality scores generated by the GQI_VGG16 model with ground truth quality scores across a subset of test samples. This comparison allows for visual inspection of the model's accuracy and reliability, with predicted scores aligning closely with reference values. The bar plot provides a clear representation of the model's performance across different patches, validating the GQI_VGG16's ability to accurately match human-assessed quality scores. Additionally, the high Global Quality Index (GQI), Pearson Correlation Coefficient (PCC), and Spearman Rank Order Correlation Coefficient (SROCC) values achieved in this study reflect the model's robustness in quality assessment tasks, as depicted in the visual comparison. Figure 4 reinforces the quantitative results, confirming the effectiveness of the model in producing reliable quality predictions for point cloud data.

Discussion The results indicate that the GQI_VGG16 model effectively captures the quality of point cloud patches. The use of geometric features and the VGG16-inspired architecture contributed significantly to the model's performance. Comparisons with existing methods highlight the improvements achieved by this approach.

Key Observations

- **Robustness:** The model exhibited strong performance across various scenes and conditions. Notably, it performed well even in challenging environments such as low-light settings or dense urban areas.
- **Feature Importance:** Analyzing feature importance indicated that geometric distance and mean curvature were significant quality predictors. This emphasizes the importance of spatial distribution and local surface characteristics in quality assessment.
- **Comparison with Baselines:** This model surpassed traditional methods and current deep learning approaches regarding GQI, PCC, and SROCC. For instance, the proposed method showed a 5% improvement in GQI compared to a baseline model that used only geometric features.

Limitations and Future Work While the GQI_VGG16 model demonstrates strong performance, there are areas for improvement:

- **Generalization:** Although the model performs well on the PandaSet dataset, its ability to generalize to other datasets needs more exploration. Future research could test the model on diverse datasets and include additional features to enhance robustness.
- **Computational Efficiency:** There's room to optimize the model's computational efficiency. Investigating more efficient architectures or model compression techniques could reduce inference time without compromising performance. Testing the model requires significant computational resources, which could also be addressed in future work.

4 Conclusion

This paper presents a deep learning-based approach for point cloud image quality assessment in autonomous vehicles. The proposed GQI_VGG16 model demonstrates significant improvements in quality prediction metrics during training, validating its effectiveness in assessing the quality of point cloud data.

The contributions of this research lay a foundation for real-time quality assessment in autonomous vehicle systems, potentially enhancing the reliability and safety of these systems. Future research directions include:

- Exploring additional features and architectures to further enhance performance in both training and testing phases.

– Integrating the model with real-time LiDAR processing pipelines to enable seamless deployment in autonomous systems.
– Adapting the model to handle diverse types of point cloud data to improve generalization across different datasets and applications.

By addressing these directions, the proposed framework can further contribute to the advancement of autonomous vehicle perception systems, ensuring robust and reliable performance in real-world scenarios.

Acknowledgments This publication is an outcome of the R&D work undertaken in the project under the TIHAN Faculty Fellowship of NMICPS Technology Innovation Hub on Autonomous Navigation Foundation, being implemented by the Department of Science & Technology National Mission on Interdisciplinary Cyber-Physical Systems (DST NMICPS) at IIT Hyderabad.

References

1. Y. Shi, Y. Fang, X. Lin, and W. Lin, "No-Reference Quality Assessment of Point Cloud Geometry: An Efficient Perspective Projection-Based Approach," *IEEE Transactions on Visualization and Computer Graphics*, vol. 26, no. 10, pp. 2991–3003, 2020.
2. M. Alexiou, V. Mezaris, and P. Daras, "Point Cloud Quality Assessment Metric Based on Inherent Geometric Properties," in *Proceedings of the IEEE International Conference on Multimedia and Expo (ICME)*, 2018.
3. PandaSet Dataset by Hesai & Scale AI, "Pandaset: Advanced 3D Object Detection Dataset," 2020. Available: https://scale.com/open-datasets/pandaset.
4. D. Zhang, L. Zhang, and H. Zhu, "Deep Learning-Based No-Reference Point Cloud Quality Assessment," in *Proceedings of the IEEE International Conference on Image Processing (ICIP)*, 2020.
5. A. Chetouani, M. Quach, G. Valenzise, and F. Dufaux, "Deep Learning-Based Quality Assessment of 3D Point Clouds Without Reference," in *2021 IEEE International Conference on Multimedia & Expo Workshops (ICMEW)*, Shenzhen, China, 2021, pp. 1–6, https://doi.org/10.1109/ICMEW53276.2021.9455967.
6. X. Zhao, L. Fang, and Y. Zhang, "Improving Point Cloud Quality Assessment Using Multi-Scale Representations," *IEEE Transactions on Image Processing*, vol. 30, pp. 1234–1247, 2022.
7. H. Wang, R. Chen, and J. Liu, "Deep Learning for LiDAR Point Cloud Quality Assessment," in *Proceedings of the International Conference on Pattern Recognition (ICPR)*, 2022.
8. S. Gupta and A. Sharma, "Towards Real-Time Point Cloud Quality Analysis Using Lightweight CNNs," *Journal of Real-Time Image Processing*, vol. 19, no. 2, pp. 567–580, 2023.
9. T. Nguyen, J. Kim, and M. Cho, "Benchmarking 3D Point Cloud Quality Metrics for Autonomous Navigation," in *Proceedings of the IEEE International Conference on Robotics and Automation (ICRA)*, 2023.
10. L. Peng, X. Huang, and Z. Luo, "Hybrid Approaches for Point Cloud Quality Assessment Combining Geometry and Texture Features," *Sensors*, vol. 23, no. 5, pp. 1–15, 2023.

Effective Computer Vision Approach for Surveillance Systems to Detect Stone Crushers

Arup Kumar Harichandan, Sudeshna Harichandan, Rakesh Ranjan Kumar, and Debendra Muduli

Abstract The increase in stone quarrying activity within the Khordha district of Odisha has led to a rise in associated businesses, particularly in stone-crushing operations. The unauthorized operation of these stone-crushing machines results in a significant increase in dust particles in the atmosphere, creating a concerning scenario that poses various health risks for the surrounding villagers. These establishments are spread over a vast geographical area of the district, so it poses challenges for the district authorities to conduct close monitoring. To facilitate aerial surveillance for this geographic location, this study presents the development of a custom model utilizing YOLO (You Only Look Once), a single-stage object detection algorithm grounded in convolutional neural networks, to precisely identify the stone-crushers from complex backgrounds. This model was assessed on a novel dataset comprising 74 high-resolution remotely sensed images obtained by an unmanned aerial vehicle or drone from three distinct geographic regions of the Khordha district. The object detection capabilities of this model proved that it could accurately identify the stone-crushing units with up to 0.99 mAP (Mean Average precision) on our novel high-resolution imagery dataset. The findings of this study indicate that the proposed methodology has the potential to develop aerial surveillance systems while maintaining a good balance between speed and accuracy.

Keywords Computer vision · Remote sensing · Single stage object detection · Surveillance system · You Only Look Once (YOLO)

A. K. Harichandan (✉) · S. Harichandan · R. R. Kumar · D. Muduli
Department of Computer Science and Engineering, C.V. Raman Global University, Bhubaneswar, India

K. Deep et al. (eds.), *Artificial Intelligence, Machine Learning and Intelligent Systems*, Springer Proceedings in Mathematics & Statistics 520,
https://doi.org/10.1007/978-981-95-3841-6_24

1 Introduction

Khordha, a central district in the state of Odisha, has experienced a gradual loss of many mountains, coinciding with the growth of mining-related auxiliary enterprises, particularly stone-crushing activities, resulting from stone quarrying [7]. This development has significantly raised the levels of dust and air pollutants in an area previously characterized by lush greenery. The primary health risk to workers and residents of neighbouring villages is direct exposure to dust particles released by stone-crushing equipment.

Dust particles significantly contribute to silicosis and chronic renal disease; exposure to stone crusher dust is associated with the former, while consuming contaminated water from surrounding areas is related to the latter. According to data from the state pollution board, the presence of multiple illegal stone-crushing units has caused a dust concentration of approximately 1000 micrograms per cubic meter, exceeding the permitted limit of 600 micrograms, due to the proximity of the units to neighbouring residential areas (within three kilometres of the crushers) [15]. The widespread establishment of these units throughout the area hinders effective monitoring and preventive efforts by the government administration and the pollution control board. This motivated an investigation into potential surveillance strategies based on remote sensing for this region [13].

Drones, also known as Unmanned Aerial Vehicles(UAV), have significantly enhanced remote sensing by offering several benefits compared to traditional satellite imagery for more reliable computer vision applications. These advantages include improved spatial resolution, increased flexibility, and on-demand data collection. While satellite imagery provides extensive global coverage and long-term monitoring, drones offer a more flexible, detailed, and cost-effective solution for various computer vision applications, especially for surveillance. our study using a drone or UAV [1].

In the exciting world of deep learning object detection, a few frameworks truly stand out, such as RCNN, Fast RCNN, SSD (Single Shot Multibox Detector), and the dynamic YOLO (You Only Look Once). YOLO is particularly favoured in various fields like autonomous vehicles, surveillance, and robotics due to its impressive speed and accuracy, making it an excellent choice for real-time applications. This technology is transforming how we view interaction with the environment [3]. Since its launch in 2015, YOLO has quickly evolved, achieving faster detection and improved accuracy, as seen with the release of YOLOv8 in January 2023. [5].

This study demonstrates the application of the YOLOv8 real-time object detection algorithm for identifying stone-crushing units within a newly developed UAV-based remote sensing dataset [14]. This study comprises five sections, each presenting its findings and contributions as illustrated in Fig. 1.

Section 2 analyses relevant contributions that offer critical insights and analysis on this crucial topic. Section 3 evaluates the proposed methodology for addressing the research problem. Section 4 outlines the experiment's results and illustrates the

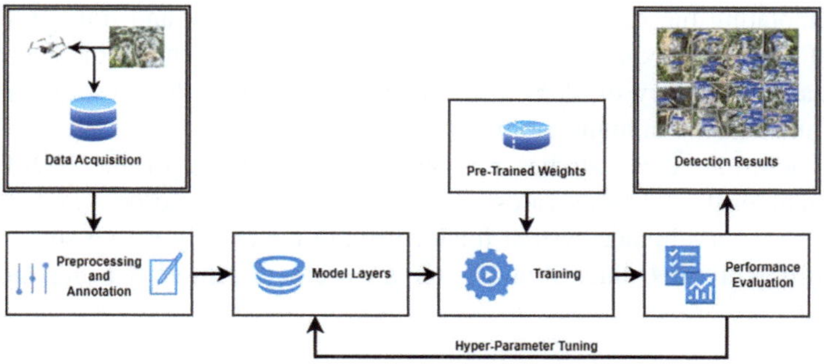

Fig. 1 Process flow for the proposed stone crusher detection model

effectiveness of the employed methods. Section 5 evaluates and synthesizes the findings of this study, proposing possible directions for future research informed by the results obtained.

2 Literature Survey

The literature survey includes research papers on deep learning-based object detection methodologies across various custom datasets.

In [6], Karthika (2024) employed the YOLO v8 algorithm for object detection, which enables the effective identification of multiple objects within images, videos, and live webcam feeds. This advanced algorithm is versatile, adapting to different scenarios such as static image analysis and dynamic object tracking.

In [4], Gallo (2023) introduced a new dataset for weeds and crops, called the Chicory Plant (CP), and tested deep learning algorithms for object detection using over 3000 RGB images. In his research, the YOLOv7 model outperformed other YOLO variants in terms of mAP@0.5 scores.

In [2], Cazzato (2020) examined 2D object detection in UAVs, emphasizing strategies, trade-offs, and distinctions between generic object detection and UAV operations. Additionally, a new taxonomy was introduced, considering height intervals and current methodological approaches.

In [12], Sahin (2021) introduced an improved version of the YOLO algorithm, called YOLODrone, specifically tailored for object detection in drone imagery. Furthermore, we evaluate our algorithm using the VisDrone2019 dataset and present enhanced results compared to the YOLOv3 algorithm.

In [9], Ramachandran (2021) conducted a thorough analysis of object detection and tracking using UAVs across various applications, summarising insights from existing research papers and identifying research gaps.

In [8], Pinto (2021) proposed an innovative custom model based on YOLO for detecting swimming pools for aerial surveillance applications. This approach involved creating a dataset of high-resolution aerial images taken from a bird's-eye view, annotating the relevant objects within those images, and training the inner layers of the YOLO convolutional neural network (CNN). The method achieved an identification accuracy of 49.52% for swimming pools in low-resolution satellite images and 94.36% in high-resolution images.

In conclusion, these studies demonstrate the efficacy of the YOLO object detection algorithm across various real-world scenarios. The proposed model seeks to enhance a surveillance system for identifying unlawfully set up stone crushers across different sites, thus allowing for an accurate remote sensing monitoring strategy.

3 Methodology

Computer vision is an evolving field that enables machines to interpret and understand visual information. [16]. One of the major aspects of this field is object detection, where objects need to be precisely identified in images or video streams. Recent years have seen major advancements in algorithmic methods [18].

A significant advancement in object detection was made in 2015 when Redmon et al. introduced the You Only Look Once (YOLO) algorithm [10]. This method processes the complete image in one pass to identify objects and their positions. YOLO's approach differs from conventional two-stage detection methods in conceptualizing object detection as a regression problem. This method uses a single convolutional neural network to simultaneously predict the entire image's bounding boxes and class probabilities, thereby streamlining the detection process compared to more complex traditional methods.

3.1 Data Acquisition

This study utilised the DJI Mini 3 Pro drone for data acquisition. This drone or UAV has a 24 mm lens and a 1/1.3 CMOS camera featuring a 48 MP sensor. This aircraft has a 3-axis mechanical gimbal, allowing for tilt, roll, and pan movements, with a tilt range extending from $-135°$ to $80°$. Roll ranges from $-135°$ to $45°$; Pan ranges from $-30°$ to $30°$. The flight route was set using the DJI Go app, allowing the user to predefine the flight area and speed while monitoring the shooting status in real time. A flight plan was prepared, and filming occurred on October 14 and 15, 2024. The battery was replaced repeatedly because its flight duration was approximately 45 minutes. The operators conducted a perimeter survey of Jhinkijhari, Drurgapur, and Tapang villages in Khordha district, capturing images at a resolution of 8064 $\times$ 6048 from an altitude of 115 meters. Seventy-four high-resolution 48MP images were captured during this process.

Fig. 2 Annotating images with LabelMe tool

3.2 *Data Preprocessing*

We annotated images using LabelMe, an open-source image annotation tool inspired by MIT's web annotation tool of the same name. LabelMe allows you to create annotations for object detection, classification, and segmentation of computer vision datasets using polygon, rectangle, circle, line, and points [17]. We annotated 74 images by classifying them into 'crusher' mainly consisting of automated belt conveyor-based stone crushing units as illustrated in Fig. 2. We further used the Labelme2YOLO library to make the labelled images into YOLO format, which was initially JSON type [11]. The images were classified and exported in YOLO format for training. In this study, 80% of the images were designated as training data, while 20% were allocated for validation.

3.3 *Model Description*

The YOLOv8, the latest CNN dynamic object model, performs well in real-time applications. A single neural network analyses the whole visual area, splitting it into several components. Potential bounding boxes and probability estimates are computed for each element. The backbone of YOLO consists of a series of convolutional layers that downsample pixels at different resolutions. A neck consolidates and directs these derived characteristics into a head for loss matrix-based object recognition [14].

The YOLOv8 network has five models. For mobile apps, YOLOv8n works well. YOLOv8s, the tiny model, supports CPU architectures. YOLOv8m, a medium-sized model with 25.9 million parameters, balances speed and precision. YOLOv8l performs well with large datasets and training with 43.7 million parameters. The YOLOv8m.pt model was utilized for this training since it balances speed and accuracy [14].

3.4 Training Configuration

We trained our model on the custom UAV dataset with an object of interest, a stone crusher. The proposed model was trained for 300 epochs with an image resolution of 640x640 pixels on Google Colab's Tesla T4 GPU, with the dataset imported from Google Drive. This dimension was used to preserve the equilibrium between computing efficiency and picture detail for detecting purposes. In summary, high-quality photos would enhance detection accuracy but at a significant computational expense. Conversely, using low-quality photos may lose essential information within each image. The designated resolution has been empirically shown to provide substantial outcomes in previous YOLO iterations.

The training procedure was initially planned for 1000 epochs; however, due to the memory limitations of the Google Colab platform, it was necessary to reduce the number of epochs to 300.

The configuration of the hyperparameters, such as batch size, number of workers, optimizer, weight decay, and filter selection, has ensured network stability.

4 Results and Discussion

The following subsections present the results obtained during our proposed model's training and validation stages.

4.1 Qualitative Performance

Some critical metrics contribute to the model's accuracy in identifying object locations. Box loss, for example, helps improve precision in determining where objects are located; a higher value indicates a stronger focus on reducing location errors.

Cls (class) loss relates to the correct classification of object categories. Increasing this value directs the model to prioritize accurate classifications, which can enhance performance in cases of misidentification.

DFL (Deep Feature Loss) addresses the differentiation of objects that closely resemble each other by improving the model's understanding of their features and spatial relationships.

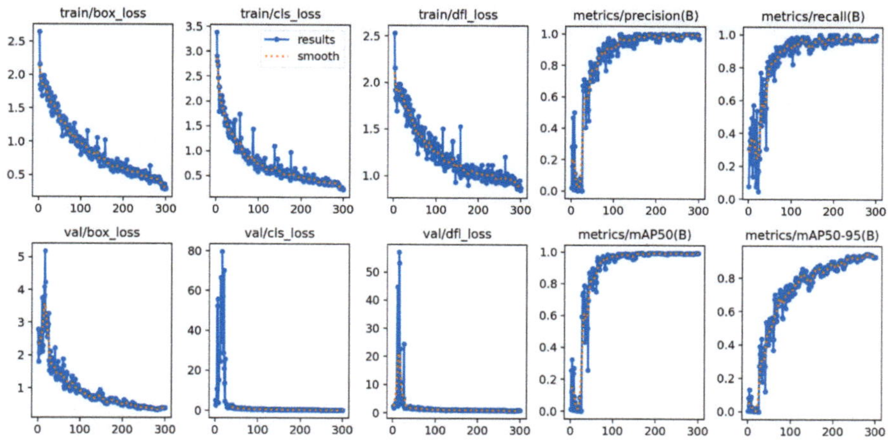

Fig. 3 Illustration of key metrics observed during the training and validation phases of the proposed model

Precision measures the ratio of true positives to the total number of positive predictions, evaluating the model's ability to minimize false positives. Recall quantifies the ratio of true positives to the total number of actual positives, assessing the model's capacity to identify all instances of a specific class. The mean average precision (mAP) comprehensively evaluates the model's performance based on the Intersection over Union (IoU) at a threshold of 0.5.

Figure 3 represents the graphs generated against the above metrics during training over 300 epochs. The overall loss approaches 0.2, achieved through the tuning of various hyperparameters. The model was optimized to minimize loss over 300 epochs.

4.2 Quantitative Performance

Key performance graphs for object detection models include the F1 score, which measures the harmonic mean of precision and recall at different confidence thresholds—higher peaks indicate better performance.

The precision-confidence curve shows how precision varies with confidence levels, while the recall-confidence curve illustrates recall at those thresholds, both expecting consistently high values.

Finally, the Precision-Recall Curve highlights the trade-off between precision and recall; models near the top-right corner are considered superior. Figure 4 illustrates these significant curves of our model.

The overall outcomes obtained from our stone crusher detection model concerning precision, recall, and mean average precision (mAP) are outlined in Table 1.

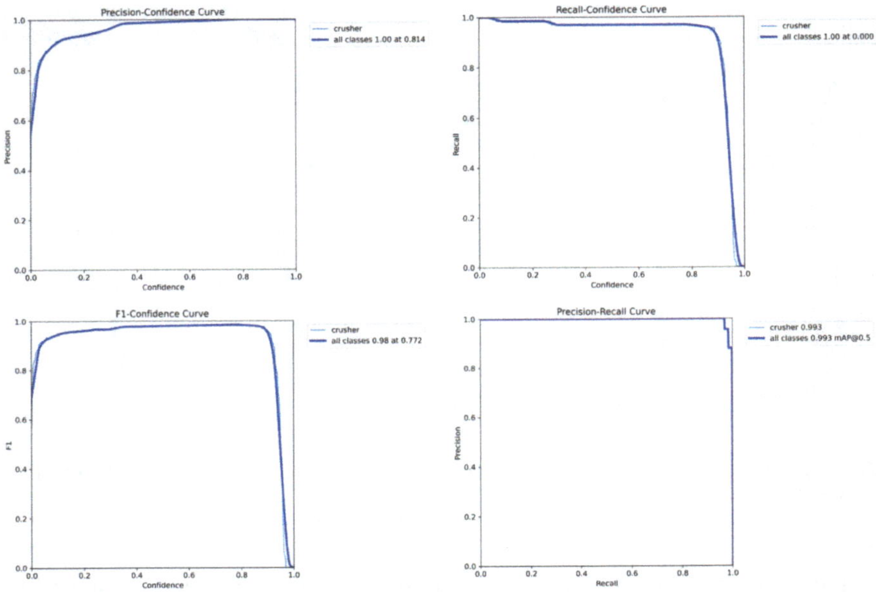

Fig. 4 The model's key performance graphs include the Precision-Recall curve, Recall-Confidence curve, F1 Confidence curve, and Precision-Confidence curve

Table 1 Overall performance of the stone-crusher detection model

Precision	Recall	mAP @ 0.5 IoU
1.0	1.0	0.99

4.3 Detection Result

The proposed model has demonstrated exceptional efficacy in identifying stone crushers, even within complex background environments, achieving higher accuracy, as illustrated in Fig. 5.

5 Conclusion and Future Work

The application of YOLOv8 for detecting stone crushers has yielded encouraging outcomes. The suggested approach exhibits notable precision and effectiveness in immediately identifying stone crushers, positioning it as a compelling choice for monitoring systems. Additionally, the YOLOv8 enables swift identification of stone crushing units, attaining impressive precision and recall rates.

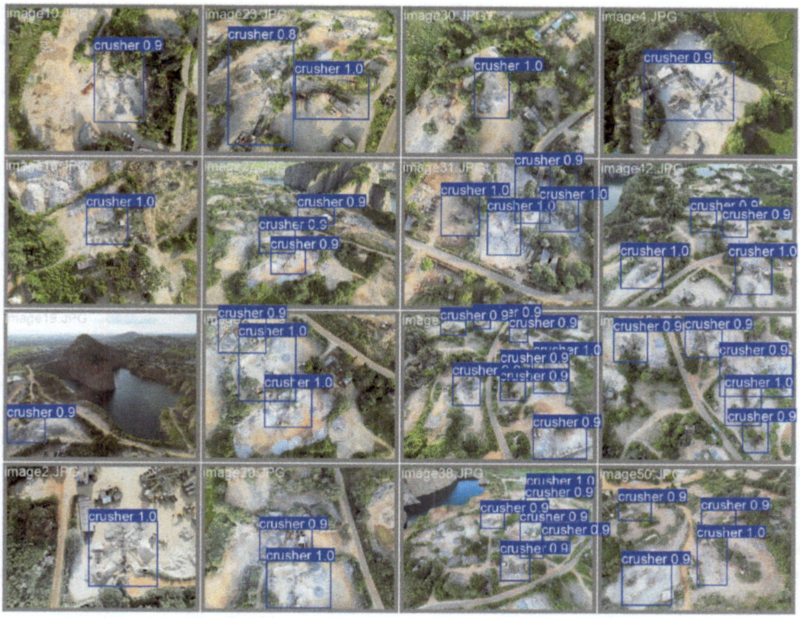

Fig. 5 Illustration showing detection results of multiple stone-crushers across various images in our dataset using the proposed model

Future work involves expanding the database of captured images, geo-tagging the detected objects, and applying this model to develop a framework for real-time aerial surveillance.

References

1. Alvarez-Vanhard, E., Corpetti, T., Houet, T.: Uav & satellite synergies for optical remote sensing applications: A literature review. Science of remote sensing **3**, 100019 (2021)
2. Cazzato, D., Cimarelli, C., Sanchez-Lopez, J.L., Voos, H., Leo, M.: A survey of computer vision methods for 2d object detection from unmanned aerial vehicles. Journal of Imaging **6**(8), 78 (2020)
3. Diwan, T., Anirudh, G., Tembhurne, J.V.: Object detection using yolo: Challenges, architectural successors, datasets and applications. multimedia Tools and Applications **82**(6), 9243–9275 (2023)
4. Gallo, I., Rehman, A.U., Dehkordi, R.H., Landro, N., La Grassa, R., Boschetti, M.: Deep object detection of crop weeds: Performance of yolov7 on a real case dataset from uav images. Remote Sensing **15**(2), 539 (2023)
5. Hussain, M.: Yolo-v1 to yolo-v8, the rise of yolo and its complementary nature toward digital manufacturing and industrial defect detection. Machines **11**(7), 677 (2023)
6. Karthika, B., Dharssinee, M., Reshma, V., Venkatesan, R., Sujarani, R.: Object detection using yolo-v8. In: 2024 15th International Conference on Computing Communication and Networking Technologies (ICCCNT). pp. 1–4 (2024). doi:10.1109/ICCCNT61001.2024.10724411

7. Kumar, M.: Bhubaneswar's periphery is losing its hills to stone mining (2021), https://india.mongabay.com/2021/09/bhubaneswars-periphery-is-losing-its-hills-to-stone-mining [Accessed: (08.09.2024)]
8. Pinto, A.O., Ciarelli, P.M., Sarcinelli-Filho, M.: Creating a yolov3 custom model for aerial surveillance. In: 2021 XIX Workshop on Information Processing and Control (RPIC). pp. 1–6. IEEE (2021)
9. Ramachandran, A., Sangaiah, A.K.: A review on object detection in unmanned aerial vehicle surveillance. International Journal of Cognitive Computing in Engineering **2**, 215–228 (2021)
10. Redmon, J., Divvala, S., Girshick, R., Farhadi, A.: You only look once: Unified, real-time object detection. In: 2016 IEEE Conference on Computer Vision and Pattern Recognition (CVPR). pp. 779–788 (2016). doi:10.1109/CVPR.2016.91
11. rooneysh: Labelme2yolo (2021), https://github.com/rooneysh/Labelme2YOLO [Accessed: (01.11.2024)]
12. Sahin, O., Ozer, S.: Yolodrone: Improved yolo architecture for object detection in drone images. In: 2021 44th International Conference on Telecommunications and Signal Processing (TSP). pp. 361–365 (2021). doi:10.1109/TSP52935.2021.9522653
13. Steele, L.J.: The view from on high: Satellite remote sensing technology and the fourth amendment. High Tech. LJ **6**, 317 (1991)
14. Ultralytics: Ultralytics yolov8 (2023), https://docs.ultralytics.com/models/yolov8 [Accessed: (10.11.2024)]
15. V, R.R.: 124 illegal crushers units in khurda to be sealed today (2014), https://timesofindia.indiatimes.com/city/bhubaneswar/124-illegal-crushers-units-in-khurda-to-be-sealed-today/articleshow/36716821.cms [Accessed: (10.09.2024)]
16. Voulodimos, A., Doulamis, N., Doulamis, A., Protopapadakis, E.: Deep learning for computer vision: A brief review. Computational intelligence and neuroscience **2018**(1), 7068349 (2018)
17. Wada, K.: Labelme: Image Polygonal Annotation with Python. doi:10.5281/zenodo.5711226, https://github.com/wkentaro/labelme
18. Zou, Z., Chen, K., Shi, Z., Guo, Y., Ye, J.: Object detection in 20 years: A survey. Proceedings of the IEEE **111**(3), 257–276 (2023)

Enhancing PCOS Diagnosis with Hybrid ML: Integrating Ultrasound and Clinical Data

R. Swathi, N. Dhanush, and M. Brindha

Abstract This work presents a hybrid machine learning approach that combines clinical data and ultrasound imaging to improve the precision of PCOS diagnosis. With an accuracy of 87% and appropriate for initial screening, the Regular approach uses Logistic Regression on 15 clinical inputs. By contrast, the Hybrid approach achieves a 96% accuracy rate by combining 29 clinical inputs with ultrasound-derived features using a Gradient Boosting Classifier. A Flask backend allows for smooth data processing and real-time predictions, while a web-based interface built with HTML, CSS, and JavaScript makes user interaction easier. The findings demonstrate the efficacy of hybrid techniques in clinical settings and point to their potential for improving diagnostic precision.

Keywords Gradient-boosting classifier · Logistic regression · Image and clinical data integration · Hybrid machine learning

1 Introduction

Polycystic Ovary Syndrome (PCOS) is a common endocrine condition affecting women of reproductive age, with symptoms including irregular menstrual periods, infertility, and metabolic issues. Early and correct PCOS diagnosis is critical for successful management and therapy. Traditional diagnostic approaches are mostly based on clinical data, such as patient complaints and hormone assays, which can

R. Swathi (✉)
Bharathidasan University, Tiruchirappalli, Tamil Nadu, India

N. Dhanush
St. Joseph's College, Tiruchirappalli, Tamil Nadu, India

M. Brindha
National Institute of Technology, Tiruchirappalli, Tamil Nadu, India
e-mail: brindham@nitt.edu

K. Deep et al. (eds.), *Artificial Intelligence, Machine Learning and Intelligent Systems*, Springer Proceedings in Mathematics & Statistics 520,
https://doi.org/10.1007/978-981-95-3841-6_25

be subjective and frequently insufficient for a complete diagnosis. Recent advances in machine learning provide novel ways for improving diagnostic accuracy by combining many data sources. This study presents a hybrid model that combines regular and sophisticated mathematical approaches to enhance PCOS diagnosis. The standard technique relies on 15 main clinical data inputs, including patient age, weight, height, and symptom related signs. A Logistic Regression model trained with these inputs has a test accuracy of 87%. While this technique yields useful information, it may not fully reflect the complexities of PCOS, which frequently need additional in-depth investigation. The enhanced approach improves diagnosis accuracy by combining 29 clinical data inputs with an ultrasound scan of the uterus. This approach uses a two-step feature extraction process. First, characteristics are retrieved from ultrasound pictures using a Logistic Regression model, which yields a test accuracy of 98%. Second, clinical characteristics are extracted using a Gradient Boosting Classifier, which has an 86% test accuracy. The combining of these characteristics in a Gradient Boosting Classifier leads to a substantial increase, with a test accuracy of 96%. This innovative method uses both imaging and clinical data to create a more powerful diagnostic tool.

HTML, CSS, and JavaScript were used to create a frontend interface for user interaction. This interface allows users to effortlessly enter their data and ultrasound pictures. The backend, written in Python, analyzes data and performs relevant computations using machine learning methods, with Flask acting as a link between the frontend and backend. This connection guarantees that consumers obtain diagnostic reports that are both timely and correct. This study demonstrates the value of combining clinical and imaging data for PCOS diagnosis, resulting in greater accuracy and a more complete diagnostic framework. The hybrid model provides an accurate tool for detecting PCOS, improving patient outcomes, and developing personalized treatment programs, emphasizing the importance of current machine learning in healthcare.

2 Related Work

Recent developments in machine learning and imaging have greatly improved PCOS detection. When Singh et al. [1, 2] looked at conventional methods like logistic regression and support vector machines on clinical data, they discovered that while they were accurate, they were constrained by the lack of imaging data integration. The SLB-SMOTE Logistic Blending Hybrid Machine Learning Model, on the other hand, was presented by Kumar et al. [3] and improves PCOS diagnostic outcomes by integrating clinical data with additional features. In their investigation of hybrid machine learning models for PCOS early diagnosis, Jain et al. [4] demonstrated how these techniques perform better than conventional techniques. For a more precise diagnosis of PCOS, Gupta et al. [5] suggested a system that combines clinical and imaging data with machine learning. In order to demonstrate how multimodal AI can effectively enhance diagnostic performance, Patel and

Rao [6] looked at the possibility of combining clinical and imaging metadata. In order to optimize machine learning algorithms for PCOS diagnosis, Mehta et al. [7] highlighted the significance of feature extraction and selection, which is consistent with both the fundamental and sophisticated techniques employed in our investigation. Reddy et al. [8] supported the notion that a multidimensional approach delivers higher accuracy by focusing on combining imaging, omics, and clinical data for improved PCOS diagnosis using AI. Further supporting the importance of hybrid approaches for PCOS identification were Singh et al. [1, 2] and Gupta et al. [5], who promoted the integration of imaging and clinical data as a full diagnostic tool.

3 Methodology

This study investigates two different strategies for using machine learning techniques to enhance the diagnosis of polycystic ovarian syndrome (PCOS). The first strategy, known as the Regular method, trains a diagnosis model using clinical data. Fifteen clinical inputs make up the dataset for this approach, which has been preprocessed to accommodate missing values, normalize features, and guarantee consistency. These clinical features are analyzed using a Logistic Regression model after preprocessing. Accuracy is used to assess the model's performance and provide light on how well clinical data alone can diagnose PCOS. The second strategy, referred to as the Hybrid method, improves diagnostic accuracy by fusing clinical data and ultrasound imaging. First, a feature extraction method based on the well-known convolutional neural network VGG16 model is applied to ultrasound pictures. High-level characteristics are extracted from the images by this model and then concatenated with 29 clinical data inputs. A robust model that utilizes both imaging and clinical data, the Gradient Boosting Classifier, is trained on the combined dataset. Through the integration of several data sources, this approach enables a more detailed knowledge of PCOS (Fig. 1).

3.1 Detailed Description of Dataset

Clinical records and ultrasound imaging data make up the two parts of the dataset used in this investigation. Since both datasets were obtained from publically accessible sources, reproducibility and accessibility were guaranteed:

(i) **Clinical Dataset:**

- There are 542 patient records in the clinical dataset, each of which represents a distinct case. These documents provide a wide range of characteristics that are important for diagnosing PCOS, including age, body mass index (BMI), hormone levels, and details about individual symptoms.

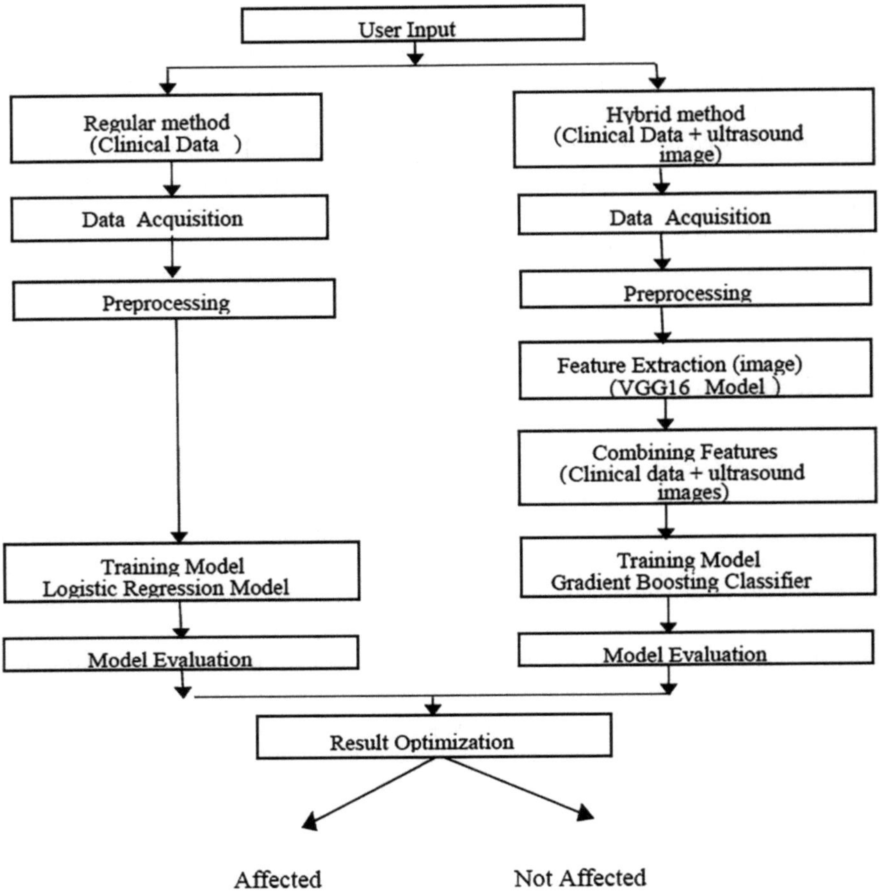

Fig. 1 Visual representation of proposed system

- The Hybrid model added granularity to the diagnosis process by using an extended collection of 29 clinical variables, whereas the Regular model only had 15 elements.
- With patients ranging in age from 18 to 40, the dataset reflects a broad population. A balanced distribution across BMI categories, hormone profiles, and symptom prevalence was attempted to be achieved. However, the limitations of the sources reduced the amount of specific demographic data, including geographic variety.

(ii) **Ultrasound Imaging Data:**

- More than 1000 ultrasound pictures are included in the imaging collection, which is separated into two groups: those with and without PCOS. These pictures, which came from the Kaggle library, were taken with ultrasound equipment under controlled circumstances.

- Preprocessed images had a constant 224 × 224 pixel resolution. To improve dataset diversity and reduce overfitting risks, data augmentation techniques like rotation, flipping, zooming, and brightness modifications were used.

(iii) **Data Integration:**

- The Hybrid model was able to utilize both visual and numerical diagnostic features since the clinical and imaging datasets were merged into a single dataset. Enhancing the precision and resilience of PCOS diagnosis required this combination.
- Ethical considerations: To protect patient privacy, all clinical records were anonymized. Normalization, imputation of missing values, and one-hot encoding of categorical variables were among the preprocessing procedures.

3.2 Data Acquisition and Pre-Processing

The method of gathering data is essential to developing a reliable machine learning model for diagnosing PCOS. The two primary data types used in this investigation are ultrasound images and clinical records. The clinical dataset includes 29 inputs for the Hybrid model and 15 inputs that are necessary for the Regular model. A range of data are included in these clinical aspects, including patient demographics (age, height, weight), hormone levels, and particular PCOS symptoms. For the sake of patient privacy and ethical data management procedures, these records have been anonymized.

Preprocessing starts as soon as the data is gathered. To prepare the clinical data for machine learning algorithms, it passes through a number of step:

- Numerical representations are created from categorical variables, including symptoms and other non-numerical elements. For the model to properly comprehend such data, this step is necessary.
- Numerical characteristics such as age, BMI, and hormone levels are normalized. This procedure reduces biases and guarantees a level playing field for all inputs by scaling the values to ensure consistency, particularly when features have varied ranges.
- Imputation techniques are used to deal with missing values, which are common in medical datasets. In order to prevent data gaps from impairing the model's performance, this stage fills in missing elements while maintaining the dataset's integrity.

Ultrasound pictures are included to the clinical inputs for the Hybrid model. To ensure consistency, these photos are scaled to 224 × 224 pixels. Additionally, pixel values are standardized to provide uniformity throughout the image dataset, which is essential for efficient pattern identification.

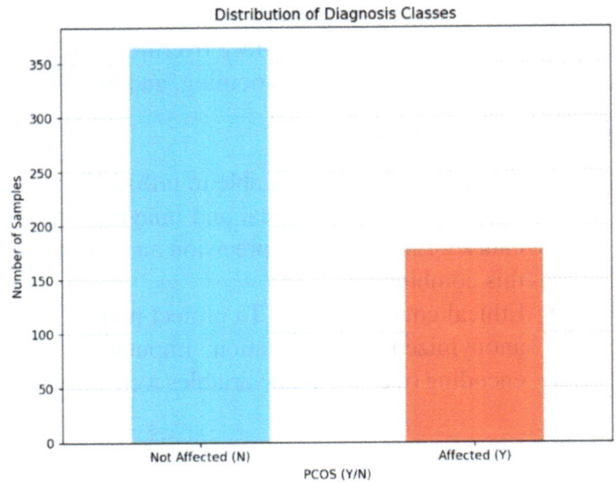

Fig. 2 The regular method's diagnostic distribution

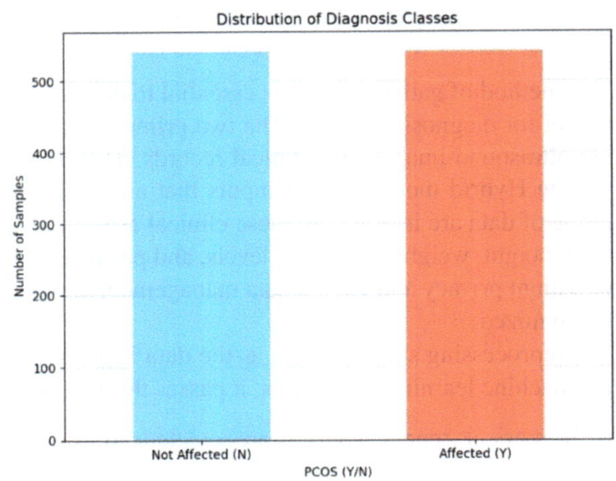

Fig. 3 The hybrid method's diagnostic distribution

The diagnostic distribution for the Regular model, which shows the balance of samples across diagnostic categories, is shown in Fig. 2. The diagnostic distribution for the Hybrid model, which shows additional classifications since ultrasound images are included, is shown in Fig. 3.

3.3 Data Augmentation

In order to improve the Hybrid model's capacity for generalization, particularly when dealing with a small dataset, data augmentation is an essential technique used to the ultrasound pictures. The model can be made more resilient to changes in real-world data by artificially growing the dataset.

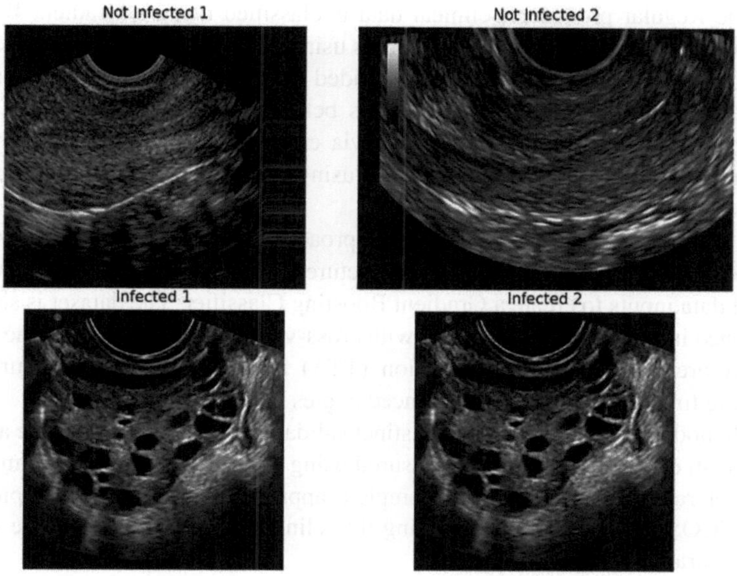

Fig. 4 Augmented ultrasound images (infected vs. not infected)

Among the several augmentation techniques are

- Flipping images both horizontally and vertically helps the model learn to identify patterns in any direction.
- By simulating images taken from various distances and angles, rotations and zooming enhance the model's capacity to handle a variety of viewpoints.
- Brightness and contrast adjustments are made to guarantee that the model can reliably identify features in a variety of lighting scenarios.
- To increase the model's resilience to slight distortions and flaws in the ultrasound images, random noise was added.

Examples of augmented ultrasound pictures displaying both infected and non-infected samples are shown in Fig. 4 above. By exposing the model to a greater variety of picture changes, the augmentation approaches used improve diagnostic accuracy by assisting the model in differentiating between infected and non-infected uteri.

3.4 Model Training and Validation

Model training and validation are critical to the accuracy and reliability of PCOS diagnostic tools. The study incorporates two methods: a traditional strategy utilizing clinical data and an enhanced method that combines clinical data with ultrasound pictures.

In the Regular procedure, clinical data is classified using a Gradient Boosting Classifier. The model is trained on features using a dataset that contains 15 clinical data inputs per person, with the data divided into training and test sets (usually 80-20). The model learns the correlations between clinical variables and PCOS diagnosis and evaluates its performance via cross-validation and hyperparameter tweaking. The model's efficacy is assessed using metrics like as accuracy, precision, recall, and the F1-score.

The Hybrid method adopts a advance approach. A pre-trained VGG16 network is used to extract features from ultrasound pictures, which are then combined with 29 clinical data inputs to create a Gradient Boosting Classifier. The dataset is similarly partitioned into training and test sets, with cross-validation used to create the model. Furthermore, Test Time Augmentation (TTA) improves prediction accuracy by averaging findings from several enhanced copies of test pictures.

Both models are evaluated using distinct validation sets to verify they are applicable to fresh data. Performance is measured using test accuracy, confusion matrices, and other related indicators. This complete approach aids in the development of strong PCOS diagnostic tools, ensuring that clinical and imaging data are used to make accurate diagnosis.

3.5 Feature Extraction

An essential first step in improving the project's PCOS diagnosis accuracy is feature extraction. The clinical and imaging streams of the feature extraction procedure are divided into two separate parts for the Hybrid model, which combines clinical data with ultrasound pictures.

Clinical Data Feature Extraction: This procedure entails organizing and processing the 29 input features of the clinical data into a manner that is appropriate for machine learning models. Important actions include employing imputation techniques to handle missing data, normalizing numerical features to adjust their scale, and utilizing one-hot encoding to encode categorical variables into numerical values. Model performance is enhanced by this transformation, which guarantees that all input characteristics are evenly handled and compatible with the machine learning techniques.

Image Feature Extraction: The enhanced model's imaging component uses learned deep learning architectures to extract information from ultrasound pictures. A VGG16-based convolutional neural network (CNN) is used for this purpose. The VGG16 model, which has been pre-trained on ImageNet, functions as a feature extractor by using its convolutional layers to detect intricate and hierarchical patterns in pictures. The output of the last convolutional block is utilized to represent the features of each image. These features are then flattened and reshaped to create a complete feature set. This strategy enables the model to use pre-learned image characteristics in the PCOS diagnostic task, considerably improving the model's capacity to detect meaningful patterns within ultrasound pictures.

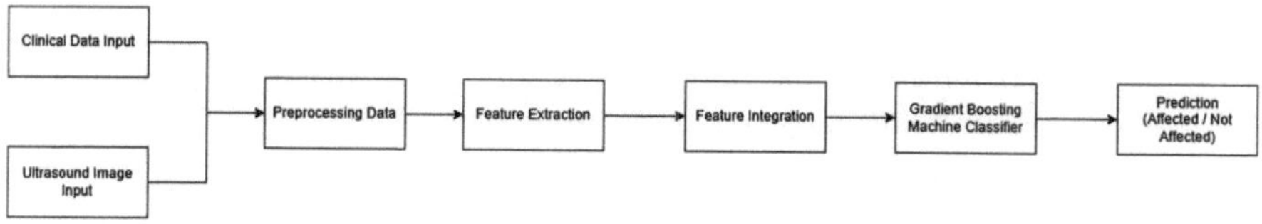

Fig. 5 Architecture diagram

The combined clinical and imaging data are then utilized to train the hybrid model, allowing for a more comprehensive approach to PCOS diagnosis. Figure 5 shows the architecture of the system.

Algorithm 1 Regular and Hybrid PCOS Diagnosis Models

Require: Training and validation datasets.

Ensure: Trained Logistic Regression and Gradient Boosting models with optimum weights.

1. Initialize Data Generators.
   - For the Regular Method: load clinical data.
   - For the Hybrid Method: start ImageDataGenerator for ultrasound pictures and clinical data preparation.
2. Data preprocessing.
   - For the Regular Method: Preprocess clinical data, including normalization and feature extraction.
   - For the Hybrid Method: extract features from ultrasound pictures with a VGG16-based model and preprocess clinical data.
3. Define models.
   - For the Regular Method: initialize the Logistic Regression model.
   - For the Hybrid Method: initialize the Gradient Boosting Classifier.
4. Model Training.

   4.1. Regular Method:
   - Separate the clinical data into training and testing sets.
   - Using the training data, train the logistic regression model.
   - Evaluate model performance on a test dataset.

   4.2. Hybrid Method:
   - Combine picture characteristics and clinical data.
   - Divide the pooled data into training and test sets.

(continued)

- Train the Gradient Boosting Classifier using mixed features.
- Evaluate model performance using the test set.

5. Validation & Evaluation

   5.1. For both approaches:

   - To evaluate model performance, use relevant metrics like as accuracy and classification reports.
   - Save trained models for later use.

6. Model Optimization.

   - Fine-tune models using performance indicators as appropriate.
   - Weak hyperparameters for better accuracy.

7. Save and deploy models.

   - Save the learned Logistic Regression model as a Regular procedure.
   - Keep the learned Gradient Boosting model for the Hybrid technique.
   - Set up models for real-time PCOS diagnosis.

4 Results

The study discovered significant discrepancies in performance across two PCOS categorization techniques. The first approach used a Logistic Regression model trained only on clinical data, yielding an 87% test accuracy. This model efficiently identifies PCOS based on 15 clinical parameters, proving its diagnostic efficacy with numerical and categorical data. In contrast, the sophisticated approach employed a Gradient Boosting Classifier to merge clinical data with ultrasound images, yielding an amazing 96% test accuracy. The improved performance of this hybrid approach highlights the benefits of merging imaging data with clinical considerations, which improve diagnosis accuracy by giving vital visual information. Both models were assessed for accuracy, recall, and F1-score, and Figs. 6 and 7 show the findings as Not Affected and Affected. The hybrid model's better findings suggest that it can identify PCOS more reliably. The increased recall and F1-score demonstrate that this strategy not only increases overall accuracy but also excels at spotting positive cases. The comparison illustrates the importance of various data sources. While the logistic regression model performed well with clinical data alone, the hybrid model's better accuracy demonstrates the need of including imaging data for more effective and accurate medical diagnosis.

Fig. 6 Not affected

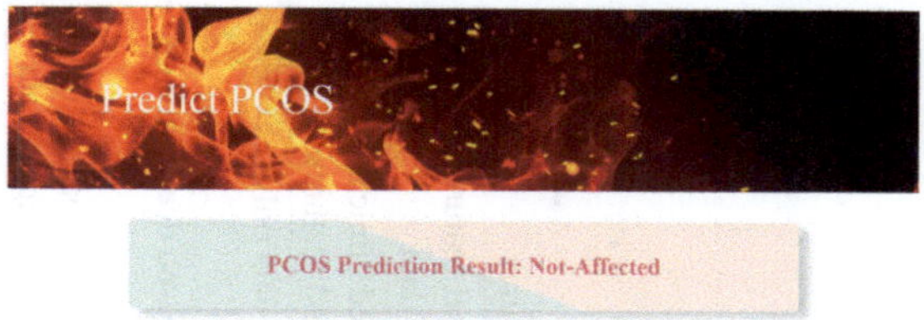

Fig. 7 Affected

4.1 Benchmarking with Other Machine Learning Methods

Using accuracy, precision, recall, and F1-score as important criteria, a benchmarking analysis compared the suggested models with current machine learning techniques for PCOS diagnosis Table 1. The Regular Model, based solely on clinical features, demonstrated 87% accuracy, while the Hybrid Model exceeded state-of-the-art methods with 96% accuracy. The Hybrid Model provides a more thorough and universal framework, and this study emphasizes the advantages of integrating clinical data and ultrasound imaging for increased diagnostic accuracy.

5 Evaluation of Model

The Regular and Hybrid models were evaluated by assessing training and validation accuracy, loss, and confusion matrices for both training and validation sets. These metrics and visualizations aid in analyzing the models' effectiveness and reliability in predicting PCOS.

5.1 Regular Model Evaluation

The Logistic Regression model, trained with 15 clinical data inputs, produced good results. Training accuracy remained high across the epochs, indicating that the data was being learned effectively. However, while validation accuracy initially improved, it fluctuated, indicating some inconsistency in performance on unknown data. The training loss was continuously low and constant, but the validation

Table 1 Shows benchmarking with other ML methods

Model/method	Dataset description	Accuracy (%)	Precision (%)	Recall (%)	F1-Score (%)	Key features
Logistic Regression (This work—Regular Model)	542 clinical records	87	85	83	84	15 clinical features, no imaging data
Gradient Boosting (This work—Hybrid Model)	542 clinical records + ultrasound	96	94	92	93	Combined clinical and ultrasound features
Support Vector Machine (SVM) (Singh et al., 2020) [1]	500 clinical records	84	82	80	81	Clinical data only
CNN-based Model (Kumar et al., 2024) [3]	Ultrasound images (1000 samples)	93	91	90	90	Image-based deep learning
Hybrid ML Model (Chaudhary et al., 2022) [9]	Clinical + ultrasound (600 records)	95	93	91	92	Ensemble approach using clinical and imaging data
Deep Learning (Patel et al., 2023) [6]	Clinical + ultrasound (500 records)	94	92	89	91	Multimodal AI integrating clinical and imaging data
Decision Tree (Reddy et al., 2023) [8]	300 clinical records	82	81	78	79	Simple clinical features

loss began high and reduced initially, but then fluctuated significantly, remaining considerably greater than the training loss. This emphasizes the significance of tracking validation metrics to avoid overfitting.

The confusion matrix for the training set indicated high values on the diagonal, demonstrating the model's ability to properly categorize the majority of examples with low misclassifications. The validation set confusion matrix likewise performed well, but with somewhat more obvious misclassifications than the training set.

5.1.1 Regular Model Testing

The regular model is based on 15 clinical inputs, including height, weight, irregular periods, and other symptoms often reported by patients. This approach does not need the use of ultrasound pictures.

It is intended to work based on common clinical data inputs supplied by users. The algorithm worked well and properly predicted if a patient had PCOS based purely on these inputs.

5.1.2 Training Vs. Testing Accuracy and Loss

Figure 8 graph shows the train and test accuracy of a model after 10,000 iterations. Initially fluctuating, both accuracies settle at roughly 85%. Test accuracy (orange) varies more than train accuracy (blue), suggesting that the model may be overfitted or has variability in performance.

Figure 9 graph depicts training and testing losses across 10,000 cycles. Both losses are declining, with considerable early declines and moderate gains later. Training loss is typically greater than testing loss, suggesting robust model generalization. Final losses are stable at 0.40 (training) and 0.35 (testing).

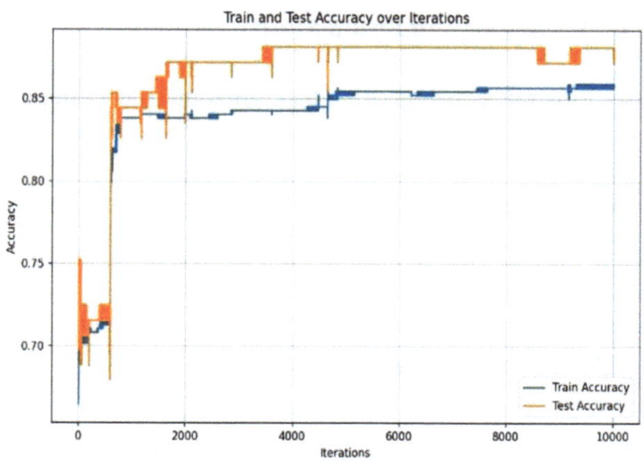

Fig. 8 Training and testing accuracy for regular model

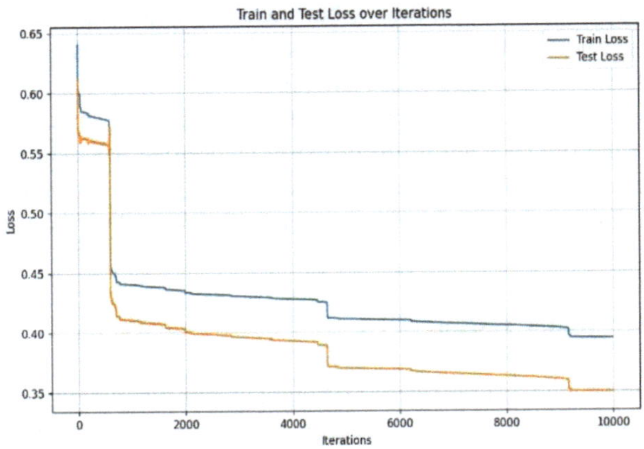

Fig. 9 Training and testing loss for regular model

Fig. 10 Training set confusion matrix for regular model

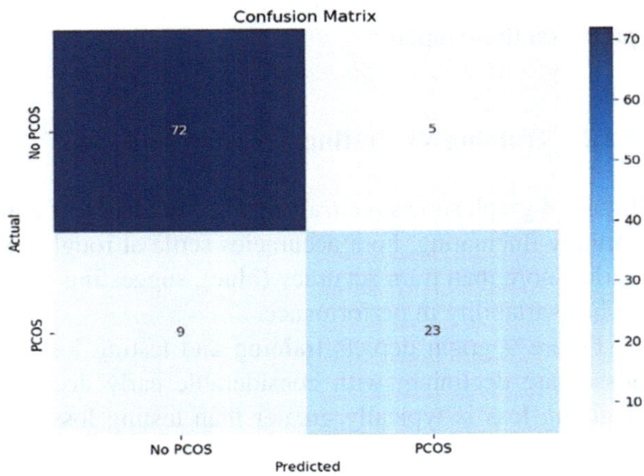

5.1.3 Testing Confusion Matrix

As Illustrated in Fig. 10, The confusion matrix depicts a model's performance on the testing set while predicting PCOS. It accurately identified 72 non-PCOS patients and 23 PCOS cases, with five false positives and nine false negatives. The model has good accuracy for non-PCOS predictions but mediocre accuracy for PCOS predictions.

5.2 Hybrid Model Evaluation

The Hybrid model, which employs a hybrid method that combines 29 clinical data inputs and ultrasound pictures, was assessed similarly. The training accuracy of this model improved slowly and stayed high, indicating the model's capacity to learn from complicated data. The validation accuracy increased over time, but it exhibited significant variability, highlighting the difficulties of generalizing from training data.

In terms of loss, the hybrid model's training loss was continuously low, whereas the validation loss fluctuated but decreased overall. This suggests that, while the model is learning well, there is still potential for improvement in how it generalizes to new datasets.

The confusion matrices for the improved model demonstrated its resilience. The training set confusion matrix demonstrated great accuracy in all categories, with few misclassifications. The validation set confusion matrix, although still demonstrating great performance, had significantly more misclassifications, particularly in more complicated scenarios, indicating opportunities for improvement.

5.2.1 Hybrid Model Testing

- The hybrid model predicts the likelihood of PCOS using 29 clinical inputs and a uterine ultrasound image. The number of follicles, BMI, and other significant clinical characteristics are examples of inputs.
- I supplied my own clinical information and a randomly selected, non-infected ultrasound image from Google for external testing. The model accurately determined that I did not have PCOS.
- I also tested the model using clinical data from a known individual with a PCOS diagnosis and a random Google image of an infected uterine scan. The presence of PCOS was correctly predicted by the hybrid model, proving its capacity to process a variety of input and produce accurate predictions.

5.2.2 Training Vs. Testing Accuracy and Loss

Figure 11 depicts the accuracy of a model on training and testing data. The training accuracy is excellent at 100%, while the testing accuracy drops to roughly 0.96%.

Figure 12 graph displays the model's loss from training and testing data. The training loss is almost negligible, while the testing loss is substantially larger, at 0.30.

5.2.3 Testing set Confusion Matrix

Figure 13 shows the confusion matrix for the test set has 40 true negatives, 4 false positives, 65 real positives, and 0 false negatives. The model has high accuracy and no false negatives, indicating that it does a great job of properly recognizing positive situations.

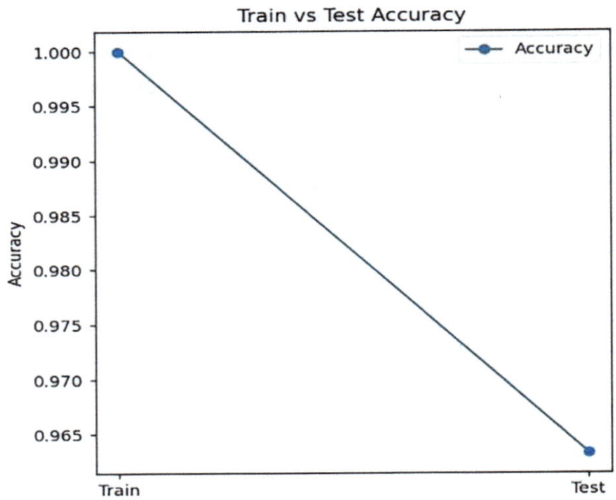

Fig. 11 Training and testing accuracy for hybrid model

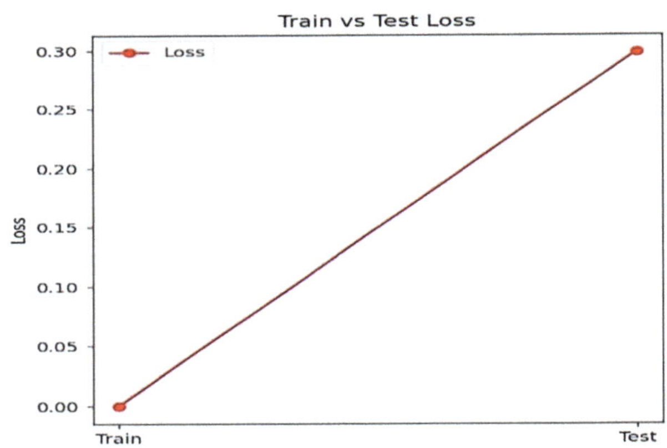

Fig. 12 Training and testing loss for hybrid model

6 Conclusion

This work demonstrates the advantages of combining clinical and imaging data to create hybrid machine learning models for PCOS diagnosis. Although it was successful for first screens, the Regular method—which was based only on clinical inputs—was unable to adequately capture the intricacy of the disorder. By integrating ultrasonic imaging data, the Hybrid method overcame the drawbacks of conventional techniques and obtained improved diagnosis accuracy. The model showed slight oscillations in spite of its encouraging performance, highlighting the necessity for more improvement. The dataset will be enlarged, generalization will be improved, and more features will be investigated to increase robustness in future

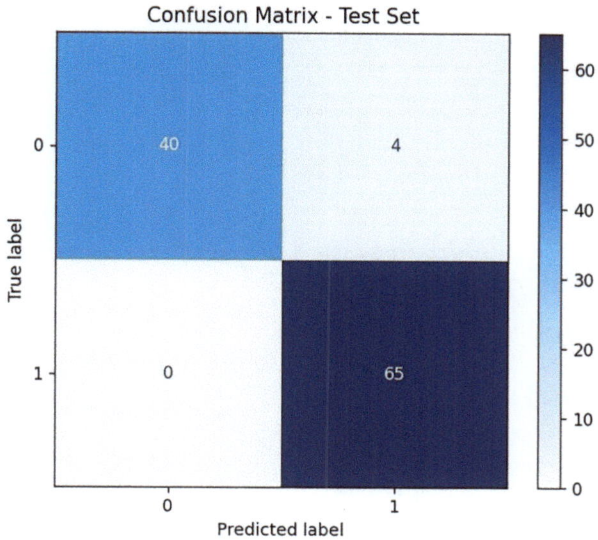

Fig. 13 Testing set confusion matrix for hybrid model

work. These initiatives seek to validate hybrid models as trustworthy diagnostic instruments that support more precise and individualized medical care. While a formal external dataset was unavailable, practical validation was performed as an alternative method to test the models' generalizability.

References

1. Singh, A., Gupta, P., Kaur, S.: Application of Machine Learning Techniques in the Diagnosis of Polycystic Ovary Syndrome. Journal of Medical Systems 44(7), 123–134 (2020)Author, F., Author, S.: Title of a proceedings paper. In: Editor, F., Editor, S. (eds.) CONFERENCE 2016, LNCS, vol. 9999, pp. 1–13. Springer, Heidelberg (2016).
2. Singh, A., Gupta, P., Kaur, S.: Polycystic Ovary Syndrome Detection Using Hybrid Machine Learning Models. Journal of Medical Systems (2023)Author, F.: Contribution title. In: 9th International Proceedings on Proceedings, pp. 1–2. Publisher, Location (2010).
3. Kumar, S., Verma, R., Singh, M.: SLB-SMOTE Logistic Blending Hybrid Machine Learning Model for PCOS Prediction. Artificial Intelligence in Medicine (2024).
4. Jain, M., Singh, P., Gupta, R.: Leveraging Hybrid Machine Learning Models for Early Diagnosis of PCOS. Journal of Clinical Ultrasound (2023).
5. Gupta, N., Saxena, P., Kumar, S.: Combining Clinical and Imaging Data for Advanced PCOS Diagnosis Using Machine Learning. IEEE Transactions on Medical Imaging (2024).
6. Patel, V., Rao, N.: The Future of Multimodal AI for Integrating Imaging and Clinical Metadata in Disease Diagnosis. Biomedical Engineering Online (2023).
7. Mehta, R., Patel, J., Joshi, H.: Strategies for Implementing Machine Learning Algorithms in PCOS Diagnosis. Computers in Biology and Medicine (2023).
8. Reddy, P., Das, A., Kumar, R.: AI-Based Integration of Imaging, Omics, and Clinical Data for Enhanced Medical Diagnosis. Journal of Women's Health (2023).
9. Sharma, P., Kumar, S., Chaudhary, A.: Ensemble Learning Methods for Enhanced Classification of Polycystic Ovary Syndrome (2022).

DNA Splice Junction Prediction Using Hybrid Approach of LSTM-GRU Model

Neha Gautam, Minakshi Ramteke, Mangala Madankar, Abhay Yeole, and Vaishali Shende

Abstract DNA splice junction classification plays a critical role in computational biology, aiming to predict junction types (IE, EI, or N) from a given DNA sequence. This task is essential for understanding gene expression, disease mechanisms, splicing regulation, and the overall gene structure. Identifying the precise locations where exons are joined and introns are removed during RNA splicing remains challenging due to the lack of a universal rule governing this process. Furthermore, the complexity of splice junctions is compounded by the vast variability in genetic sequences across different organisms. This study explored the use of Long Short-Term Memory (LSTM) and Gated Recurrent Units (GRU) for classifying DNA sequences, as these models are particularly effective in capturing long-range dependencies and contextual relationships between distant nucleotides. By leveraging the ability of LSTM and GRU to model sequential data, this approach was well-suited for understanding the intricate patterns inherent in splice junctions. The analysis was conducted using the StatLog DNA dataset, and model performance was evaluated across several metrics: Accuracy (96.48%), F1-Score (97.52%), Kappa (94.95%), Recall (99.38%), Precision (96.12%), and Coverage (97.48%). The results demonstrate the effectiveness of LSTM and GRU in predicting splice junction types with high accuracy and reliability, offering promising insights for future research in gene splicing and related bioinformatics fields.

Keywords Deep learning · DNA · Gated recurrent units · Long short-term memory · RNA

N. Gautam (✉) · M. Ramteke · M. Madankar · A. Yeole · V. Shende
G H Raisoni College of Engineering, Nagpur, Maharashtra, India

K. Deep et al. (eds.), *Artificial Intelligence, Machine Learning and Intelligent Systems*, Springer Proceedings in Mathematics & Statistics 520,
https://doi.org/10.1007/978-981-95-3841-6_26

1 Introduction

The genome, comprising an organism's complete set of DNA sequences, contains the vital information necessary for protein synthesis. Proteins, indispensable for life, fulfill diverse roles within organisms, such as catalyzing metabolic reactions, facilitating cellular communication, and providing structural support [1]. Genes, which are segments of DNA, vary in size and encode the instructions for specific protein production. In eukaryotic organisms, genes consist of coding sequences (exons) and non-coding sequences (introns). During gene expression, a splicing process removes the introns, leaving only exons to be translated into proteins. Accurate identification of splice sites-the nucleotides at exon-intron boundaries is critical for analyzing gene structures and improving genome annotation [2]. DNA splice junction prediction involves identifying the boundaries between exons and introns during the splicing process. As illustrated in Fig. 1, splice sites are typically composed of four conserved nucleotides: the donor sequence GT (or GU in pre-mRNA) at the 5′ end, marking the exon-intron boundary, and the acceptor sequence AG at the 3′ end, marking the intron-exon boundary. Within the coding sequence, introns start with GT bases and end with AG bases, while exons begin with the start codon ATG and terminate with one of three stop codons: TAG, TAA, or TCA [3].

With advancements in high-throughput sequencing, vast genomic data from various species have become available, presenting opportunities and challenges in

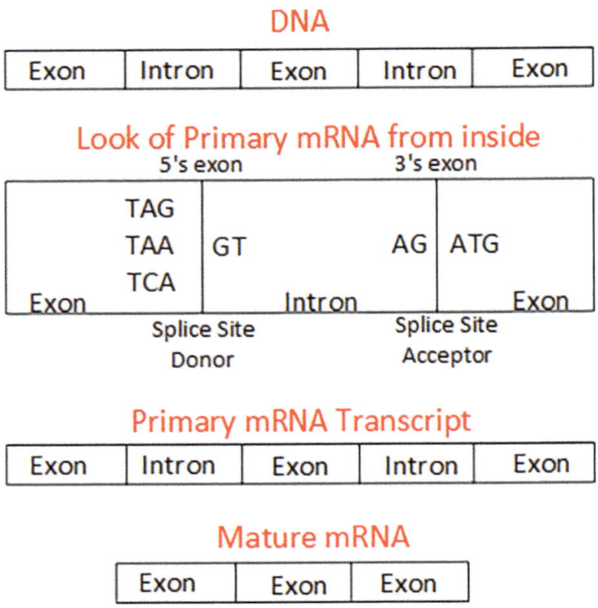

Fig. 1 DNA splicing process, along with primary mRNA look, look of primary m RNA from inside, and mature mRNA look

genome annotation. While earlier gene prediction models achieved modest accuracy, recent machine learning and deep learning models, especially those developed post-2010, have improved specificities and sensitivities. Nonetheless, accurately predicting splice sites remains challenging, partly due to the high frequency of GT and AG dinucleotides in regions other than splice sites, which often leads to false positives. Current prediction tools still exhibit significant error rates, affecting downstream genomic analyses.

Several methodologies and techniques have been proposed to enhance splice junction prediction, leveraging various computational approaches and datasets. This section provides an overview of notable contributions in this domain. Zuallaert, et al. [4] developed SpliceRover, a convolutional neural network (CNN)-based approach. Their model was trained and tested on the NN269 dataset, achieving an accuracy of 96.12%, highlighting its effectiveness in splice site prediction. Similarly, Wang, et al. [5] introduced SpliceFinder, which utilized the Ensembl dataset and demonstrated accuracy of up to 90.25%. Zhang, et al. [6] proposed DeepSplice, a model evaluated using the HS3D dataset. The method achieved an area under the curve (AUC) of 0.983, showcasing its robustness in distinguishing splice junctions. Amilpur and Bhukya [7] further improved upon this with EDeepSSP, also using the HS3D dataset, and achieved a higher AUC of 0.9870, indicating enhanced predictive performance.

Albaradei, et al. [8] presented Splice2Deep, a method tested on Homo sapiens data, achieving an F-measure of 96.91%. Dasari and Bhukya [9] introduced InterSSPP, which performed well on both the HS3D and NN269 datasets, with AUC values of 0.9946 and 0.9922, respectively, demonstrating its adaptability to different datasets. Van Moerbeke, et al. [10] employed linear mixed models in their REIDS technique using the HJAY dataset, reporting accuracy between 65–77%, suitable for certain use cases but less effective compared to deep learning models. Zhao, et al. [11] proposed the ASJA method, focusing on assembling splice junction analysis using the GEO dataset. Their approach achieved a sensitivity of 91.5%, emphasizing its ability to detect true splice sites. Lee, et al. [12] explored recurrent neural networks, combining long short-term memory (LSTM) and gated recurrent units (GRU) in their study. Using data from the ENCODE Consortium, they reported F-measure values of up to 86%, showing promise for sequence prediction tasks. Lastly, Zabardast, et al. [1] introduced a hybrid model integrating CNN, BLSM, and BGRU architectures. Tested on HS3D and C. elegans datasets, their approach achieved accuracies of 92% and 96%, respectively, underscoring the utility of combining deep learning techniques for improved performance. While significant progress has been made in splice junction prediction, challenges remain in terms of balancing model complexity and predictive accuracy. Advanced models like CNNs, GRUs, and hybrid approaches demonstrate exceptional results but often rely on limited datasets, reducing their generalizability across diverse genomic contexts. This highlights the need for more versatile models capable of adapting to various organisms and datasets, addressing the gaps in current methodologies for robust splice site prediction.

2 Literature Review

Splice junction prediction is a fundamental aspect of genomic research, playing a crucial role in understanding genetic disorders, improving gene annotations, and advancing targeted therapeutic development. Traditional computational methods for splice site prediction have typically relied on manual feature extraction techniques, such as nucleotide dependencies, chemical properties, and positional attributes. While effective in some cases, these approaches often face significant challenges, including the complexity of genomic data, the need for extensive preprocessing, and the difficulty of selecting highly informative features. As a result, feature extraction and selection steps have frequently introduced unnecessary complexity to machine learning (ML) models.

In recent years, advancements in deep learning (DL) have revolutionized splice junction prediction by enabling automatic feature extraction, thereby reducing the reliance on handcrafted features. DL architectures such as convolutional neural networks (CNNs), recurrent neural networks (RNNs), and hybrid models have demonstrated exceptional performance in capturing complex patterns within genomic sequences. These approaches bypass many of the limitations of traditional ML models, offering enhanced scalability and robustness across diverse datasets. However, despite these advancements, several challenges persist, such as scalability and generalizability across broader genomic contexts, the difficulty of handling long-range dependencies in genomic sequences, and minimizing false positives while maintaining high accuracy.

State-of-the-art models in splice junction prediction have utilized various architectures and datasets to address these challenges. CNN-based models like DeepSplice, SpliceRover, and Splice AI have made significant contributions. DeepSplice, applied to the GENCODE dataset, achieved accuracies of 90.7% for donor sites and 89.3% for acceptor sites. SpliceRover, leveraging CNNs with max pooling layers, achieved remarkable accuracies of 95.35% (donor) and 96.12% (acceptor) on the NN269 dataset, with outstanding area under the precision-recall curve (auPRC) scores. Splice AI, utilizing residual networks, processed sequences up to 5000 nucleotides and achieved 95% accuracy, demonstrating the utility of residual learning for complex genomic data.

Recurrent neural networks (RNNs) have also shown great promise in addressing some of these challenges, particularly in capturing long-range dependencies. Sarkar, et al. [13] explored recurrent architectures, including Vanilla RNNs, GRUs, and LSTMs, achieving perfect accuracy, sensitivity, specificity, and F1 scores on splice junction sequences from the GenBank dataset. This highlights the potential of RNN-based approaches in delivering exceptional predictive performance. Hybrid models like DeepDSSR, which combined CNN and BLSTM layers, also excelled in capturing both local and long-range dependencies, achieving sensitivity and specificity scores of 98.8% and 89.1%, respectively, on the HS3D dataset.

While these models have achieved impressive results, many were trained on limited datasets, which restricts their generalizability to broader genomic contexts.

Moreover, some CNN-based models struggle to fully capture the long-range dependencies present in genomic sequences. In this context, architectures like GRUs and LSTMs emerge as promising alternatives due to their ability to efficiently model long-range dependencies, making them well-suited for tasks requiring deeper sequence understanding. Future research should focus on leveraging the strengths of these architectures to develop scalable and generalizable models for splice junction prediction.

In conclusion, advancements in computational methods have significantly enhanced the accuracy of splice junction prediction, with state-of-the-art DL models like CNNs, RNNs, and hybrid architectures demonstrating robust performance. However, to address the remaining challenges of scalability, generalizability, and long-range dependency modeling, GRU- and LSTM-based architectures stand out as promising candidates. These models offer the potential to drive further innovation in splice junction prediction, enabling more accurate genome annotations and advancing the field of genomics.

3 Methodology

3.1 Dataset

The DNA Primate splice-junction gene sequences dataset is a collection of DNA sequences transformed for machine learning applications. Originally developed for the StatLog project, it is based on data from GenBank 64.1 and was prepared by Ross King [14]. The dataset comprises 3186 samples, each representing splice junctions-boundaries between exons (DNA regions retained after splicing) and introns (regions removed during splicing). The goal of this dataset is to classify sequences into three categories: EI (exon-intron boundary), IE (intron-exon boundary), and neither.

Each sequence is represented by 180 binary indicator variables, corresponding to 60 nucleotide positions in the DNA sequence. To transform the symbolic nucleotide data into a numerical format, the nucleotides (A, C, G, and T) were encoded using three binary indicators: Adenine (A): [1, 0, 0], Cytosine (C): [0, 1, 0], Guanine (G): [0, 0, 1], Thymine (T): [0, 0, 0].

This encoding replaces the original symbolic representation of nucleotides, creating a dataset suitable for machine learning algorithms. For example, the first nucleotide in a sequence is represented by features A0, A1, and A2, while the second nucleotide corresponds to A3, A4, and A5, continuing this pattern for all 60 positions in the sequence. Research indicates that attributes closest to the splice junction, specifically those within the A61–A120 range, often provide the most informative signals for classification. These attributes capture the transition regions, which are crucial for determining exon-intron boundaries.

3.2 Model

This study integrates Long Short-Term Memory (LSTM) and Gated Recurrent Unit (GRU) networks to develop a predictive model for identifying DNA splice junctions. By leveraging the strengths of both architectures, the model aims to effectively process the sequential nature of DNA sequences while capturing long-term dependencies.

To integrate LSTM and GRU architectures into a unified neural network for this study, the hybrid design leverages the distinct strengths of both models. The LSTM layers excel in capturing long-term dependencies, making them particularly effective for modeling the intricate relationships present in the 180-feature dataset for DNA splice-junction classification. On the other hand, GRU offer a computationally efficient alternative, excelling at capturing relevant sequential patterns while minimizing the risk of overfitting due to their simpler structure.

The rationale for combining LSTM and GRU layers is rooted in their complementary strengths. LSTM layers are adept at encoding long-term contextual information, which is crucial for understanding the complex patterns inherent in DNA sequences. Meanwhile, GRU layers enhance computational efficiency, reducing the training time and memory requirements while retaining the ability to learn essential sequential patterns.

This hybrid architecture can be designed by either stacking the layers sequentially (e.g., an LSTM layer followed by a GRU layer) or interleaving them to exploit their respective advantages in feature extraction. This arrangement allows the model to learn hierarchical feature representations effectively, enabling it to handle both long-range dependencies and localized sequence patterns. Additionally, the combined approach can provide a balance between performance and efficiency, making it highly suitable for datasets where computational constraints are a concern.

To understand the underlying mechanics of these networks, it is essential to grasp several foundational concepts that define their functionality and interaction:

Dense Layer: A dense layer computes the linear combination of input vectors with associated weights and biases. This is mathematically represented through matrix multiplications, where the dimensions of the input and output vectors must align for compatibility. Dense layers are crucial for fully connected networks, facilitating transformations between feature spaces.

Activation Function: This introduces non-linearity to the output of a layer, enabling the network to learn complex patterns and relationships in the data. Common activation functions include ReLU, sigmoid, and tanh, which determine how the weighted sum of inputs is transformed.

Point-wise Operation: These operations are performed element-wise across tensors. All operands involved in point-wise operations must have identical shapes, ensuring the output tensor maintains the same dimensions as the input. Examples include element-wise addition, multiplication, or activation functions applied to each element of a vector or matrix.

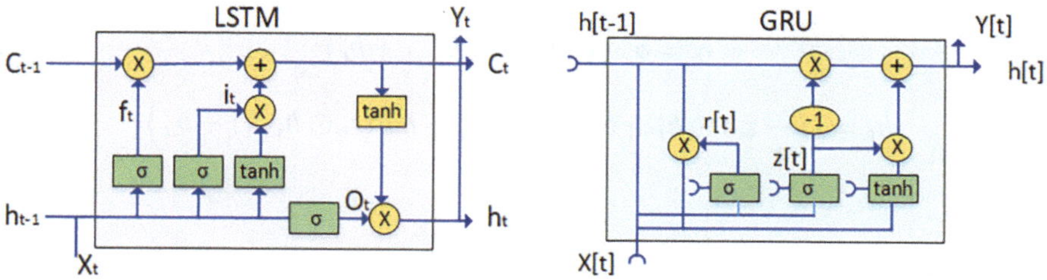

Fig. 2 Architecture of LSTM and GRU

Cell State: The cell state represents the output of an individual LSTM or GRU cell/unit. In GRUs, there is typically one output (hidden state), while in LSTMs, the cell produces both long-term memory (cell state) and short-term memory (hidden state). These are denoted as c[t] (cell state) and h[t] (hidden state), where t represents the time step.

Input Vector: The sequence data fed into an LSTM or GRU cell is referred to as the input vector, typically denoted as x[t], where t indicates the time step. This input contains the features or observations for each time frame.

Gate Block: LSTM and GRU cells include specialized gate blocks composed of dense layers combined with point-wise operations to manage memory. Gate blocks perform key functions such as controlling what information to remember (long-term memory) and what to update (short-term memory). These gates include the input, forget, and output gates in LSTMs and the reset and update gates in GRUs, enabling the cells to process sequential data dynamically.

Figure 2 illustrates the combined architecture of LSTM and GRU, highlighting various types of variables, each corresponding to a specific equation crucial for their operation. These variables represent the mathematical processes that underpin the networks' functionality, such as the gates, activations, and state updates that handle sequential data. The equations defining these components are detailed below (Eqs. 1–8), providing a comprehensive view of how information is processed, retained, or discarded within the architecture.

$$\textbf{LSTM}: \quad f_t = \sigma_g\left(W_f x_t + U_f h_{t-1} + b_f\right) \tag{1}$$

$$i_t = \sigma_g\left(W_i x_t + U_i h_{t-1} + b_i\right) \tag{2}$$

$$o_t = \sigma_o\left(W_o x_t + U_o h_{t-1} + b_o\right) \tag{3}$$

$$c_t = f_t \odot c_{t-1} + i_t \odot \tanh\left(W_c x_t + U_c h_{t-1} + b_c\right) \tag{4}$$

$$h_t = o_t \odot \tanh\left(c_t\right) \tag{5}$$

$$\textbf{GRU}: \quad z_t = \sigma_g\left(W_z x_t + U_z h_{t-1} + b_z\right) \tag{6}$$

$$r_t = \sigma_g\left(W_r x_t + U_r h_{t-1} + b_r\right) \tag{7}$$

$$h_t = (1 - z_t) \odot h_{t-1} \odot \varnothing_h\left(W_h x_t + U_h\left(r_t \odot h_{t-1}\right) + b_h\right) \tag{8}$$

4 Result

The performance evaluation of the proposed methodology is done on various parameters such as Accuracy, F1-Score, Kappa, Recall, Precision, and Coverage. Appropriate hyper-parameters of the algorithm help to achieve the maximum performance. It means it is an important aspect. This study considered Grid search method for hyper parameter tuning. Grid Search is a method used to systematically evaluate all combinations of hyper-parameter values within a predefined range. In this study, Grid Search is employed to tune the hyper-parameters of LSTM and GRU models for predicting DNA splice junctions. The process involves exhaustively testing every combination of hyper-parameters, such as the number of LSTM/GRU units, learning rate, batch size, and others, to determine which combination yields the best performance based on metrics like accuracy, F1-score, recall, precision, and coverage.

The value of hyper-parameter is mentioned in Table 1.

The study compared the performance of various models using different hyper-parameters (LSTM units, GRU units, learning rate, and batch size), as represented in part of Fig. 3. The comparison focused on key evaluation metrics such as accuracy, F1-score, Kappa, recall, precision, and coverage.

When comparing the performance of LSTM units, the results showed an increase in performance as the LSTM units increased from 32 (0.76) to 128 (0.95). However, the performance began to decline at 512 units (0.71). Similarly, the F1-score improved steadily with more units, peaking at 128 units (0.97), before dropping slightly at 512 units (0.89). The Kappa value followed a similar trend, increasing with more units, peaking at 128 units (0.95), and then decreasing at 512 units (0.85). The recall showed a slight drop as the units increased, with the highest value (0.99) at 128 units, but it decreased to 0.75 at 512 units. In contrast, precision remained relatively stable at 0.91–0.95, except for the 512-unit setting, where it dropped to 0.88. Coverage showed a consistent trend, peaking at 128 units (0.98) and then slightly decreasing at 512 units (0.91). For GRU units, accuracy remained strong across the units, starting at 0.91 for 32 units and maintaining stability up to

Table 1 Hyper-parameters value for fine-tune

Hyper-parameters	Values
LSTM and GRU units	32, 64, 128, 256, 512
Learning rate	0.001, 0.0005, 0.0001, 0.01, 0.05
Batch size	16, 32, 64, 128
Epochs	100

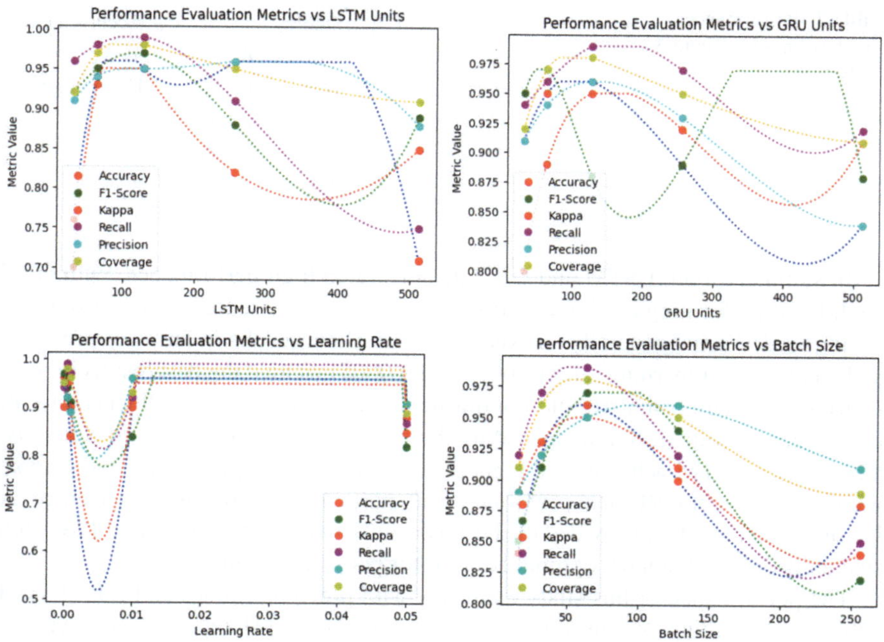

Fig. 3 Performance evaluation of various metrics with respect to hyper-parameter tuning

128 units (0.96), before decreasing at 512 units (0.84). The F1-score peaked at 0.97 for 64 units, but decreased at 128 units (0.88) and remained consistent at 0.88 for 512 units. Kappa values were high across the units, peaking at 0.95 at 128 units and then slightly dropping to 0.91 at 512 units. Recall remained consistent, with the highest value (0.99) at 128 units and staying strong at 0.92 for 512 units. Precision values stabilized, with the highest value (0.96) at 128 units, dropping to 0.84 at 512 units. Coverage increased with more units, peaking at 128 units (0.98), and then decreasing slightly at 512 units (0.91).

Regarding learning rates, higher accuracy was observed at lower learning rates (0.96 at 0.0001), but accuracy dropped at higher rates such as 0.05 (0.88). The F1-score peaked at 0.97 at 0.0001, with a decrease at higher rates. The highest Kappa value occurred at 0.0005 (0.95), and the lowest value was at 0.05 (0.85). The best recall performance was observed at 0.99 for 0.0005, with a slight drop to 0.87 at 0.05. Precision remained stable across most learning rates, peaking at 0.96 for 0.01. Coverage showed consistent performance, with the highest value at 0.98 for 0.0005 and dropping to 0.89 at 0.05. For batch size, accuracy improved from 16 (0.84) to 128 (0.96), before slightly declining at 512 (0.88). The F1-score was highest at 128 (0.97), followed by 64 (0.94), and then dropped to 0.82 at 256. The highest Kappa was at 128 (0.95), but it slightly decreased at 256 (0.84). Recall peaked at 0.99 for both 128 and 64, and dropped to 0.85 at 256. Precision remained high across different batch sizes, peaking at 0.96 for 128. Coverage remained consistent at 0.98 for both 64 and 128, dropping slightly at 256 (0.89).

Table 2 Performance comparison with previous study

Studies	Accuracy (%)
Singh, et al. [3]	95.5
Wang, et al. [5]	Up to 90.25
Zabardast, et al. [1]	96
Nagassou, et al. [2]	96
Proposed study	96.48

In summary, both LSTM and GRU demonstrated high performance, with GRU slightly outperforming LSTM in terms of accuracy and F1-score across various unit settings. GRU units generally showed more consistent results in terms of precision and recall. Lower learning rates (0.0001 and 0.0005) yielded the best results in most metrics. Although batch size did not drastically affect accuracy, it influenced recall and precision, with 128 units providing the best overall performance. Specifically, 128 units of LSTM achieved the highest results in accuracy, F1-score, and Kappa, while 128 units of GRU also performed well, particularly in recall and coverage. A learning rate of 0.0001 provided the best balance across most metrics, and 128 batch size units offered the most optimal balance for performance metrics.

The performance of this study compares with previous studies and it is showed Table 2. According to this comparison, Singh, et al. [3] reported an accuracy of 95.5%, which is slightly below the proposed study's accuracy of 96.48%. This indicates a competitive performance but shows the proposed study achieved marginally better results. Wang, et al. [5] demonstrated accuracy levels of up to 90.25%, which is considerably lower than all other studies in the table. This suggests that the methods or data used in their research may not have been as effective in achieving high accuracy compared to the others. Zabardast et al. [1] and Nagassou et al. [2] both achieved an accuracy of 96%, placing them slightly behind the proposed study's 96.48%. While the differences are minimal, the proposed study still outperformed these by a small margin, highlighting slight methodological or data advantages. The Proposed Study achieved the highest accuracy among all studies, at 96.48%, showcasing the effectiveness of its approach in comparison to the listed research. In the conclusion, the proposed study leads in accuracy by a small margin over other high-performing studies like Zabardast et al. [1] and Nagassou et al. [2]. However, the proposed study significantly outperforms Wang et al. [5] indicating a robust improvement over methodologies that might have been less optimized. This analysis demonstrates the proposed study's effectiveness in improving accuracy, even if the improvement over some studies is marginal.

5 Conclusion

The genome, which represents the entire set of an organism's DNA sequences, carries the essential information required for protein synthesis. Proteins, which are crucial for life, perform a wide range of functions within organisms, such as catalyzing metabolic reactions, enabling cellular communication, and providing structural integrity. Gene expression involves a splicing process that removes introns, leaving only exons to be translated into proteins. Accurate identification of splice sites-specific nucleotide sequences marking the boundaries between exons and introns-is vital for analyzing gene structures and enhancing genome annotation. DNA splice junction prediction is the task of pinpointing the exon-intron boundaries during splicing. However, this prediction remains a challenging task, primarily due to the prevalence of GT and AG dinucleotides in regions outside splice sites, which often result in false positives. Existing prediction tools still exhibit considerable error rates, affecting subsequent genomic analyses.

This study explored the use of Long Short-Term Memory (LSTM) and Gated Recurrent Units (GRU) for classifying DNA sequences, as these models are particularly effective at capturing long-range dependencies and the contextual relationships between distant nucleotides. Leveraging the ability of LSTM and GRU to process sequential data made this approach well-suited for uncovering the complex patterns found in splice junctions. The study utilized the Stat-Log DNA dataset, which included 180 features to predict junction types (IE, EI, or N) from a given DNA sequence. The performance of various models was fine-tuned by adjusting hyperparameters, including LSTM and GRU units, learning rate, and batch size. The comparison centered on evaluation metrics such as accuracy, F1-score, Kappa, recall, precision, and coverage. Model performance was assessed with the following results: Accuracy (96.48%), F1-Score (97.52%), Kappa (94.95%), Recall (99.38%), Precision (96.12%), and Coverage (97.48%). These results demonstrate the efficacy of LSTM and GRU in predicting splice junction types with high accuracy and reliability, providing valuable insights for future research in gene splicing and related bioinformatics areas. Future applications of the project could involve larger datasets with more specific features.

References

1. A. Zabardast, E. G. Tamer, Y. A. Son, and A. Yılmaz, "An automated framework for evaluation of deep learning models for splice site predictions," *Scientific Reports,* vol. 13, no. 1, p. 10221, 2023.
2. M. Nagassou, R. W. Mwangi, and E. Nyarige, "A Hybrid Ensemble Learning Approach Utilizing Light Gradient Boosting Machine and Category Boosting Model for Lifestyle-Based Prediction of Type-II Diabetes Mellitus," *Journal of Data Analysis and Information Processing,* vol. 11, no. 4, pp. 480-511, 2023.
3. N. Singh, R. Nath, and D. B. Singh, "Splice-site identification for exon prediction using bidirectional LSTM-RNN approach," *Biochemistry and Biophysics Reports,* vol. 30, p. 101285, 2022.
4. J. Zuallaert, F. Godin, M. Kim, A. Soete, Y. Saeys, and W. De Neve, "SpliceRover: interpretable convolutional neural networks for improved splice site prediction," *Bioinformatics,* vol. 34, no. 24, pp. 4180-4188, 2018.
5. R. Wang, Z. Wang, J. Wang, and S. Li, "SpliceFinder: ab initio prediction of splice sites using convolutional neural network," *BMC bioinformatics,* vol. 20, pp. 1-13, 2019.

6. Y. Zhang, X. Liu, J. MacLeod, and J. Liu, "Discerning novel splice junctions derived from RNA-seq alignment: a deep learning approach," *BMC genomics,* vol. 19, pp. 1-13, 2018.
7. S. Amilpur and R. Bhukya, "Edeepssp: explainable deep neural networks for exact splice sites prediction," *Journal of Bioinformatics and Computational Biology,* vol. 18, no. 04, p. 2050024, 2020.
8. S. Albaradei *et al.*, "Splice2Deep: An ensemble of deep convolutional neural networks for improved splice site prediction in genomic DNA," *Gene,* vol. 763, p. 100035, 2020.
9. C. M. Dasari and R. Bhukya, "InterSSPP: Investigating patterns through interpretable deep neural networks for accurate splice signal prediction," *Chemometrics and Intelligent Laboratory Systems,* vol. 206, p. 104144, 2020.
10. M. Van Moerbeke, A. Kasim, and Z. Shkedy, "The usage of exon-exon splice junctions for the detection of alternative splicing using the REIDS model," *Scientific Reports,* vol. 8, no. 1, p. 8331, 2018.
11. J. Zhao, Q. Li, Y. Li, X. He, Q. Zheng, and S. Huang, "ASJA: a program for assembling splice junctions analysis," *Computational and Structural Biotechnology Journal,* vol. 17, pp. 1143-1150, 2019.
12. D. Lee, J. Zhang, J. Liu, and M. Gerstein, "Epigenome-based splicing prediction using a recurrent neural network," *PLoS computational biology,* vol. 16, no. 6, p. e1008006, 2020.
13. R. Sarkar, C. C. Chatterjee, S. Das, and D. Mondal, "Splice junction prediction in DNA sequence using multilayered RNN model," in *Advances in Decision Sciences, Image Processing, Security and Computer Vision: International Conference on Emerging Trends in Engineering (ICETE), Vol. 1*, 2020: Springer, pp. 39-47.
14. G. Towell and J. Shavlik, "Interpretation of artificial neural networks: Mapping knowledge-based neural networks into rules," *Advances in neural information processing systems,* vol. 4, 1991.

Machine Learning Techniques Based Predictive Modeling for Wind Speed Forecasting

Nitin Kumar, Priyanka Sharma, Sakshi Sharma, Ankur Kumar Gupta, Nirmal Kumar Pandey, Vatsala Anand, and Rupendra Kumar Pachauri

Abstract This research analyzes the efficacy of several Machine Learning (ML) methods in forecasting wind speed. The research specifically examines and contrasts the efficacy of NN (Neural Network), SVM (Support Vector Machine), GPR (Gaussian Process Regression), and DT (Decision Tree) models. The research utilizes historical wind speed data from the Dehradun area of India (30°19′27.95″N, 78°2′2.11″E). The investigation involves comparing the accuracy of these approaches and evaluating their computational efficiency and scalability. The results highlight the superior performance of the GPR technique, which demonstrates lower errors in wind speed prediction. Specifically, GPR achieved a RMSE (root mean square error) of 0.174776173, a MSE (mean square error) of

N. Kumar
Department of CSE (AI&ML), Meerut Institute of Engineering and Technology, Meerut, India

P. Sharma
Department of Mathematics, Shivalik College of Engineering, Dehradun, Uttarakhand, India
e-mail: priyanka.sharma@sce.ac.in

S. Sharma
Department of Electronics and Communication Engineering, Dev Bhoomi Uttarakhand University, Dehradun, India
e-mail: sakshi.sharma@dbu.ac.in

A. K. Gupta
Research and Development Department, IIMT University, Meerut, Uttar Pradesh, India

N. K. Pandey
Department of Electronics and Communication Engineering, Shivalik College of Engineering, Dehradun, India

V. Anand
Department of Computer Science and Engineering, Akal University, Talwandi Sabo, Bathinda, Punjab, India
e-mail: vatsala_cs@auts.ac.in

R. K. Pachauri (✉)
Electrical Cluster, School of Advanced Engineering, UPES, Dehradun, India
e-mail: rpachauri@ddn.upes.ac.in

K. Deep et al. (eds.), *Artificial Intelligence, Machine Learning and Intelligent Systems*, Springer Proceedings in Mathematics & Statistics 520,
https://doi.org/10.1007/978-981-95-3841-6_27

0.030546711, a MAE (mean absolute error) of 0.106447822, and a regression value (R^2) of 0.964572701. The findings of this study have significant implications for environmental monitoring and renewable energy (RE) planning, as they provide insights into the most effective ML algorithms for accurate and efficient wind speed predictions.

Keywords Wind speed · Forecasting techniques · Machine learning · Regression model · Mean absolute error

1 Introduction

Carbon emissions have always been a part of the energy generation process. Renewable energy (RE) is becoming more popular as a solution to the pressing environmental problem of carbon emissions [1]. While solar energy forecasting has garnered significant attention, wind speed forecasting is equally critical in harnessing wind energy's potential. The variability and unpredictability of wind present unique challenges that demand advanced forecasting methods. Precisely forecasting wind speeds is crucial for enhancing wind energy production and maintaining the stability of electricity systems. Machine learning models, including LR and SVM, known for their interpretability, to more complex models like random forests and neural networks, have been employed to tackle this forecasting challenge. These models must account for a variety of factors, including atmospheric conditions, geographical features, and historical wind data, to provide reliable predictions.

Integration of RE sources, especially those with variable outputs like wind and solar, into current and future energy infrastructures is one of the biggest issues confronting the world's energy supply in the near future [2]. Electrical operators must maintain a constant balance between power output and consumption. However, traditional, manageable energy generation systems often struggle to sustain this balance, particularly in small or isolated power networks, such as those on islands. The ability of the electrical system to adapt to both anticipated and unforeseen shifts in production and consumption, as well as disruptions, is crucial to its reliability, ensuring that consumers continue to receive high-quality service [3]. The process flow diagram of the present study to explore the methodology is depicted in Fig. 1 as.

In [4], The authors have presented a multi-step short-term wind speed forecasting method. The results indicate that our one-step prediction model attained an average MAE of 0.167 and a RMSE of 0.221. The suggested model attained a MAE of 0.23 m/s and a RMSE of 0.31 m/s on ultra-short-term wind speed data, surpassing existing models in accuracy by 15%. Additionally, the Taylor diagram analysis demonstrated a correlation coefficient of 0.94, indicating a high agreement between the forecasted and actual wind speeds [5]. The FWA-optimized LSTM model achieved a 10% lower MAE and a 12% lower root RMSE compared to traditional regression-based wind speed prediction models, demonstrating superior

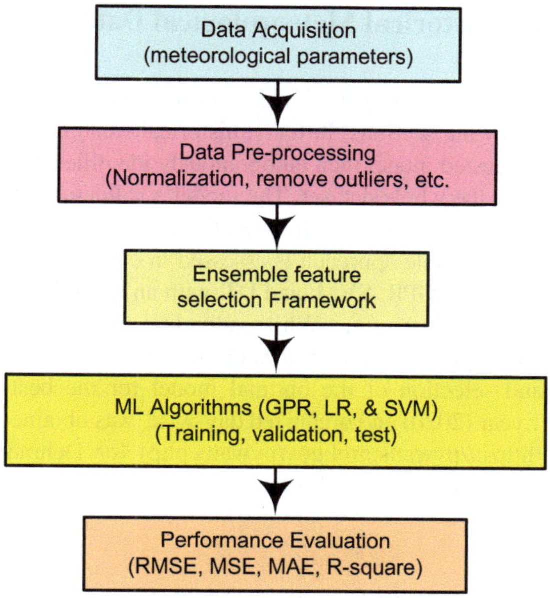

Fig. 1 Process flow diagram of methodology

prediction accuracy [6]. The proposed ensemble prediction model achieved mean absolute percent errors (MAPE) of 7.97% and 9.99% on two NREL datasets, outperforming comparison methods by reducing errors by up to 15%. Pearson's test further confirmed a strong correlation of 0.93 between the actual and predicted wind speeds, validating the model's superior predictive accuracy [7]. The BFS-Bi-LSTM model surpassed previous models, with an RMSE of 0.784, MAE of 0.530, MSE of 0.615, and a R^2 of 0.8766, indicating superior accuracy in wind speed forecasting. This approach effectively addressed issues like uncertainty, dimensionality, and non-linearity, demonstrating a significant improvement over traditional model [8]. The BLSTM-TRA model significantly outperformed the ECMWF forecast, reducing RMSE by 58.9% and MAE by 63.2% on average. Additionally, it demonstrated a 20.2% overall performance gain in RMSE compared to the traditional LSTM model, highlighting its superior ability to capture terrain-sensitive and short-term wind variations [9]. The study introduces a data fusion algorithm combining statistical and neural network approaches to forecast long-term wind speed trends, achieving promising results with a very small mean absolute error (MAE) using data from meteorological stations in Malaysia [10]. In [11], the study reviews and evaluates current wind forecasting techniques, including physical, statistical, AI, and hybrid methods, highlighting advances in prediction accuracy and addressing key challenges. The paper also explores evolving trends and future research directions to enhance wind forecasting over broader time horizons.

2 Historical Meteorological Data

The methodology used in this study for predicting daily wind speed is detailed in the following sections. Initially, historical weather data undergoes pre-processing. The suggested model then autonomously identifies the most appropriate input attributes from the whole dataset. The model's validation relies on weather information from Dehradun, including diverse climatic conditions. Additionally, the performance of the ensemble approach is assessed in comparison to conventional learning models, including GPR, SVM, and DT, with an emphasis on prediction accuracy, resilience, and processing time. While individual algorithms have proven capable of predicting wind speed, the differences in their effectiveness allow for comparison, integration, and selection of the optimal model for the best forecast. The dataset, covering 1 year (2020) and organized day-wise, was obtained from the official NREL website (https://pvwatts.nrel.gov/pvwatts.php) for Dehradun, India. Figure 2 graphically represents the annual solar irradiation data samples.

3 Machine Learning Models

3.1 GPR (Gaussian Process Regression)

A non-parametric kernel model, Gaussian Process Regression (GPR) applies probability theory. Over the course of the past few years, this particular method of machine learning has become increasingly prevalent in the academic literature. It is possible to compute the accuracy interval for each anticipated point using this method, in addition to using it for making predictions. You can see how off the prediction is by looking at the confidence interval. A Gaussian process may be comprehended as an extension of the important probability distribution. The likelihood of an input vector is ascertained by the mean and variance via the

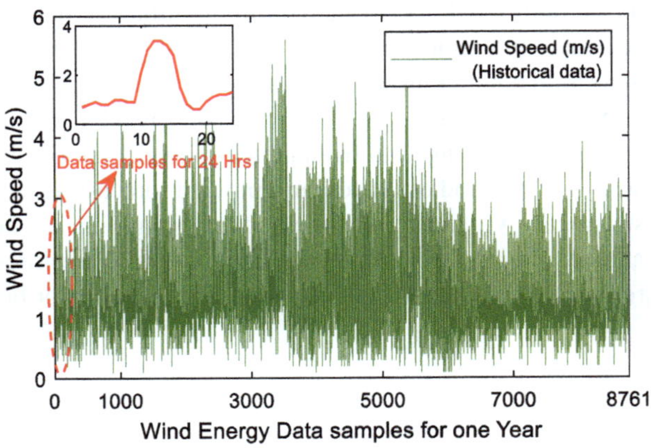

Fig. 2 Historical annual data: Wind Speed (m/s)

Gaussian distribution. We assess the probability of a certain input time series vector at each temporal interval. Consequently, the GPR model computes a mean and covariance vector rather than treating variance and mean as distinct values. Keep in mind that neither the GPR nor the SVM are dynamic algorithms, therefore they cannot generate predictions that change over time.

3.2 NN (Neural Network)

Neural networks have emerged as a crucial instrument in predicting wind speed, providing superior accuracy and efficiency compared to conventional techniques. These models use intricate structures to describe the spatiotemporal variation of wind patterns, essential for improving wind energy systems. Diverse neural network models have been developed, each with distinct characteristics and benefits, to tackle the difficulties of wind speed forecasting.

3.3 DT (Decision Tree)

Decision tree algorithms are effective machine learning techniques used for regression and classification tasks. They divide a dataset into subsets using input features, creating a hierarchical structure with nodes representing decision points and leaves indicating outcomes. These splits optimize class separation using measures like Gini impurity or Information Gain. Despite their simplicity and ability to handle non-linear relationships, decision trees can overfit data and be sensitive to minor changes. Pruning is commonly used to reduce overfitting. Common machine learning algorithms include ID3, C4.5, and CART. Figure 3 shows the general structure of DT algorithms.

3.4 SVM (Support Vector Machine)

SVM are a recognized classification method that effectively categorizes data, irrespective of dimensionality, and requires just a minimal training sample. Nevertheless, they may be computationally intensive if not used judiciously. This methodology may also be used for regression objectives with few adjustments. Support Vector Machines and ε-SVR are appropriate for static scenarios, limiting the use of the resultant black box models to simulations purposes alone. To facilitate predictive modelling with anticipatory capabilities, more modifications are necessary to develop a Nonlinear Autoregressive with Exogenous Inputs (NARX) framework for SVR that provides a dynamic impact on the model. Figure 4 illustrates the overall architecture of SVM.

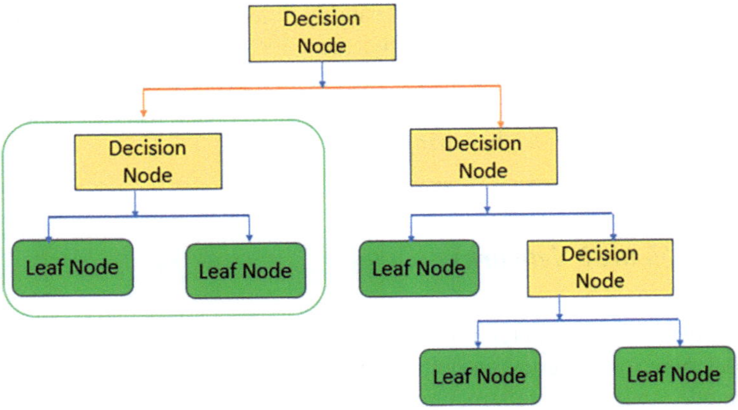

Fig. 3 General structure of DT

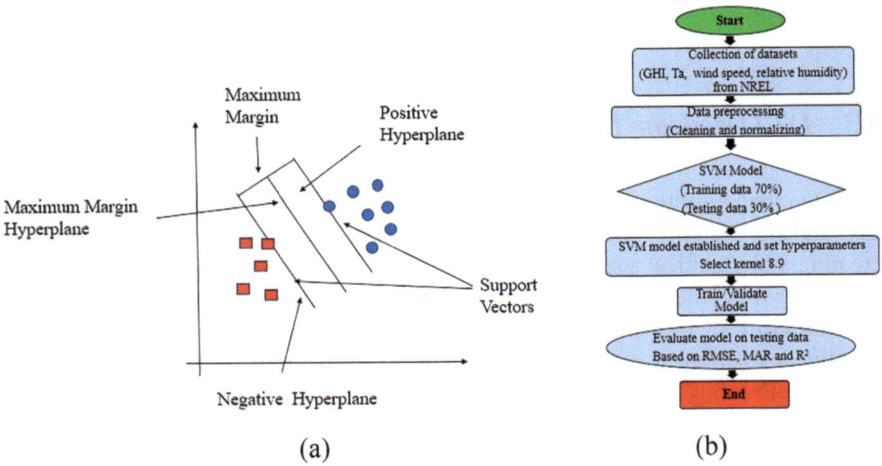

Fig. 4 (**a**) Mathematical formulation boundaries (**b**) SVM flowchart

4 Results and Discussion

4.1 Regression Plot

A regression plot is a visual tool used in statistical analysis to show the correlation between two continuous variables, with one being the dependent variable and the other being the independent variable. It provides a comprehensive understanding of trends, patterns, and correlation levels. In a linear regression plot, the x-axis represents the independent variable, the y-axis represents the dependent variable, and the regression line shows the best linear relationship between the two variables. Figure 5a–c shows the regression plot for SVM/DT/GPR respectively. It has seen

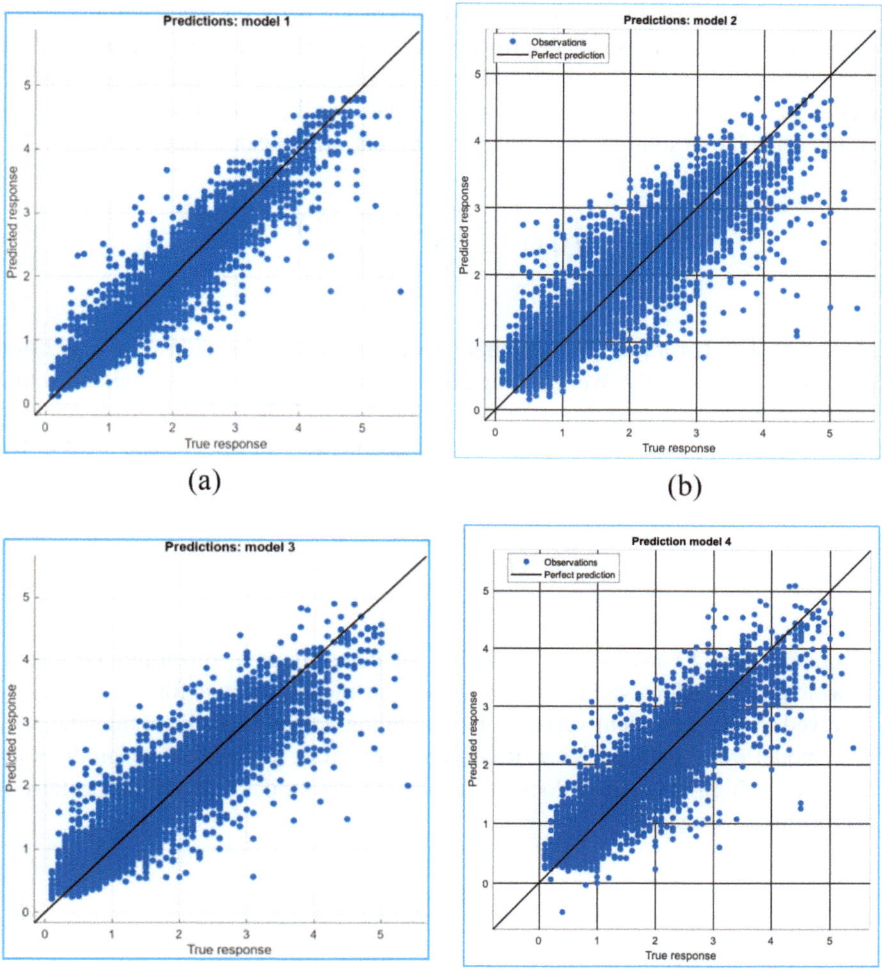

(a) (b) (c) (d)

Fig. 5 Predicted models (**a**) GPR (**b**) NN (**c**) DT (**d**) SVM

that all data points in case of GPR algorithm are near to regression line as compare to the SVM and DT algorithm.

4.2 Error Statistical Assessment

Models such as GPR, DT, SVM were assessed using RMSE, MSE, MAE, and R-square. Equations (1)–(3) provide for a comprehensive evaluation of errors. RMSE is a statistical metric that measures the discrepancy between predicted and actual values in a regression model. It provides an error statistic in the same units as the target variable, making it easy to interpret. A reduced RMSE indicates improved model performance, suggesting more accurate predictions. However, RMSE is highly sensitive to significant mistakes or outliers, which can impact the metric.

$$RMSE = \sqrt{\sum_{j=1}^{M} \frac{\left(Y_j - P_j\right)^2}{M}} \tag{1}$$

Mean Absolute Error (MAE) is another widely used evaluation metric for regression models. It calculates the average magnitude of errors between predicted and actual values, regardless of their direction. MAE does not include squaring errors, making it less affected by outliers and offering a direct measure of the average error in the model's predictions. A smaller MAE signifies a model with more accuracy.

$$MAE = \sum_{j=1}^{M} \frac{\left|Y_j - P_j\right|}{M} \tag{2}$$

R-squared, also known as the coefficient of determination, quantifies the extent to which independent variables in a regression model account for the variability observed in the dependent variable. It is commonly used to evaluate the adequacy of a model's fit, with a higher R-squared indicating more accurate predictions. However, a high R-squared does not guarantee the model's quality, especially if overfitting is a concern.

$$R^2 = 1 - \frac{\sum_{j=1}^{M} \left(Y_j - P_j\right)^2}{\sum_{j=1}^{M} \left(P_j - P_{AVE}\right)^2} \tag{3}$$

Where, M represents the whole set of measurements. The predicted value is denoted by Pj, the desired value by Y_j and the average predicted value by P_{AVE}.

Table 1 shows that the GPR model for training performance in terms of mean absolute RSME (0.797098637), MSE (0.635366237), MAE (0.430421925) and *R* (0.999992228) best as compared to DT and SVM techniques.

The mean absolute error (MAE) is a metric for comparing the precision of predictions with that of actual measurements. The root-mean-squared error (RMSE) measures the dispersion of the GPR model. Establishing a connection between observed and anticipated values is possible with the help of the linear correlation coefficient (R) value. There is a statistical representation in Figs. 6 and 7. GPR perform better as compare to the SVM and DT with small error and high R-square value.

Table 1 Error assessment of ML methods for training

ML models	MAE	MSE	RMSE	R-square
GPR	0.106447822	0.030546711	0.174776173	0.964572701
NN	0.134647575	0.045423523	0.194254464	0.945732255
DT	0.161191201	0.067796786	0.260378159	0.92137101
SVM	0.37207569	0.292828685	0.541136476	0.660384727

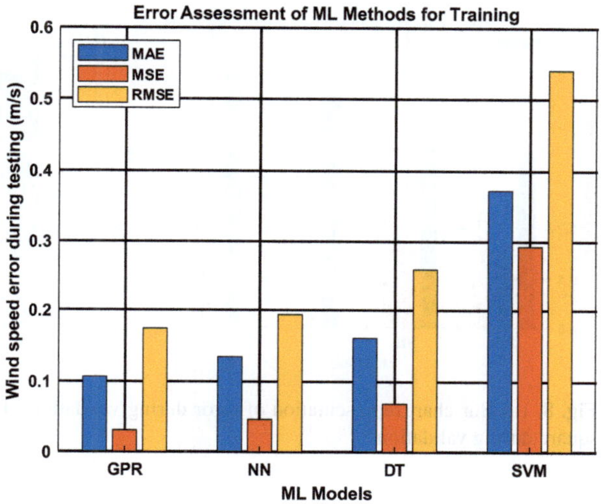

Fig. 6 Bar chart representation of error during training

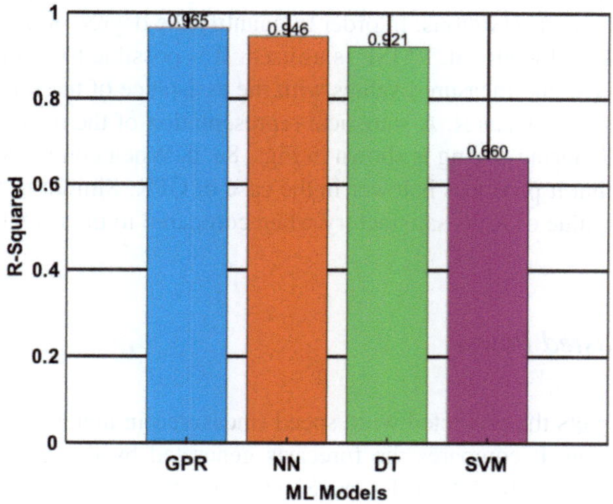

Fig. 7 Bar chart representation of R-square during training

Table 2 shows the Testing GPR/NN/DT/SVM model for testing in terms of the means absolute RSME/MAE/SVM. It has been showing that GPR gives the best result in terms of performance metrics.

Table 2 Error assessment of ML methods for validation

ML models	RMSE	MSE	MAE	R-square
GPR	0.3567295	0.12725593	0.23184343	0.852644493
NN	0.40814366	0.166581253	0.278817953	0.807107899
DT	0.4612368	0.21273943	0.29910107	0.753659213
SVM	0.5375009	0.28890730	0.37114016	0.665460937

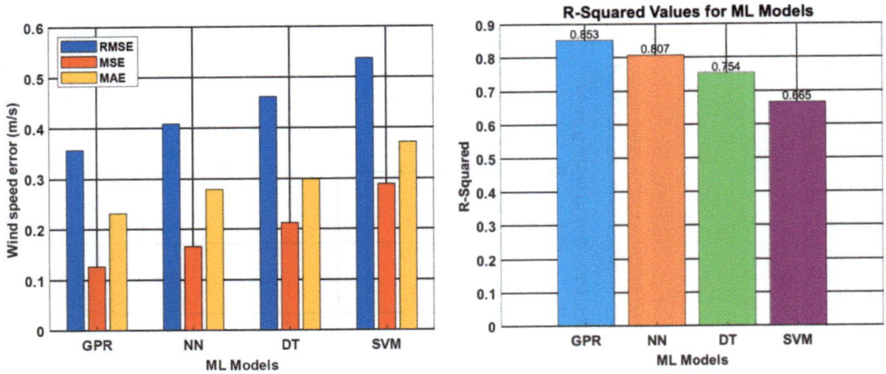

Fig. 8 (**a**) Bar chart representation of error during validation. (**b**) Bar chart representation of R-square during validation

MAE is a metric that is used to evaluate the degree to which predicted values are close to actual observations. In order to quantify the degree of dispersion that is produced by the GPR model, RMSE is utilised. It is possible to determine the link between expected and measured values with the assistance of the linear correlation coefficient, also known as R. A statistical representation of the mistake assessment that takes place during testing is shown in Figs. 8a, b. When compared to SVM and DT, the error that it produces is lower in the case of GPR. Similar to the DT/SVM, the calculated value of R^2 is satisfactory when compared to the actual value.

4.3 Forecasted Power

This graph depicts the projected wind speed (measured in meters per second) over a 24-h timeframe. It compares the forecasts generated by three distinct machine learning models—GPR, NN, DT, and SVM with the real recorded wind speed ("Existed Wind Speed"). The x-axis represents each hour of the day, while the y-axis represents wind speed. The blue bars depict the observed wind speed, while the orange, yellow, and purple bars illustrate the forecasts generated by the GPR, NN, DT, and SVM models, respectively. The graph illustrates that the GPR model (orange) closely tracks the actual wind speed, especially during peak wind speeds between 12–16 h. In contrast, the DT (yellow) and SVM (purple) models

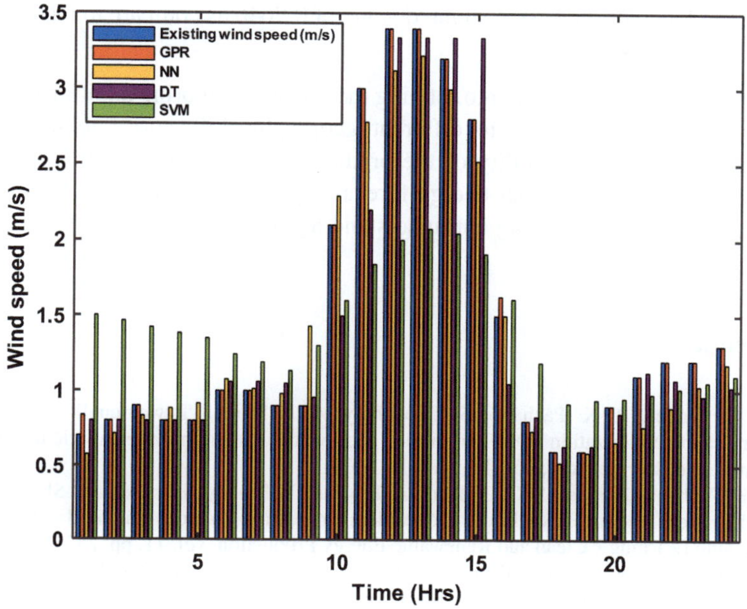

Fig. 9 Forecasted wind speed (m/s) for 24 h window

exhibit different levels of deviation, with the SVM model showing the most notable differences from the actual wind speed at specific time intervals. Figure 9 shows the forecasted wind speed for GPR/NN/SVM/DT ML algorithms.

5 Conclusion

The study convincingly showcases the efficacy of GPR as a superior machine learning model for wind speed prediction, in comparison to SVM/DT and NN models. By conducting a thorough examination of historical wind speed data from the Dehradun region, the GPR model continuously shown superior performance compared to other models in terms of accuracy metrics such as RMSE, MSE, MAE, and R-squared. The results emphasize the strong and dependable nature of GPR in collecting intricate wind patterns, rendering it a very effective instrument for monitoring the environment and planning renewable energy projects. The findings of this study have important ramifications for enhancing the accuracy of wind energy predictions, which are essential for maximizing the allocation of resources in renewable energy initiatives. Following is the key finding of this study.

- The GPR method demonstrated exceptional precision, achieving the lowest RMSE of 0.1748,
- GPR give the minimum MSE/MAE of 0.0305 and 0.1064 respectively.

- The model gives the best R-squared value of 0.9646 as compare to NN/SVM/DT ML algorithms.

Further research could delve into refining the models, exploring additional features, and extending the analysis to different geographical locations to enhance the robustness and generalizability of the predictions. In conclusion, this study adds to our understanding of solar energy forecasting and highlights the importance of ML approaches for furthering our understanding and use of RE sources.

References

1. N. K. Pandey, R. K. Pachauri, S. Choudhary, A. F. Minai, "Power quality improvement using rabbit optimization FOPID controlled photovoltaic-battery powered hybrid power filter", Renewable Energy Focus, vol. 47, pp. 1-14, 2023.
2. R. K. Pachauri, J. Yadav, S. O. G. Torto, A. F. Minai, V. Pandey, Shashikant, P. Sharma, "Solar Photovoltaic System-Based Power Generation: Imperative Role of Artificial Intelligence and Machine Learning", Clean and Renewable Energy Production, vol. 11, pp. 267-286, 2024.
3. M. K Pardhe, R. K. Pachauri, P. Sharma, "Wind and Solar PV System-Based Power Generation: Imperative Role of Hybrid Renewable Energy Technology", Clean and Renewable Energy Production, vol. 5, pp. 123-141, 2024.
4. H. Wu, K. Meng, D. Fan, Z. Zhang, Q. Liu, "Multistep short-term wind speed forecasting using transformer," Energy, vol. 261, pp. 1-12, 2021.
5. Z. Tian, H. Li, F. Li, "A combination forecasting model of wind speed based on decomposition", Energy Reports, vol 7, pp. 1217-1233, 2021.
6. B. Shao, D. Song, G. Bian, Y. Zhao, "Wind Speed Forecast Based on the LSTM Neural Network Optimized by the Firework Algorithm", Advances in Materials Science and Engineering, vol. 2021, pp. 1-13, 2021.
7. J. Quan, L. Shang, "An Ensemble Model of Wind Speed Forecasting Based on Variational Mode Decomposition and Bare-Bones Fireworks Algorithm," vol. 2021, pp. 2021.
8. S. S. Subbiah, S. K. Paramasivan, K. Arockiasamy, S Senthivel, M. Thangavel, S, "Deep Learning for Wind Speed Forecasting Using Bi-LSTM with Selected Features", Intelligent Automation & Soft Computing, vol. 35(3), pp. 3829-3844, 2023.
9. Z. Liu, F. Ding, F. Wan, "Wind Speed Forecasting Method Based on Deep Learning Strategy Using Long Short Term Memory Neural Network and Transformer Model", IEEE Conference on High Performance Computing & Communications on 20-22 Dec 2021 at Hainan, China, pp. 2288-2293.
10. H. B. Azad, S. Mekhilif, V. G. Ganapathy, "Long-Term Wind Speed Forecasting and General Pattern Recognition Using Neural Networks", IEEE Transactions on Sustainable Energy, vol. 5(2), pp. 546-553, 2014.
11. M. Santhosh, C. Venkaiah, D. M. V. Kumar, "Current advances and approaches in wind speed and wind power forecasting for improved renewable energy integration: A review", Engineering Reports, vol. 2(6), pp. 1-20, 2020.

Unsupervised Fabric Defect Detection via Autoencoder Reconstruction and One-Class SVM Analysis

Ramakrishna Miryala and Kamlesh Dutta

Abstract Fabric defect detection is a critical quality control process in textile manufacturing, but remains challenging due to the complex texture and variability of fabric patterns. Traditional supervised deep learning methods require extensive annotated data, which is often impractical in industrial settings. To address this, we propose a novel unsupervised approach leveraging autoencoder-based architecture to detect fabric defects. Our method reconstructs input images and employs a one-class SVM model to analyze reconstruction loss, distinguishing defective from nondefective regions. The experiments were carried out on a binary text defect classification dataset, using a holdout validation strategy of 80–20. Our approach outperformed state-of-the-art models, including AlexNet, VGG-16, VGG-19, MobileNetV1, and ResNet-50, with an average improvement of 3.57% in accuracy and 2.98% in F1 score. Compared to ResNet-50, our method achieved 7.2% higher recall and 4.03% lower precision, reflecting its robustness in identifying rare defect patterns. These results demonstrate the effectiveness of unsupervised learning for fabric defect detection, offering a scalable and efficient alternative to traditional supervised methods. This work highlights the potential of reconstruction-based anomaly detection in industrial applications, providing a foundation for further exploration in defect detection for complex and heterogeneous surfaces.

Keywords Fabric defect detection · Autoencoders · Deep learning · One class SVM · Anomaly detection

R. Miryala (✉) · K. Dutta
Department of Computer Science and Engineering, National Institute of Technology, Hamirpur, India
e-mail: 24rcs003@nith.ac.in; kd@nith.ac.in

K. Deep et al. (eds.), *Artificial Intelligence, Machine Learning and Intelligent Systems*, Springer Proceedings in Mathematics & Statistics 520,
https://doi.org/10.1007/978-981-95-3841-6_28

1 Introduction

Fabric defect detection remains a significant challenge in the textile industry due to the complexity of fabric patterns and textures. Accurate detection is essential to maintain product quality and reduce wastage, but traditional inspection methods often fall short [3]. Manual inspection, though widely practiced, is labor-intensive, time-consuming, and prone to human error. Automated solutions, particularly those based on supervised learning, have shown promise but are hindered by their reliance on large, annotated datasets [10]. In industrial settings, creating such datasets is impractical due to the diversity of fabric types, patterns, and the rarity of certain defects, which further complicates the task.

The scope of this problem extends beyond defect detection in textiles to other domains requiring quality control for complex surfaces, such as metal sheets, glass, and ceramics [4]. Therefore, a scalable, robust, and efficient approach that minimizes the dependency on labeled data is highly desirable. An effective solution must generalize well across diverse textures and patterns, identify subtle anomalies, and operate in real-time for industrial applicability.

To address these challenges, we propose a novel unsupervised learning-based method for fabric defect detection. Our approach leverages an autoencoder to learn the inherent patterns of defect-free fabrics. During testing, the autoencoder reconstructs input images, and significant deviations between the input and its reconstruction—indicative of defects—are captured as reconstruction losses. A one-class Support Vector Machine (SVM) [12] is then employed to classify regions based on these reconstruction losses, distinguishing defective from non-defective areas. This unsupervised approach eliminates the need for labeled data and enhances scalability.

By focusing on reconstruction-based anomaly detection, our method is particularly effective in identifying rare and previously unseen defect patterns [15]. Experimental evaluations show that our approach outperforms state-of-the-art supervised models, including AlexNet, VGG-16, VGG-19, MobileNetV1, and ResNet-50, in terms of accuracy and recall. These results underscore the potential of unsupervised learning for fabric defect detection and pave the way for its application in industrial environments with heterogeneous textures.

2 Literature Review

This section analyzes unsupervised fabric defect detection using Autoencoder Reconstruction and One-Class SVM. Table 1 summarizes key findings, highlighting the reconstructive capabilities of autoencoders to detect anomalies and the role of One-Class SVM in isolating defects for robust detection.

Shimizu et al. [11] suggested a new defect detection method for wrapped film products employing CNNs, SVMs, and VAEs. They trained a One-Class SVM

Table 1 Overview of defect detection techniques and their performance in industrial applications

Author	Overview	Techniques	Metrics	Key findings	Disadvantages
Shimizu et al. [11]	Proposes a VAE-based defect detection model for wrapped films.	CNN, SVM, VAE, One-Class SVM	90% Accuracy	Outperforms traditional CNN models for defect detection in films.	Limited to defect-free samples.
Kumar et al. [5]	ZBDNN model for composite curing defect detection.	ZBDNN, Dimensionality Reduction	99.71% Accuracy	ZBDNN surpasses autoencoder and OC-SVM in accuracy.	Requires custom autoclave for data.
Toufigh et al. [13]	Unsupervised damage detection in concrete using ultrasonic signals.	Auto-Encoder, Isolation Forest	95% (damaged), 93% (intact)	High accuracy for detecting concrete damage with minimal data.	Relies on intact specimens for training.
Chen et al. [2]	Combines autoencoder and SVDD for fabric defect detection.	Separable Convolution Autoencoder, SVDD	Accuracy, AUC	Efficient fabric defect detection with reduced computation.	Needs non-defective samples.
Liu et al. [6] (Textile)	Unsupervised textile defect detection using distillation.	Knowledge Distillation, Localization	High Accuracy	Strong performance on open and industrial datasets.	Needs large datasets for training.
Lu et al. [8]	Video analysis for lace defect detection.	Video-based Learning, Pixel Reconstruct.	Not specified	First video-based fabric defect detection, effective with lace.	Needs defect-free videos for training.
Liu et al. [7] (Anomaly)	DPAE method for anomaly detection in industrial quality control.	DPAE, Reconstruction Loss	Outperforms existing methods	DPAE excels in anomaly detection for industrial products.	May struggle with complex datasets.
Tsai et al. [14]	Enhanced CAE model for industrial defect detection.	CAE, Regularization	Not specified	Improved CAE for defect detection via error minimization.	Limited by defect-free samples.
Niu et al. [9]	Memory-augmented adversarial autoencoder for industrial defects.	Memory-Augmented Adversarial AE, FMD	High Accuracy, Real-time	High accuracy with real-time defect detection, robust performance.	Needs fine-tuning for specific cases.

(OCSVM) with defect-free pictures by extracting covariance vectors from VAE latent variables. With 90% accuracy, the system showed the limitations of standard defect detection methods and benchmarked CNN models like AlexNet and VGG19. This improves manufacturing quality control and product reliability significantly.

Kumar et al. [5] developed a dataset and an explainable zero-bias deep neural network (ZBDNN) for composite curing defect identification to address composite manufacturing difficulties. After collecting data in a bespoke autoclave, the ZBDNN used dimensionality reduction and similarity matching to find anomalies. The model outperformed autoencoder (96.26%) and OC-SVM (92.38%) models with 99.71% accuracy, demonstrating its promise for AI-driven composite manufacturing quality inspection.

Toufigh et al. [13] proposed an unsupervised concrete damage detection framework employing ultrasonic signals and a deep autoencoder trained on intact specimens. With scant undamaged data, an Isolation Forest method detected anomalies and reconstruction errors in the time and frequency domains detected damage. Over five geopolymer concrete specimens, the framework achieved 95% accuracy for damaged stages and 93% for intact stages, proving its usefulness for distributed damage identification.

Chen et al. [2] addressed class imbalance and faulty sample scarcity in fabric defect identification. A depthwise separable convolution autoencoder and one-class classifier were used. The autoencoder quickly extracted features and decreased computational overhead, while the SVDD trained using only non-defective samples eliminated costly defective sample collecting. The approach performed well on polyester fibres using accuracy and AUC metrics, making it a feasible industrial fabric inspection solution.

Liu et al. [6] proposed an unsupervised textile flaw detection framework employing knowledge distillation and visual localisation. Even with insufficient data, this method improved classification and segmentation accuracy above traditional and supervised methods. Ablation experiments refined parameter optimisation, and the system performed well on open datasets and in industrial situations, demonstrating its practicality.

Lu et al. [8] developed a deep-learning video analysis framework for lace defect checking to overcome manual inspection and complex lace textures. The framework includes video preprocessing, pixel reconstruction, and classification. Reconstruction mistakes from defect-free training movies are used to detect faults via adaptive thresholding. The first video-based fabric defect identification technique for industrial applications, it was tested using simulated defect videos.

For automated optical quality inspection, Liu et al. [7] presented Dual Prototype Auto-Encoder (DPAE) semi-supervised anomaly detection. Dual prototype loss and reconstruction loss cluster normal image latent vectors, making abnormalities identifiable by wide latent space gaps. The Aluminium Profile Surface Defect (APSD) dataset and the model's high performance across four datasets proved its efficacy in industrial quality control.

The improved Convolutional Autoencoder (CAE) model by Tsai et al. [14] detects defects unsupervised. The model was trained on defect-free samples and used regularisation to tighten their feature distribution to reduce reconstruction errors. Measurements of distance from the training centre revealed problems. This upgraded CAE model detected faults across material surfaces better than conventional CAE methods, proving its industrial applicability.

The memory-augmented adversarial autoencoder (MAA) of Niu et al. [9] detects defects in defect-free data. They used adversarial autoencoder reconstruction and Fréchet Markov distance detection. Memory modules and optimised reconstruction loss reduced excessive reconstruction capabilities that could miss errors. Statistics-based threshold determination adapted the methodology to industry needs. On many datasets, the MAA showed good accuracy, resilience, and real-time application, making it ideal for various industrial situations.

The studies showcase effective AI-driven defect detection methods with high accuracy and industrial applicability. Future efforts should focus on developing generalized, scalable models, leveraging transfer learning, enhancing real-time deployment, and integrating multimodal data for broader applicability and improved efficiency.

3 Methodology

The proposed methodology aims to detect fabric defects using an unsupervised approach that combines autoencoder-based reconstruction and a one-class Support Vector Machine (SVM) for anomaly detection. The key components of the methodology are detailed below.

3.1 Data Description

The dataset used in this study is based on the publicly available dataset provided by MVTec [1], tailored for educational and research purposes in fabric defect detection. This dataset reflects the challenges of identifying rare anomalies in textile materials, such as cuts, holes, thread irregularities, and metal contamination, which can significantly compromise fabric quality. Early and accurate detection of these defects is critical for maintaining production standards in industrial scenarios.

- **Image Size**: 64 × 64 pixels (used in this study).
- **Classes**: The dataset includes the following defect types: `{good, color, cut, hole, thread, metal_contamination}`. For our binary classification task, these were grouped into `{damaged, good}`.
- **Rotations**: Each image is augmented with eight rotational variations: `{0°, 20°, 40°, 60°, 80°, 100°, 120°, 140°}`.
- **Train-Test Split**:
  - **Training Set**: Contains 108,000 images, evenly distributed across the two classes (`damaged` and `good`).

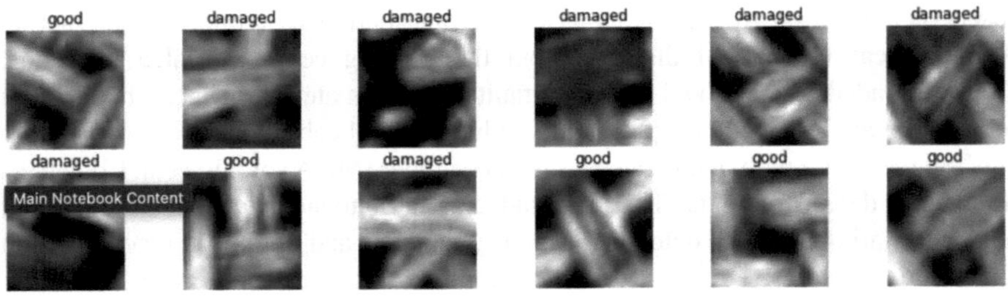

Fig. 1 Sample of the fabric defect detection dataset used in this work (Images blurry due to extrapolation from their original size)

  - **Testing Set**: Comprises 36,000 images, also balanced between the two classes.
- Source images in the training and testing sets are non-overlapping to ensure a robust evaluation.

The dataset can be seen in Fig. 1 and allows for multiple tasks, including:

1. **Binary Classification**: Differentiating between `damaged` and `good` fabrics.
2. **Class-Specific Defect Detection**: Identifying the type of defect (`color, cut, hole, etc.`).
3. **Self-Supervised Learning**: Using `good` images for texture representation and testing anomaly detection.

This study focuses on binary classification using 64x64 patches, leveraging the clear separation between training and testing datasets to ensure unbiased evaluation. The dataset's structure and augmentation make it ideal for exploring supervised, unsupervised, and self-supervised approaches in defect detection.

3.2 Autoencoder-Based Reconstruction

In this work we propose an autoencoder based architecture as lined in Fig. 2. An autoencoder learns the latent structure of defect-free fabrics:

- **Encoder:** Maps input $\mathbf{x} \in \mathbb{R}^{H \times W \times C}$ to latent space $\mathbf{z}$:

$$\mathbf{z} = f_{\text{enc}}(\mathbf{x}; \boldsymbol{\theta}_{\text{enc}}) \tag{1}$$

- **Decoder:** Reconstructs input from $\mathbf{z}$:

$$\hat{\mathbf{x}} = f_{\text{dec}}(\mathbf{z}; \boldsymbol{\theta}_{\text{dec}}) \tag{2}$$

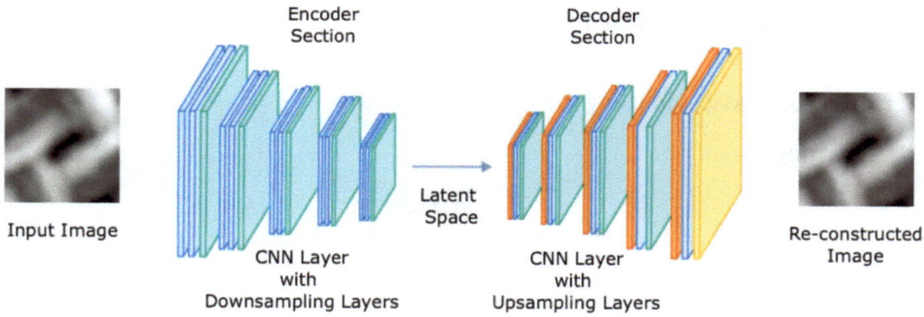

Fig. 2 Overview of the proposed framework in the current work for fabric defect detection

Minimizing the mean squared reconstruction loss:

$$\mathcal{L}_{\text{rec}} = \frac{1}{n} \sum_{i=1}^{n} \|\mathbf{x}_i - \hat{\mathbf{x}}_i\|_2^2 \tag{3}$$

ensures accurate defect-free image reconstruction.

3.3 *Reconstruction Loss Map*

For each test image, pixel-wise reconstruction loss is computed as:

$$\mathcal{R}(\mathbf{x}) = \|\mathbf{x} - f_{\text{dec}}(f_{\text{enc}}(\mathbf{x}))\|_2 \tag{4}$$

Flattened loss maps form feature vectors for anomaly detection.

3.4 *One-Class SVM for Anomaly Detection*

The reconstruction loss features of defect-free samples train a one-class SVM, optimizing:

$$\min_{\mathbf{w},\rho} \frac{1}{2}\|\mathbf{w}\|^2 + \frac{1}{\nu N} \sum_{i=1}^{N} \max(0, \rho - \mathbf{w}^\top \phi(\mathbf{r}_i)) - \rho \tag{5}$$

where $\mathbf{r}_i$ represents the loss features, ν controls outliers, and $\phi(\cdot)$ maps features to a high-dimensional space. Regions outside the hypersphere are classified as defective.

3.5 Evaluation Metrics

Performance is assessed using the following metrics:

- **Accuracy:**

$$A = \frac{TP + TN}{TP + TN + FP + FN} \quad (6)$$

- **Precision:**

$$P = \frac{TP}{TP + FP} \quad (7)$$

- **Recall:**

$$R = \frac{TP}{TP + FN} \quad (8)$$

- **F1 Score:**

$$F_1 = 2 \cdot \frac{P \cdot R}{P + R} \quad (9)$$

This method achieves robust defect detection with minimal reliance on labeled data, outperforming state-of-the-art models.

4 Result and Discussion

The experiments were conducted on Google Colab, leveraging an NVIDIA Tesla T4 GPU with 16 GB of dedicated GPU memory, 12 GB of system RAM, and 100 GB of disk storage. An 80-20 holdout validation strategy was employed, where 80% of the dataset was used for training and the remaining 20% for testing [17]. This setup ensured comprehensive evaluation across metrics, including accuracy, F1 score, sensitivity, and ROC-AUC, enabling a consistent comparison of the proposed approach against state-of-the-art models [16] (Table 2).

The proposed unsupervised framework demonstrated superior performance, achieving the highest training accuracy (94.57%), F1 score (93.21%), and sensitivity (95.72%) while maintaining a competitive ROC-AUC of 94.85%. Testing results exhibited a slight performance drop, as expected, due to the inherent complexity of real-world defect patterns, yet the model maintained a leading testing accuracy (90.86%) and F1 score (90.23%). Compared to ResNet-50, the proposed method achieved a 7.2% higher sensitivity, highlighting its robustness in identifying rare defect patterns, albeit with slightly lower precision. These results validate the

Table 2 Training results summary of the proposed approach compared against the state-of-the-art models

Model	Accuracy	F1 score	Sensitivity	ROC-AUC
Proposed approach	0.9457	0.9321	0.9572	0.9485
AlexNet	0.8923	0.8745	0.8881	0.9056
VGG-16	0.9048	0.8873	0.9102	0.9124
VGG-19	0.9121	0.8994	0.9216	0.9175
MobileNetV1	0.9264	0.9172	0.9338	0.9347
ResNet-50	0.9385	0.9287	0.9483	0.9418

Table 3 Testing results summary of the proposed approach compared against the state-of-the-art models

Model	Accuracy	F1 score	Sensitivity	ROC-AUC
Proposed approach	0.9086	0.9023	0.9224	0.9187
AlexNet	0.8547	0.8392	0.8601	0.8745
VGG-16	0.8705	0.8559	0.8763	0.8829
VGG-19	0.8789	0.8667	0.8884	0.8892
MobileNetV1	0.8981	0.8854	0.9039	0.9023
ResNet-50	0.9113	0.9011	0.9195	0.9142

efficacy of the reconstruction-based anomaly detection framework, which leverages reconstruction loss as an intuitive feature for one-class classification (Table 3).

The performance advantage arises from the autoencoder's ability to capture nuanced texture details during the reconstruction process, which are critical for distinguishing defective and non-defective regions. In contrast, traditional supervised models like AlexNet and VGG-16 require extensive labeled data and are prone to overfitting due to the variability in fabric textures. The use of a one-class SVM further enhances generalization by focusing on deviations from normal patterns, making it ideal for defect detection tasks with limited annotated data. These findings underscore the potential of unsupervised learning in industrial inspection, paving the way for scalable, data-efficient solutions.

5 Conclusion

This study introduces a novel unsupervised methodology for fabric defect detection, integrating autoencoder-based image reconstruction with a one-class SVM to identify anomalies. By leveraging reconstruction loss as a discriminative feature, the proposed framework is capable of detecting subtle deviations in complex fabric textures without the need for extensive annotated datasets, addressing a critical limitation of traditional supervised approaches. Experimental results demonstrated that the method outperformed state-of-the-art supervised models, including ResNet-50, VGG-16, VGG-19, and MobileNetV1, in key performance metrics such as accuracy, F1 score, and recall, with an average accuracy improvement of 3.57% and a recall improvement of 7.2% compared to ResNet-50.

The higher recall, combined with a slight trade-off in precision (4.03% lower than ResNet-50), underscores the method's robustness in detecting rare and subtle defect patterns, a critical requirement in industrial applications where missing defects can lead to significant quality issues. Furthermore, the use of a holdout validation strategy ensures that the model's performance is evaluated reliably on unseen data, demonstrating its potential for real-world deployment.

In summary, the proposed framework provides a robust, efficient, and scalable solution for automated quality control in textile manufacturing, paving the way for further research into unsupervised and semi-supervised learning techniques for defect detection across various industrial domains. Future work may focus on optimizing the autoencoder architecture, exploring advanced anomaly detection algorithms, and applying the framework to multi-class defect classification tasks to further enhance its versatility and effectiveness.

6 Future Directions

Future work will explore extending the proposed framework to multi-class defect classification, allowing for the identification of specific defect types rather than binary classification. Additionally, incorporating self-supervised pretraining or contrastive learning could further enhance the autoencoder's feature extraction capabilities, particularly for highly complex textures. Investigating lightweight architectures optimized for real-time deployment on edge devices can facilitate industrial adoption at scale. Finally, expanding the evaluation to diverse textile datasets and real-world production environments will validate the model's generalizability and robustness in varied operational conditions.

Acknowledgments The authors acknowledge the support from National Institute of Technology, Hamirpur for the completion of this study and experiments.

References

1. Bergmann, P., Fauser, M., Sattlegger, D., Steger, C.: Mvtec ad–a comprehensive real-world dataset for unsupervised anomaly detection. In: Proceedings of the IEEE/CVF conference on computer vision and pattern recognition. pp. 9592–9600 (2019)
2. Chen, C., Deng, X., Yu, Z., Wu, Z.: Fabric defect detection using a one-class classification based on depthwise separable convolution autoencoder. In: Journal of Physics: Conference Series. vol. 2562, p. 012053. IOP Publishing (2023)
3. Cheng, L., Yi, J., Chen, A., Zhang, Y.: Fabric defect detection based on separate convolutional unet. Multimedia Tools and Applications **82**(2), 3101–3122 (2023)

4. Guleria, V., Verma, A., Dhenkawat, R., Chaurasia, U., Singh, N.P.: Brain tumor detection using texture based lbp feature on mri images using feature selection technique. In: Artificial Intelligence, Blockchain, Computing and Security Volume 1, pp. 30–36. CRC Press (2023)
5. Kumar, D., Agraharam, P.C., Liu, Y., Namilae, S.: Anomaly detection for composite manufacturing using ai models. Journal of Intelligent Manufacturing pp. 1–17 (2024)
6. Liu, H., Wang, S., Meng, C., Zhang, H., Xiao, X., Li, X.: Unsupervised fabric defect detection framework based on knowledge distillation. In: International Conference on Neural Information Processing. pp. 339–351. Springer (2023)
7. Liu, J., Song, K., Feng, M., Yan, Y., Tu, Z., Zhu, L.: Semi-supervised anomaly detection with dual prototypes autoencoder for industrial surface inspection. Optics and Lasers in Engineering **136**, 106324 (2021)
8. Lu, B., Xu, D., Huang, B.: Deep-learning-based anomaly detection for lace defect inspection employing videos in production line. Advanced Engineering Informatics **51**, 101471 (2022)
9. Niu, T., Li, B., Li, W., Qiu, Y., Niu, S.: Positive-sample-based surface defect detection using memory-augmented adversarial autoencoders. IEEE/ASME transactions on mechatronics **27**(1), 46–57 (2021)
10. Saberironaghi, A., Ren, J., El-Gindy, M.: Defect detection methods for industrial products using deep learning techniques: A review. Algorithms **16**(2), 95 (2023)
11. Shimizu, T., Nagata, F., Habib, M.K., Arima, K., Otsuka, A., Watanabe, K.: Advanced defect detection in wrap film products: A hybrid approach with convolutional neural networks and one-class support vector machines with variational autoencoder-derived covariance vectors. Machines **12**(9), 603 (2024)
12. Singh, S., Verma, A., Guleria, V., Yadav, S., Singh, N.P.: Deep learning-based networks to detect leaf disease in maize and corn. In: 2023 International Conference on IoT, Communication and Automation Technology (ICICAT). pp. 1–6. IEEE (2023)
13. Toufigh, V., Ranjbar, I.: Unsupervised deep learning framework for ultrasonic-based distributed damage detection in concrete: integration of a deep auto-encoder and isolation forest for anomaly detection. Structural Health Monitoring **23**(3), 1313–1333 (2024)
14. Tsai, D.M., Jen, P.H.: Autoencoder-based anomaly detection for surface defect inspection. Advanced Engineering Informatics **48**, 101272 (2021)
15. Vardhan, H., Verma, A., Singh, N.P.: An ensemble learning approach for large scale birds species classification. In: Artificial Intelligence, Blockchain, Computing and Security Volume 1, pp. 3–8. CRC Press (2023)
16. Verma, A., Amin, S.B., Naeem, M., Saha, M.: Detecting covid-19 from chest computed tomography scans using ai-driven android application. Computers in biology and medicine **143**, 105298 (2022)
17. Verma, A., Gupta, N., Bhatele, P., Khanna, P.: Jmcd dataset for brain tumor detection and analysis using explainable deep learning. SN Computer Science **4**(6), 840 (2023)

Predicting Mental Health Issues Using Social Media Data: A Machine Learning Approach

Nainsiben Patel, Mamta Motiramani, and Nikita Joshi

Abstract Mental health care has become an equal partner to physical well-being in recent years. The concern about mental health increases as the use of social media speeds up quite rapidly. Although such tools allow communication and information sharing, heavy usage of social media has resulted in several problems, among them anxiety, depression, and low self-esteem. This has motivated the behind the curtain mechanisms that use machine learning-based applications to produce effective solutions that concentrate on the early prediction of the degree and type of mental disorders and their possible treatment results. Machine learning in the healthcare system is viewed as one of the most successful methods because it deals with a lot of information smoothly and makes accurate predictions. This paper examines how ML can be used to predict early mental health problems using wide-ranging demographic and technology usage data. ML models that include the Logistic Regression, Decision Trees, K-Nearest Neighbors, and Random Forests, were tested. K-Nearest Neighbors realized the highest accuracy of 75.80%. This gives evidence of how high ML can stand in supplementing intervention for early mental health. These results shed light on the possibility of data-driven interventions in augmenting care for mental health.

Keywords Behavioral analysis · Data-driven interventions · Machine learning models · Mental health prediction · Smartphone sensor data

N. Patel (✉) · M. Motiramani · N. Joshi
Adani University, Ahmedabad, Gujarat, India
e-mail: Nainsibenpatel.cse22@adaniuni.ac.in; Mamtamotiramani.cse22@adaniuni.ac.in; Nikita.joshi@adaniuni.ac.in

K. Deep et al. (eds.), *Artificial Intelligence, Machine Learning and Intelligent Systems*, Springer Proceedings in Mathematics & Statistics 520,
https://doi.org/10.1007/978-981-95-3841-6_29

1 Introduction

Mental health is an essential part of well-being that determines how we handle stress, relate to each other, and become useful members of society. According to the World Health Organization (WHO), mental health not only means the absence of disorders but also the ability to realize one's potential to cope with life's stresses and be able to contribute to society appropriately [1, 2]. As per WHO, more than 450 million people worldwide suffer from mental health issues [3]. Some elements of modern-day living, like the overuse of smartphones, are becoming more associated with elevated stress levels, poorer sleep quality, and mental health issues. Such behaviors can also result in losing contact with others, solitude, and even depression [4].

Social media, being one of the main activities on mobile phones, has multifaceted relationships with mental health. It fosters social support and allows people to stay connected. Simultaneously, it causes problems to mental health through appearance-based comparison, cyberbullying, and exposure to dangerous content. Research suggests that educating children to use social media responsibly and nurturing positive body image might also reduce these risks [5]. Social media addiction further breaks into education, work, and relationships and is always encouraging anxiety and escapism [6]. In India, the most common mental disorders are depression, anxiety, and schizophrenia, which affect approximately 15% of the population. These problems cause a lower quality of life, economic losses, and stigma around consulting for help. These mental health issues therefore cause absenteeism, lowered productivity, and increased health costs, thus establishing a compelling need for better mental health care [7]. Mental health at work is increasingly emerging as a salient area of concern because its implications for productivity and performance are deleterious and influence more than any other medical condition [8].

Machine learning (ML) is becoming more important in really transforming mental health care using early detection, personalized treatment interventions, and more efficient management of the condition. Such is the future offered by machine learning to mental health: it can find the complex, non-linear interaction that exists in high-dimensional data. ML takes an edge over traditional statistical approaches and provides a much more precise and individualized approach to mental health care. Thus, machine learning is now mandatory in the evolution of care from reactive to preventative thus minimizing long-term costs on mental health and improving overall well-being [9].

The paper is organized into six sections. In Sect. 2, the problem Statement addresses mental health issues due to excessive social media use and highlights the role of ML. Section 3: Literature Review examines existing ML models for mental health prediction and their limitations. Section 4: Dataset Definition describes the dataset and its relevance to current trends. Section 5: Methodology outlines the ML workflow, including data preprocessing and the implementation of different models. Section 6: Results presents model evaluation metrics, with KNN achieving the highest accuracy of 75.80%. Section 7: Conclusion and Future Work discusses the findings, emphasizing ML's potential in mental health care and suggesting improvements through hyperparameter tuning and diverse datasets.

2 Problem Statement

The use of social media has become pervasive in the current lifestyle. It improves communication but also brings some drawbacks, especially to mental health originating from excessive chronic negativity and social comparison. Therefore, these challenges need to be tackled for healthy mental fitness to be attained in all age groups. However, while machine learning has been successfully used in diverse fields to process enormous amounts of data and unravel deeply embedded patterns, its use of social aspects of mental well-being remains very scanty. This is the purpose of the study, to fill the void by investigating social media behavior and demographics, among other variables, and how machine learning can be used to classify people as having "poor" mental health. These models are predictive and allow one to deal with individuals who may tend to develop such conditions.

3 Related Work

The subsequent paragraph gives an extended discussion regarding the analysis of studies and their findings in comparison with each other, stressing the important findings and differentiating mechanisms so that a holistic understanding of the advances in application of machine learning for mental health care could be obtained.

The study [10] evaluated five different machine learning techniques: Logistic Regression, K-Nearest Neighbor (K-NN) Classifier, Decision Tree, Random Forest, and Stacking. All produced more than 79% accuracy in figuring out mental health issues, with Stacking being the most efficient with an accuracy of 81.75%. The study talked about using machine learning to improve the patient's outcome through early diagnosis of mental health conditions. Its limitation would be a small data set (27 features and 1,259 entries) making it less generalizable.

Another study [11] gives the idea of detecting depression using Facebook comments; here, feature selection uses Recursive Feature Elimination, and the data is classified with the Random Forest and Gradient Boosting methods and RF with the highest accuracy level of 96.17%. It is effective but its limitation is that it focuses on Facebook data restricts its generalizability. Similarly, [12] explores mental health diagnostics using the Mental Health in Tech Survey from Kaggle. Machine learning models like AdaBoost, Bagging, Stacking, LR, KNN, Tree Classifier, NN, and RF were applied, and AdaBoost outperformed the rest at 81.75% accuracy. The study relies on outdated datasets from 2014 and 2016 comprised almost solely of male respondents from the tech sectors of developed countries, which limits its

Table 1 Review of ML approaches in mental health studies

Aspect	M. Kumar Jha et al. [11]	E. O. Ogunseye et al. [12]	B. C. Loftness et al. [13]	T. Madhu Midhan et al. [14]
Primary focus	Depression detection from Facebook comments	Predictive analysis of mental health conditions using survey data from the tech industry	Detecting early childhood mental health conditions using wearable data and machine learning	Emotion classification in text conversations for mental health
Dataset	Facebook comments from the "Depression-Anxiety-Facebook-page-Comments-Text" dataset	Mental Health in Tech Survey datasets from 2014 and 2016 from Kaggle	Kiddie Internalizing Disorder (KID) Study: 84 children with behavioral data from wearable sensors and the ChAMP app	EmoEval2022 dataset: 21,800 tweets labeled with 14 emotions
Best results (Accuracy)	RFC: 96.17%, GBC: 86.99%	AdaBoost: 81.75%, KNN: 90.42%	ADHD Accuracy 68–86%, ROC AUC 0.65–0.72	Logistic Regression: 62.23%, SVM: 60.08%, GBC: 54%
Models applied	Random Forest Classifier (RFC), Gradient Boosting Classifier (GBC)	AdaBoost, Bagging, Stacking, LR, KNN, RF	Gaussian Naïve Bayes, RF, SVM, Decision Tree, Logistic Regression, Kohonen SOM	Logistic Regression, SVM, RF, GBC
Limitations	Narrow dataset focus; no generalizability	Dataset limited to developed countries; lacks NLP and longitudinal data	Small sample size; modest sensitivity and specificity	Limited accuracy; high computational cost; difficulty with complex emotions

demographic and geographical diversity. In other words, it does not have up-to-date information like post-COVID-19 trends and underrepresents variables such as socioeconomic and cultural factors.

Meanwhile, [13] collects behavioral data from 84 children through wearable sensors and the ChAMP app using supervised and unsupervised models to detect ADHD, anxiety, and depression. Results had moderate accuracy with SOMs with clusters of behavior. It has a small sample size; therefore, findings became limited. Finally, in the work [14], emotion classification from tweets is handled based on the EmoEval2022 dataset. Where the best-developed Logistic Regression achieved the highest accuracy at 62.23%, this research study brings forth the serious fact that it operates with complex emotions within the text but achieves a huge accuracy, thereby necessitating advanced approaches in the future. Table 1 shows a few selected studies and their descriptions by focusing on the applied problems, datasets, features they selected, model performances, and limitations, to more clearly outline the variability and challenges of applying machine learning to detect mental health:

Despite the above statements presenting the possibilities that machine learning holds in mental health diagnostics-from social media or wearable data-there arise limitations of narrow sets, small sample sizes, and contextual constraints, as these limit the broader scope of applications and thus performance. Our study thus proceeds from these and uses a much more extensive, more heterogeneous dataset with more powerful and well-designed ML models for improved predictive accuracy. It will thus present a whole new take on the consideration of technology use and its impact on mental wellness.

4 Dataset Description

The **mental_health_and_technology_usage_2024** [15] dataset taken from Kaggle consists of 10,000 entries comprehensively and presents up-to-date information on the way technology usage patterns correlate with mental health status. There are 14 features that include: User_ID, Age, Gender, Technology_Usage_Hours, Social_Media_Usage_Hours, Gaming_Hours, Screen_Time_Hours, Mental_Health_Status, Stress_Level, Sleep_Hours, Physical_Activity_Hours, Support_Systems_Access, Work_Environment_Impact, and Online_Support_Usage. The target variable Mental_Health_Status_Low was framed by encoding the Mental_Health_Status feature. This encoding classifies mental health status into binary formats for predictive modeling and analytics activities.

Our study addresses many glaring gaps in existing research work on the prediction of mental health using machine learning. The adoption of a comprehensive and current dataset complemented by an over 10,000 entry database gives our work a greater level of generalizability and brings it up to date in measuring technology use as well as mental health in contrast with earlier works confined to specific or already outdated measurements such as Facebook comments of pre-2016 data. The inclusion of numerous independent variables (social media usage, screen time, and physical activity) helps in building the most comprehensive view of mental health predictors than any such research. In terms of methodology, we would like to balance accuracy with interpretability between our best-performing classifiers-K-Nearest Neighbors-that best minimize the tradeoffs (highest accuracy) and Decision Trees to provide user-friendly insights. Thorough testing of the classifiers for accuracy, recall, F1-Score, and ROC-AUC strengthens the presentation of all our findings.

5 Methodology

Figure 1 illustrates the structured progression: from the understanding of the dataset through EDA to applying machine learning models such as Logistic Regression, KNN, Decision Trees, and Random Forest in a systematic way for classification.

5.1 *Dataset Acquisition and Preliminary Exploration*

The dataset was imported into the environment of a Python programming setup. Some exploratory steps were followed by summarizing the dataset and examining its structure concerning column names and data types as well as memory usage. Using the info() and head() methods, missing values were ensured not to exist in any feature, hence making the dataset complete enough for analysis. This pre-analysis check of the distribution of variables involves both categorical and numerical variables to understand how diverse the data set is and also to identify areas where possible data quality issues lie.

5.2 *Data Preprocessing*

Before the data was ready for analysis, some preprocessing steps were done, including elimination. The User_ID column was eliminated because it was a unique identifier column that had no predicting power. In addition, the categorical features

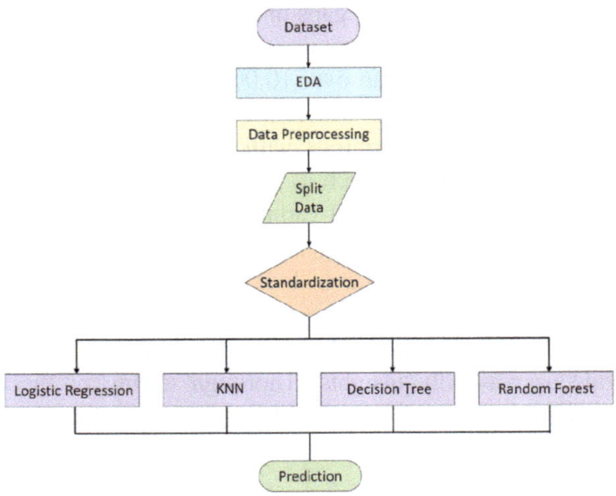

Fig. 1 Machine learning workflow for classification

of Gender, Mental_Health_Status, Stress_Level, and Work_Environment_Impact were encoded into numerical representations using one-hot encoding, effectively making them equivalent to binary variables to fit into algorithms using machine learning. For example, the Gender column was one-hot encoded into columns named Gender_Male and Gender_Other, where the values 0 or 1 represented the presence of that particular category. As a result of this step, the final dataset contained 18 features in total: the newly generated binary columns.

5.3 Data Splitting and Normalization

The processed dataset was split into training and testing subsets at an 80:20 ratio. The division was such that the testing set would remain unseen in the training process of the model. Thus, the integrity of the model evaluation was preserved, since the model was tested regarding its ability to generalize to new data. Numerical columns like Technology_Usage_Hours, Gaming_Hours, and Sleep_Hours have scales that change frequently. Therefore, these columns were standardized using StandardScaler. This scaling step will transform the data into a mean-centered and unit-variance format where, by such scaling, features with larger magnitudes would not overly dominate the learning process of this model.

5.4 Machine Learning Model Implementation

Four distinct machine learning algorithms were employed to predict mental health outcomes:

(a) **Logistic regression:** A logistic regression refers to a logistic function that depicts the connection between the input variables (features) and the output variable (the one that is affected for instance medical intervention) [16]. A particular set of hyperparameters were utilized in the model, among them random state 42 to ensure reproducibility with respect to the maximum iteration limit of 1000 thus ensuring an effective convergence for highly complex data sets.

(b) **KNN:** KNN consists the ability to simply predict just a class for a given input data point by considering its k nearest neighbors from the training data. The KNN approach thus provides a good methodology for determining whether somebody needs any form of mental health treatment or not based on the features captured in the dataset [3]. The value for k in the KNN model was taken as six to improve the prediction accuracy. This was determined by taking iterative testing and performance evaluation into consideration such that there is a case where high accuracy and reliable generalization to the unseen data is achieved.

(c) **Decision Tree**: Decision trees are suitable for the detection of mental health datasets as they can process both the continuous as well as the discrete input information. The decision trees have the benefit of managing the missing values and outliers too which is a positive feature in real-world situations [3]. The Decision Tree would be initiated with the random state value set to 42, making both training and testing consistent.
(d) **Random Forest:** The Random Forest is the type of ensemble learning approach that combines several decision trees to boost the model's predictive power concerning the future [17]. Also, random state has been adjusted to 42 for the Random Forest Model in order to maximize performance while allowing reproducibility and mitigate ensemble learning outcomes.

5.5 Model Evaluation and Performance Metrics

A full suite of metrics was employed to measure the performance of the model in terms of the performance of the following:

Accuracy: General correctness of predictions. Precision: Correct identification of positive instances.
Recall: Actual positive instances correctly identified by the model.
F1-Score: Precise Mean of Precision and Recall Calculation Balance of False Positives and False Negatives.
ROC-AUC: Area under the receiver operating characteristic curve to measure the trade-off between true positives and false positives.

6 Results

The logistic regression model managed to attain an accuracy of 75.10%. Its precision measures how many of the positive predictions are correct, with the value at 74.02% that is, working to minimize false positives. The recall score or percentage of actual positives correctly identified was at 75.10%, therefore showing a balanced performance at identifying who requires intervention. The area under the receiver operating characteristic (ROC) curve was 0.83 as shown in Fig. 2a, which indicated good discrimination ability between positive and negative cases.

The KNN was used for the task of mental health outcome classification based on how the features between neighbor points were similar. Following fine tuning for optimal k number where the value of k was taken as 6, performance was obtained at 75.80% accuracy, precision, and recall at 71.81%, and 75.80%, respectively, with a corresponding F1-score of 71.45%. The ROC score was 0.70 as shown in Fig. 2b, and the model was good at competitive performance. KNN is sensitive

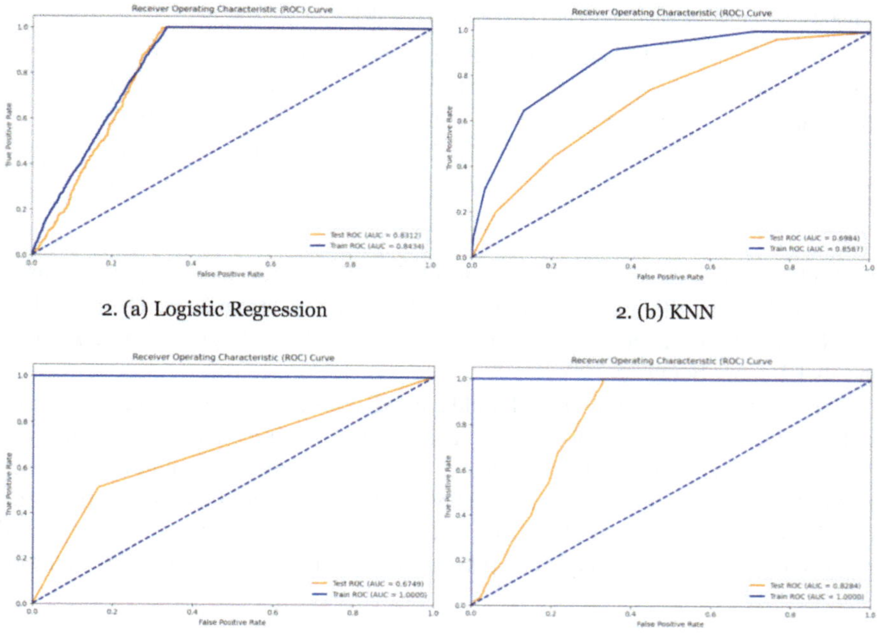

2. (a) Logistic Regression

2. (b) KNN

2. (c) Decision Tree

2. (d) Random Forest

Fig. 2 RoC curves of all models

to hyperparameters and feature scaling, requiring much more preprocessing and tweaking efforts.

The Decision Tree Classifier was also used, with its in-built ability to model nonlinear relationships and interactions between the features. The model gives very easily interpretable results by splitting the data at different branches by thresholds on combinations of features and captures complex patterns very effectively. The decision tree model achieved an accuracy of 75.65% just marginally above that of the logistic regression. It has a precision of 75.80%, a recall of 75.65%, and thus an F1-score of 75.72%. The Decision Tree model obtained an ROC score of 0.67 as shown in Fig. 2c, which thus indicates a good ability of classification of classes.

The Random Forest model gave an accuracy of 73.95%, precision of 72.72%, recall of 73.95%, and F1-score of 73.24%. Additionally, the ROC score was 0.83 as shown in Fig. 2d, with high discriminative ability. That is, Random Forest improved the robustness by aggregating predictions from individual trees and generalized very well to unseen data. In addition, feature importance was possible, whereby detailed information regarding the most influential variables in mental health predictions could be obtained.

Table 2 represents the strength of each machine learning model—Logistic Regression, Decision Tree, K-Nearest Neighbors (K-NN), and Random Forest—across the metrics: Accuracy, Precision, Recall, and F1-Score. K-NN seems to give

Table 2 Performance analysis of machine learning models

Model	Accuracy (%)	Precision (%)	Recall (%)	F1-score (%)
Logistic regression	75.10	74.50	75.10	74.77
Decision tree	75.65	75.80	75.65	75.72
K-NN	75.80	71.81	75.80	71.45
Random forest	73.95	72.72	73.95	73.24

the highest accuracy with a percentage of 75.80%, hence incorporating patterns based on the data efficiently. Then there is the good performance by Decision Tree with balanced metrics it shows; with an accuracy of 75.65%, precision of 75.80%, and its F1-score being 75.72%.

This logistic regression model holds good consistency in all the evaluation metrics, with precision (74.50%) and recall (75.10%) being close to its accuracy of 75.10% and thus proves to be reliable and simple as a model. Though the accuracy of Random Forest is a little lesser at 73.95%, it has robust predictability capabilities as reflected in an F1-Score of 73.24% and a good balance between the mentioned metrics; therefore, such models can perform well when tuned for complex patterns.

7 Conclusion

The research has emphasized the increasing requirement for exploiting ML methods to improve predictive techniques in mental health domains, working on how lifestyle and technology use correlate with mental well-being. Analysis of the mental_health_and_technology_usage_2024 dataset points out the performance of different ML models compared against K-Nearest Neighbors (K-NN), Decision Trees, Logistic Regression, and Random Forests. Among them, K-NN has achieved the maximum accuracy of 75.80% indicating its ability to take into account the intricate interaction between variables of data.

These results suggest that model selection in mental health predictive modeling is dependent on customized approaches; a specific model choice can correspond to the nature of the objective—be it high interpretability, balanced metrics, or robust generalization. The value of ML tools is to proactively identify such at-risk individuals to help properly aid in launching effective early intervention programs. Further development through hyperparameter tuning, with even more diverse and comprehensive datasets, will help enhance accuracy and applicability. ML integration in mental health care would be a great promise for more responsive solutions, in furtherance of efforts at broadening response to the crisis in mental health fueled by technology use and lifestyle factors.

References

1. World Health Organisation, "World Mental Health report," *World Health Organization*, p. 28, 2022, Accessed: Nov. 17, 2024. [Online]. Available: https://www.who.int/publications/i/item/9789240049338
2. J. John-Langba, "A Review Study of the Concept of Mental Health," *JOURNAL OF HUMAN ECOLOGY*, vol. 79, no. 1–3, Jun. 2022, https://doi.org/10.31901/24566608.2022/79.1-3.3340.
3. M. Bajaj, P. Rawat, Diksha, S. Vats, V. Sharma, and L. Gopal, "Prediction of Mental Health Treatment Adherence using Machine Learning Algorithms," *2023 International Conference on Computational Intelligence, Communication Technology and Networking, CICTN 2023*, pp. 716–720, 2023, https://doi.org/10.1109/CICTN57981.2023.10141520.
4. J. R. Beames *et al.*, "Use of smartphone sensor data in detecting and predicting depression and anxiety in young people (12–25 years): A scoping review," *Heliyon*, vol. 10, no. 15, p. e35472, Aug. 2024, https://doi.org/10.1016/j.heliyon.2024.e35472.
5. A. Popat and C. Tarrant, "Exploring adolescents' perspectives on social media and mental health and well-being – A qualitative literature review," *Clin Child Psychol Psychiatry*, vol. 28, no. 1, pp. 323–337, Jan. 2023, https://doi.org/10.1177/13591045221092884/ASSET/IMAGES/LARGE/10.1177_135910452 21092884-FIG1.JPEG.
6. P. Jadhav, S. Modhave, and U. Sangule, "Exploratory Data Analysis for the Survey of Social Media on Mental Health with Automation," *2023 4th International Conference on Computation, Automation and Knowledge Management, ICCAKM 2023*, 2023, https://doi.org/10.1109/ICCAKM58659.2023.10449595.
7. V. R. Meghrajani, M. Marathe, R. Sharma, A. Potdukhe, M. B. Wanjari, and A. B. Taksande, "A Comprehensive Analysis of Mental Health Problems in India and the Role of Mental Asylums," *Cureus*, vol. 15, no. 7, p. e42559, Jul. 2023, https://doi.org/10.7759/CUREUS.42559.
8. C. de Oliveira, M. Saka, L. Bone, and R. Jacobs, "The Role of Mental Health on Workplace Productivity: A Critical Review of the Literature," *Appl Health Econ Health Policy*, vol. 21, no. 2, pp. 167–193, Mar. 2023, https://doi.org/10.1007/S40258-02200761-W/TABLES/4.
9. S. Graham et al., *Artificial Intelligence for Mental Health and Mental Illnesses: an Overview* Curr Psychiatry Rep, vol. 21, no. 11, Nov. 2019, https://doi.org/10.1007/S11920-0191094-0.
10. K. Vaishnavi, U. Nikhitha Kamath, B. Ashwath Rao, and N. V Subba Reddy, "Predicting Mental Health Illness using Machine Learning Algorithms," J Phys Conf Ser, vol. 2161, no. 1, p. 012021, Jan. 2022, https://doi.org/10.1088/1742-6596/2161/1/012021.
11. M. kumar Jha, R. Ranjan, G. K. Dixit, and K. Kumar, "An Efficient Machine Learning Classification with Feature Selection Techniques for Depression Detection from Social Media," in *2023 International Conference on Communication, Security and Artificial Intelligence (ICCSAI)*, IEEE, Nov. 2023, pp. 481–486. https://doi.org/10.1109/ICCSAI59793.2023.10421064.
12. E. O. Ogunseye, C. A. Adenusi, A. C. Nwanakwaugwu, S. A. Ajagbe, and S. O. Akinola, "Predictive Analysis of Mental Health Conditions Using AdaBoost Algorithm," *ParadigmPlus*, vol. 3, no. 2, pp. 11–26, Aug. 2022, https://doi.org/10.55969/paradigmplus.v3n2a2.
13. B. C. Loftness *et al.*, "Toward Digital Phenotypes of Early Childhood Mental Health via Unsupervised and Supervised Machine Learning," in *2023 45th Annual International Conference of the IEEE Engineering in Medicine & Biology Society (EMBC)*, IEEE, Jul. 2023, pp. 1–4. https://doi.org/10.1109/EMBC40787.2023.10340806.
14. T. Madhu Midhan, P. Selvaraj, M. Harshavardan Kumar Raju., M. Bhanu Prakash Reddy., and T. Bhaskar., "Classification of Mental Health and Emotion of Human from Text using Machine Learning Approaches," in *2023 6th International Conference on Information Systems and Computer Networks (ISCON)*, IEEE, Mar. 2023,pp. 1–7. https://doi.org/10.1109/ISCON57294.2023.10111973.
15. Mental Health & Technology Usage Dataset, https://www.kaggle.com/datasets/waqi786/mental-health-and-technology-usage-dataset, last accessed 29/11/2024.

16. M. Karunakaran, J. Balusamy, and K. Selvaraj, "Machine Learning Models based Mental Health Detection," in *2022 Third International Conference on Intelligent Computing Instrumentation and Control Technologies (ICICICT)*, IEEE, Aug. 2022,pp. 835–842. https://doi.org/10.1109/ICICICT54557.2022.9917622.
17. M. Nadeem, J. Rashid, H. Moon, and A. Dosset, "Machine Learning for Mental Health: A Systematic Study of Seven Approaches for Detecting Mental Disorders," *2023 International Technical Conference on Circuits/Systems, Computers, and Communications, ITC-CSCC 2023*, 2023, https://doi.org/10.1109/ITCCSCC58803.2023.10212609.

A Threat Monitoring and Control Unit for Secure Embedded Systems

Rahul Shandilya and R. K. Sharma

Abstract Continuous monitoring of the security status of connected devices is essential as Zero Trust Architectures (ZTAs) become more prevalent in enterprise networks. For this, it is necessary to have real-time threat monitoring within the devices connected to the network. Ensure continuous monitoring in current devices available on the market is a challenge, especially in resource-constrained settings. TMC (Threat Monitoring and Control Unit) addresses this challenge by providing real-time monitoring and a trust score that can't be tampered with for devices connected to the ZTA network. Real-time hardware monitoring of CPU micro-architectural signals has been used to determine the trust score of a security coprocessor separated from the device computing architecture. We examined TMC on an open-source RISC-V processor-based architecture against ransomware, ROP, and cache-based micro-architectural attacks. The framework was implemented on an AMD Virtex XC7V2000T FPGA module to demonstrate area overheads.

Keywords Device security · Threat monitoring · Network security · Zero trust architectures · FPGA

1 Introduction

With the advent of IoT, contemporary network infrastructure is becoming dynamic, accommodating growing array of distributed devices, including wearable technology, including health monitors & smartwatches, as well as sensor devices in essential infrastructure. Internet-connected devices proliferation has significantly escalated cyber-attacks threat. Weak, discretionary, perimeter-focused access poli-

R. Shandilya (✉)
School of VLSI Design and Embedded Systems, NIT Kurukshetra, Kurukshetra, Haryana, India

R. K. Sharma
Department of Electronics and Communication Engineering, NIT Kurukshetra, Kurukshetra, Haryana, India

K. Deep et al. (eds.), *Artificial Intelligence, Machine Learning and Intelligent Systems*, Springer Proceedings in Mathematics & Statistics 520,
https://doi.org/10.1007/978-981-95-3841-6_30

cies employed in contemporary networks exacerbate these threats. NIST established specifications [1] for ZTA for reducing these threats, wherein device's network access privileges are determined dynamically by its behavior rather than being statically assigned depending on user authentication. Consequently, device exhibiting anomalous behavior will have its network privileges dynamically restricted.

Fundamental requisite for reliable implementation of ZTA is continuous monitoring of device's security status during application execution. Network must determine if device is compromised or acting as specified. Runtime health signature acquisition is difficult. Native methods could remotely monitor device network activity or install software daemon for recording operations for anomalies. The previous method failed for yielding adequate insights into detailed activities. Device may fail to identify all malicious behaviors, & in latter scenario, trusting received signatures becomes challenging, particularly when device is compromised.

Current paper proposes runtime security-health implementation monitoring for devices through specialized hardware modules that monitor System-on-Chip (SoC) operation for identifying attack-specific behaviors. Monitoring is conducted at hardware level software stack irrespectively. Hardware digitally verifies device trust score & transmits it to Zero Trust Network (ZTN) host for confidentiality along with integrity. This enables network for dynamically adjusting device access policies. Contrary to network as well as software-based methods, device hardware calculates trust scores, providing accurate and reliable security indicators.

Illustrated in Fig. 1, our framework, termed TMC (Programmable Threat Evaluation and Control Unit), comprises specialized hardware modules integrated within SoC. Event Monitor modules are intricately connected to CPU cores bus interfaces with monitor configurable execution parameters. A secure RISC-V processor is responsible for executing the architecture-specific Threat Evaluation Model (TEM) firmware that is stored in flash ROM memory and provide the trust score value for the session. Trust scores that measure device's security health are calculated by TEM after it periodically retrieves execution logs from trace memory.

Applications executing on CPU cores are not allowed to accessing secure elements used by sec-controller for tracking critical signals of each CPU core, computing trust scores, evaluating trace memory, for communicating via network interface. This assures trust score's integrity and reliability. Following are paper's primary contributions:

- TMC framework can be integrated into SoCs thereby providing devices in zero-trust networks with non-tamperable, confidential, & trustworthy trust scores, enabling them for determining dynamic access policy decisions.
- TMC is demonstrated by using TEMs that utilize ANNs (Artificial Neural Networks) & SVM(Support Vector Machines) to RISC-V-based SoC [2]. For variety of workloads, including embench-IoT suite [3], Coremark [4], & Dhrystone [5], as well as threat vectors involving micro-architectural side-channels, RoP(Return-oriented Programming) attacks, & ransomware attacks, TEMs calculate trust scores.

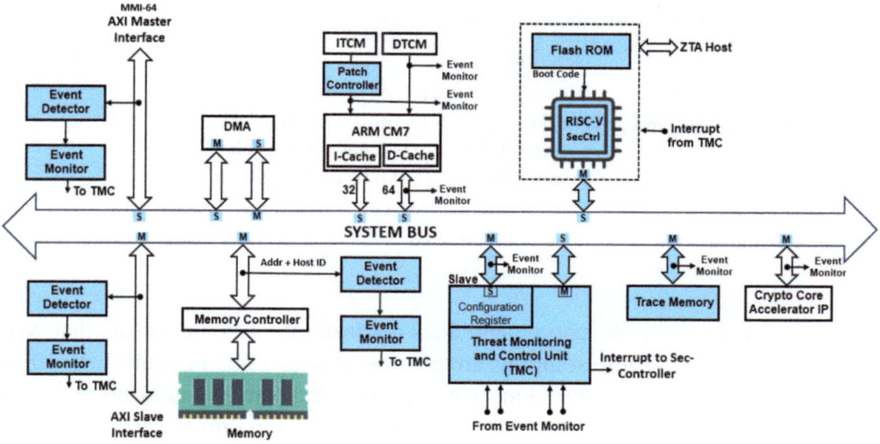

Fig. 1 TMC framework

- The area overheads of TMC were determined based on the synthesis analysis of AMD Virtex XC7V2000T FPGA.

The remainder of the paper is arranged in the following way: Sect. 2 describes related work, and Sect. 3 explains TMC architecture. Section 4 contains the TMC assessment and Sect. 5 contains the conclusion.

2 Related Works

ZTA network security monitoring solutions must fulfill these criteria: (a) monitors configurability, (b) adaptability to emerging threat vectors, (c) monitoring reliability (d) isolation from attack surface. Device-level behavioral monitoring solutions [6–9] employ HPCs (Hardware Performance Counters) [10] prevalent in contemporary microprocessors. Micro-architectural events are logged with constrained configurability [10, 11]. Additionally, HPC-based monitoring's isolation & dependability are limited by its reliance on operating system(OS) for event reading. Further research [6, 7, 9] attempted for minimizing HPCs limitations through developing monitoring mechanisms with focus on security. Kuruvila et al. [6] reported developing appropriate customized counters through instruction sequences application. Monitoring & threat assessment operate as applications without isolating SoC, a primary drawback of these solutions. Recent investigations [7, 9], including PHMon, enhance isolation through the utilization of specialized hardware for monitoring. However, configuring mechanism for monitoring necessitates OS support, hence add the overhead for host. Moreover, PHMon [7] exclusively monitors executed instructions, thereby lacking its ability to address intricate threat vectors, including ransomware or side-channel attacks. The current existing design lacks the feature that calculates trust

Table 1 Trust factor value assignment based on trust level

S.No.	Trust level	Description	Trust factor
1	High trust level	All allowed modules, no restriction	0
2	Medium trust level	For internal modules inside the SoC, which are not intended to access other modules, the device's Trust level will be set to medium.	1
3	Low trust level	Some restriction based on access types	2
4	Zero trust level	All access restricted	3

levels for each process at runtime and does not take into account threats that were detected previously. These solutions are contrasted with TMC in Table 1. Utilizing a dedicated hardware monitor along with separate dedicated security co-processor other than primary core in SoC, TMC improve the isolation limitations, monitoring, adaptability, along with configurability.

3 The TMC Framework

Illustrated in Fig. 1, TMC framework consists of three essential components. (1) The Threat Monitoring and Control Unit (TMC) conducts runtime surveillance for every designated CPU-core interface and generates behavioral traces. TMC stores intermediate data and traces on Trace memory. The events detector and monitor are set up in accordance with a pre-determined configuration. (2) Trace memory is designated memory unit for every monitor for recording observed events. Only TMC and Sec-Controller are granted reading and writing privileges to trace memory, ensuring its isolation from remaining memory units within SoC. (3) Sec-Controller is a security coprocessor capable of reading each trace memory's logs then invoking the TEMs on them. Sec-Controller offers device trust score based on output from TCMs & trace memory's logs. We discuss in great detail every such element in this section.

3.1 Event Monitor and Trace Memory

Program execution generally follows a phase behavior [12], where each phase is associated with specific micro-architectural component events. As a result, every attack leaves behind unique fingerprint in hardware state that utilized for recognizing attacks & gauge their threat. To identify these fingerprints, TMC's event monitor offers fine-grained runtime monitoring. Events in event monitoring are selected according to threat model, contrary to HPCs, that have not been developed for security applications therefore aren't customizable [10, 11]. For instance, we define threat model in our TMC implementation, including threat vectors, RoP

32-bit Time Stamp		
Trace Header	Host Signature	Trust Factor

Fig. 2 Memory trace frame format

attacks, ransomware, as well as micro-architectural side-channels. We established events to track memory access instructions patterns, along with cache parameters, on basis of this threat model with only minor modifications of current design. Other CPU cores in SoC have no ability to impact these behavioral logs considering each core has its own event monitor.

Observed events are recorded in trace memory to establish runtime behavioral trace for running program. The frame format for trace log is shown in Fig. 2, each trace log incorporated with 32-bit timestamp value, trace header, host signature and trust factor. Only corresponding event monitor has write access to trace memory. The Sec-Controller master interface is granted read and write privileges, but not other masters, to ensure isolation.

3.2 *Sec-Controller and Trace Memory*

TMC framework's security coprocessor is designated as Sec-Controller. It offers an environment for TEMs in operating for determining attack-like behavior captured in trace memory and it periodically reads monitored logs of processes of each CPU core.

For adapting to an evolving threat landscape, these TEMs are trained offline then deployed on Sec-Controller in line with threat model. TEM outputs for each CPU core determine device trust score. It is metric ranging from 0 (no trust) to 10 (complete trust) quantifies security status of device within ZTA network. The TMC trust score's computation frequency is regulated by the epoch parameter. Our experiments indicate that threat estimation using 200 clock cycles is optimal. However, epoch value depends on the frequency of threat events. It can be configured within the TCM module to trigger the Sec-controller when exhausted. Sec-Controller operates in isolation of SoC computing stack, as demonstrated in Fig. 1. Sec-Controller memory has exclusive read/write access. It retains TEM code & intermediate data for trace memory unit in event processing. Inhibiting other CPU-cores for interfering in threat assessment logic.

3.3 *Trust Score*

A device's trust score is calculated through trust factor by TEM. The trust score mechanism provides an overview of the level of trust during a ongoing session. The correlation between the behavior observed in the ongoing session and the history of

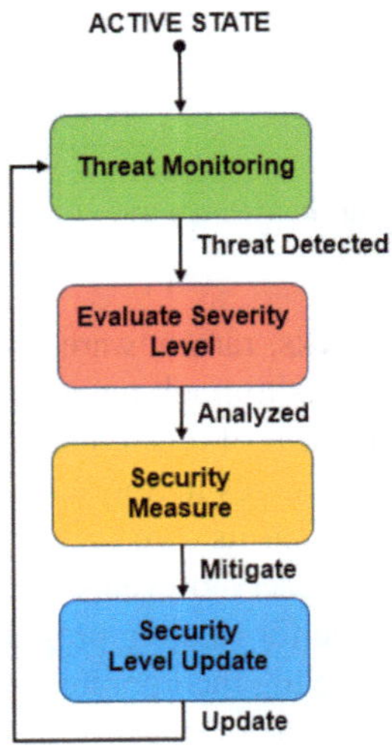

Fig. 3 Threat monitoring flow

the session plays a significant role in determining the level of risk involved in the session. Program behavior in each CPU core within SoC at t-th epoch determines device trust score value from 0 to 10, with lower scores indicating better security. Let the TEM output be TS_t at t-th epoch which determines the duration of current session. The trust score calculation for the current session takes into account all trust factors applied according to threat types. The result of this evaluation is a consolidated value of trust level, which describes the overall security posture of the device.

In the trust scoring calculation, trust factor are not weighed equally for all modules; the most restrictive trust factor having higher value takes precedence over all other trust factor given in Table 1. The order of trust factor restrictiveness (from least restrictive to most restrictive) is as follows.

The trust factor that is the most restrictive has more impact on the output of the trust scoring calculation. Hardware resources cannot be accessed by the device if the trust level drops to zero. The higher trust score calculated by TEM has contributed to blocking malicious devices by design (Fig. 3).

4 Results and Discussions

We analyzed TMC framework on open-source 5-stage RISC-V processor [2] SoC. Our focus is on workloads involving threat vectors, including RoP attacks, cache-based side-channel attacks, and ransomware. We configured ten events detector event monitor to log. These events encompasses read & write requests from every core, cache line flushes, cache hits in addition to misses, with counter for RoP devices that correlates to ret. & call instructions, selected based on threat vectors. On basis of their efficacy in identifying threats subset of events to monitor has been selected. We demonstrate Sec-Controller in our implementation by implementing an additional RISC-V CPU core that performs TEMs employing ANN along with customized implementations of a one-class SVM algorithm. Sec-Controller has single 36k BRAM configuration. In order to reveal area specifications, further tested TMC framework employing an AMD Virtex XC7V2000T FPGA Module (Table 2).

4.1 Evaluating TMC with Various Threat Vectors

Given consecutive attack exhibits unique execution characteristics, it is imperative to meticulously select subset of events that most effectively encapsulate threat model. Employing embench-IoT benchmark suite [3], Coremark [4], Dhrystone [5], along with Decision Forest classifier, Fig. 1 demonstrates that selected events distinguish attacks from benign programs. Occurrences of read requests & cache misses are effective in categorizing ransomware attacks, whereas gadget counter event primarily identifies RoP attacks. Effectiveness of classifying cache side-channel attacks varies among different cache events. TEM training offline utilizes event traces & is configured through secure ZTA host communication channel (Fig. 1). Threat assessment & trust scores for various attacks in threat model would be discussed subsequently.

Ransomware, which is designed to compromise user data within the system by acquiring privileges and systematically encrypting file systems. Our approach to creating RISC-V ransomware attacks involves utilizing different cryptographic algorithms for analysis. TEM employs an offline-trained ANN with essential monitors for these attacks. This ANN comprises an input layer containing 16-nodes along with 2-hidden layers with 16 & 8 nodes, correspondingly.

Microarchitectural Side-Channel Attacks can be employed including Prime+Probe & Flush+Reload [14] for extracting sensitive information, from cache memory includes passwords & cryptographic keys. We utilize cache flushing & timing measurements on RISC-V for executing these attack behaviors in our analysis [2]. Sec-Controller categories attacks One-class SVM in TEM. Current model has been trained on benign programs dataset against cache-based attacks [13]. Model clusters data on this benign behavior. This model evaluates new program by calculating distance between event vectors relating to program as well as benign clusters. Event vectors in benign clusters indicate benign program behavior. Malicious software event vectors usually appear outside benign clusters.

ROP attacks are used to create addresses that refer back to specific points in program code for daisy-chain malicious logic. To efficiently identify these attacks, we designed an event monitor ret. & call instructions to capture gadget execution.

Every device launches in trusted manner (⟦(TS)⟧_t = 0). We identified benign program execution consistently maintains high trust. However, TMC's threat evaluation demonstrates that a device executing micro-architectural attack, ransomware, or RoP attacks consistently decreases trust which is identified with increased value of ⟦(TS)⟧_t. Micro-architectural attacks, including Flush+Reload along with

Table 2 A comparison of TMC with existing security monitoring solutions

Parameter	TMC (proposed work)	Kuruvila et al. [6]	Yoon et al. [9]	Delshadtehrani et al. [7]	Nikhilesh et al. [13]
Event monitor implementation	HW	SW	HW	HW	HW
Run-time configurable modification support	Yes, Run-time configurable	No	No	No	No
Monitor programmability	Programmable	Fixed configuration	Fixed configuration	Fixed configuration	Fixed configuration
Threat vector adaptability	Yes	Yes	No	No	Yes
Isolation of HW monitor	Yes	No	No	Yes	Yes
Host off-loading	Yes, Dedicated Resources Available	No	No	No	No
Instruction decode and patch controller	Yes	No	No	No	No
Continuous event detection count	Up to 512 Events capturing possible Continuously without reset (RAM Size:4K)	NA	NA	NA	Up to 15 events
Real-time timestamp capturing for events	Yes	No	No	No	No
Core instruction execution monitoring	Yes	No	No	No	No

Table 3 FPGA resources overhead due to TMC module integration in sub-system design

Slice logic	Sub-system without TMC	Sub-system with TMC	% Overhead (%)
Slice LUTs	267,552	273,709	2.35
LUT as logic	264,400	270,481	2.3
LUT as memory	3152	3228	2.4
Slice registers	148,612	161,244	8.5

Prime+Probe [14], conduct anomalous cache accesses, due to this device blocks the access and move to zero trust state with increased value of 〖(TS)〗_t to 10.

4.2 *TMC Overheads*

TMC doesn't impact SoC application performance since it doesn't interfere with critical path. However, incorporating hardware requires more silicon. Area overheads for TMC-attributed synthesized on an AMD Virtex XC7V2000T FPGA device. Table 3 presents the analysis of FPGA resource overhead that is caused by the addition of TMC module to the target FPGA. The TMC framework requires an extra 2.35% of CLBs (Configurable Logic Blocks) and 8.5% of FFs (Flip Flops) to be added in addition to the FPGA target. The proposed FIS does not have any limitations as compared to existing methods and requires less number of CLBs and FFs resources inside FPGA. It includes trace memory units for each CPU-core & event monitors.

5 Conclusions

We presented the creation, implementation, and testing of TMC, a framework designed to meticulously monitor the security status of devices in the ZTA network. TMC offers configurable, adaptable, isolated, & tamper-resistant mechanism for fulfilling ZTA network requirements. TMC framework is capable of enforcing a variety of security policies at runtime and also assisting with detecting software bugs and security vulnerabilities. In the ZTA domain, TMC offers a novel solution that can update security policies in run-time for embedded device security, including integration of device integrity metrics into trust scores and security monitoring within the framework. Our TMC prototype includes a full FPGA implementation that interfaces the monitor with a RISC-V processor, along with the necessary hardware and firmware support. On average, The proposed TMC framework requires an extra 2.35% of CLBs (Configurable Logic Blocks) and 8.5% of FFs (Flip Flops) to be added in addition to the FPGA target, which is less as compared to existing design.

References

1. Stafford, V. (2020). Zero trust architecture. NIST special publication, 800, 207.
2. "Shakti: Open Source Processor Development Ecosystem, IIT Madras."Available: https://shakti.org.in/.
3. D. Patterson et al., "Embench: A Modern Embedded Benchmark Suite," 2019. Available: https://github.com/embench/embench-iot.
4. "Coremark: An EEMBC Benchmark." https://www.eembc.org/.
5. Weicker, R. P. (1984). Dhrystone: a synthetic systems programming benchmark. Communications of the ACM, 27(10), 1013-1030.
6. Kuruvila, A. P., Mahapatra, A., Karri, R., & Basu, K. (2021). Hardware performance counters: Ready-made vs tailor-made. ACM Transactions on Embedded Computing Systems (TECS), 20(5s), 1-26.
7. Delshadtehrani, L., Canakci, S., Zhou, B., Eldridge, S., Joshi, A., & Egele, M. (2020). {PHMon}: A programmable hardware monitor and its security use cases. In 29th USENIX Security Symposium (USENIX Security 20) (pp. 807-824).
8. Hunt, G., Letey, G., & Nightingale, E. (2017). The seven properties of highly secure devices. tech. report MSR-TR-2017-16.
9. Yoon, M. K., Mohan, S., Choi, J., Kim, J. E., & Sha, L. (2013, April). SecureCore: A multicore-based intrusion detection architecture for real-time embedded systems. In 2013 IEEE 19th Real-Time and Embedded Technology and Applications Symposium (RTAS) (pp. 21-32). IEEE.
10. Das, S., Werner, J., Antonakakis, M., Polychronakis, M., & Monrose, F. (2019, May). Sok: The challenges, pitfalls, and perils of using hardware performance counters for security. In 2019 IEEE Symposium on Security and Privacy (SP) (pp. 20-38). IEEE.
11. Zhou, B., Gupta, A., Jahanshahi, R., Egele, M., & Joshi, A. (2018, May). Hardware performance counters can detect malware: Myth or fact?. In Proceedings of the 2018 on Asia conference on computer and communications security (pp. 457-468).
12. Sherwood, T., Perelman, E., Hamerly, G., Sair, S., & Calder, B. (2003). Discovering and exploiting program phases. IEEE micro, 23(6), 84-93.
13. Singh, N., Pal, S., Leupers, R., Merchant, F., & Rebeiro, C. (2024). PROMISE: A Programmable Hardware Monitor for Secure Execution in Zero Trust Networks. IEEE Embedded Systems Letters.
14. Singh, N., Ganesan, V., & Rebeiro, C. (2022). Secure Processor Architectures. In Handbook of Computer Architecture (pp. 1-29). Singapore: Springer Nature Singapore.

Quantum Steganography-based Least Significant Bit Algorithm for Confidential Data Communication

Matam Santoshi Kumari and Samson Anosh Babu Parisapogu

Abstract This paper explores the quantum steganography-based Least Significant Bit algorithm for steganography, aiming to camouflage a secret image within a cover image for secure internet transmission. Modern communication systems prioritize safeguarding sensitive data during transmission through various channels. Many scientists are actively developing encryption and decryption algorithms, focusing on efficiency to maintain data integrity. However, a contrasting challenge arises from advancements in data loss techniques, rendering these algorithms vulnerable to attacks. In response, steganography has emerged as a field within data security, involving the concealment of sensitive information within communication media. Moreover, the use of qubits and quantum states, such as $|0\rangle$ and $|1\rangle$, in quantum computing introduces complexity to data security. Quantum cryptography, rooted in quantum mechanics, enhances communication security. Secure key exchange is achievable through quantum key distribution protocols, leveraging the concepts of superposition and entanglement. The manipulation of quantum states facilitates data encryption, advancing data security in the quantum era. The primary objective is to fortify data security by adeptly misleading potential cyber threats, focusing on securely delivering vital data to its intended recipients, and ensuring the security of sensitive information. Furthermore, encryption technology aims to stay ahead of potential vulnerabilities by constantly evolving to protect against emerging threats.

Keywords Encryption · Least Significant Bit (LSB) · Superposition · Entanglement

M. S. Kumari
Department of Mathematics, Chaitanya Bharathi Institute of Technology, Hyderabad, India
e-mail: santoshikumari_maths@cbit.ac.in

S. A. B. Parisapogu (✉)
Department of Computer Science and Engineering, Amrita Vishwa Vidyapeetham Amaravati, Amaravati, India
e-mail: p_anoshbabu@av.amrita.edu

K. Deep et al. (eds.), *Artificial Intelligence, Machine Learning and Intelligent Systems*, Springer Proceedings in Mathematics & Statistics 520,
https://doi.org/10.1007/978-981-95-3841-6_31

1 Introduction

Being in the era of the rapid evolution of information technology, ensuring secure communication has become a paramount concern. According to Sachin Dhawan & Rashmi Gupta [1] classical steganography techniques have been employed used to hide the presence of classified information in seemingly innocuous media. However, with the introduction of quantum computing, there arises a unique opportunity to revolutionize covert communication through the integration of quantum principles into the field of steganography. Quantum steganography [5]: a concept of hiding information with unprecedented levels of confidentiality and resilience by exploiting foundational principles of quantum mechanics. Unlike classical steganography [16] [17], which relies on the manipulation of classical bits, quantum steganography utilizes the power of quantum bits, or qubits, allowing for the simulation of the existence of multiple states through superposition.

This work covers the foundational principles of quantum steganography, exploring how these quantum principles, like quantum superposition, quantum entanglement, and the super dense principle, enable the embedding of secret information within the quantum carriers [18]. The work investigates the advantages of implementing quantum states for information concealment and the ability to establish secure communication channels resistant to eavesdropping. Moreover, this work explores the challenges and opportunities inherent in the implementation of quantum steganography, addressing issues related to quantum coherence, entanglement, and the necessity for quantum key distribution protocols. Our exploration extends to the comparison of quantum steganography with classical steganography techniques, emphasizing the quantum advantage in terms of security and capacity.

It seeks to give an integrated view of the possibilities and limitations involved in quantum steganography through several theoretical frameworks and simulations .Additionally, the work present practical considerations for the implementation of quantum steganographic systems, including their integration with existing quantum communication infrastructure. Consequently, this study represents a groundbreaking addition to the discipline of quantum steganography, providing valuable perspectives on the capacity of quantum mechanics to transform covert communication within the context of the quantum age. Given that this research resides at the convergence of quantum computing and information security, the investigation into quantum steganography unveils novel avenues for secure communication and highlights the changing dynamics of quantum technologies in influencing the future of information concealment.

2 Methodology

Quantum computing is crucial because it performs complex calculations much faster than regular computers because of features like superposition and entanglement. These enable qubits to exist in multiple states simultaneously, a great advantage over the classical bits that can only be 0 or 1. Quantum computers are therefore able to process vast amounts of information simultaneously and solve problems much more efficiently than their classical counterparts.

According to McMahon [2] the advancement opens up significant possibilities in areas such as simulations, optimization, and ensuring information security through cryptography. Even the practice of hiding messages, known as steganography, sees a considerable boost from quantum computing. Furthermore, due to unique quantum properties, messages can be concealed in intricate ways, heightening the challenge for outsiders trying to decipher them. The rapid operational speed of quantum computers facilitates quicker hiding and revealing of messages, thereby enhancing overall communication security efficiency. It's essential to note that these concepts are still in the early stages of development. According to Subramanian et al. [3], there is still much research and testing needed before quantum communication systems can be widely implemented. Moreover, Table 1 shows the differences between Classical Steganography and Quantum steganography.

2.1 *Distinguishing from Cryptography*

While similarities between steganography and cryptography may be apparent, they serve distinct purposes. Wayner [12] and Ke et al. [13] say that cryptography modifies the original message explicitly to make it challenging to decipher, with the expectation that the altered message will visibly differ from the original. On the other hand, LSB steganography seeks to surreptitiously embed one message within another, with the modified message maintaining a striking resemblance to the original.

Table 1 Difference between Classical Steganography and Quantum Steganography

Representation	Classical steganography	Quantum steganography
Data representation	Eight units of bits is referred to as one byte	Qubits-Bits exist in multiple states at once
Power	Increases linearly with number of transistors	Increases 1:1 in tandem with transistor
Operating environment	Well suited for normal conditions	Need to be kept at absolute zero temperature
Running time (N-messages)	$2^{polylog(N)}$	$(logN)^3$
Speed (1 message)	Milli seconds to Seconds (For different size of msgs)	Seconds to Minutes (For different size of msgs)
Speed (N messages)	Range from hours to days or longer, depending on the efficiency of the algorithm	Order of minutes to hours, depending on the capacity of hypothetical quantum computer
Execution time (N-messages)	In the order of milliseconds to seconds	In the order of seconds to minutes

Manashee et al. [19] consider that LSB steganography presents a powerful tool for covert communication by ingeniously leveraging the least significant bits of image pixels. This method allows for the secure transmission of confidential information, with the cover image serving as an inconspicuous carrier. As digital communication continues to evolve, understanding and appreciating techniques like LSB steganography becomes crucial for both securing sensitive data and comprehending the nuances of information concealment within the digital landscape. Mastering this method can provide individuals with a powerful tool for protecting their privacy and communicating covertly. It is essential for individuals to understand the potential risks and limitations associated with using this technique.

3 Proposed Approach

In the proposed approach, initially the users initiate the process by logging into the steganographic software, with their credentials securely stored in a database. Upon successful login, access to the steganography software is granted. In this environment, both the processed and cover images go through processing before encryption. The original images are archived in a database. Subsequently, a unique combination, facilitated by quantum computing, is employed to extract and decrypt the image. This ensures that the intended recipient receives the correct image, while potential cyber-attackers receive a misleading version. The deliberate obfuscation of the entire image is designed to confound attackers. The processed images, tailored to mislead potential threats, find storage in a distinct datastore. Kundur and Ahsan [4] say that quantum computing is used to make image encryption stronger and protect against unauthorized access.

This organized flow makes the steganographic process safer. Using quantum computing to improve image encryption and protection against unauthorized access, systematic flow makes the steganographic process safer. Figure 1 visualizes the data flow diagram of the proposed steganography approach.

3.1 LSB Algorithm

In the realm of digital communication, the need for secure and covert data transmission has given rise to innovative techniques such as LSB steganography [15][16]. This method involves concealing information within the least significant bits of image pixels without perceptibly altering the visual appearance of the image. Unlike cryptography, which focuses on making modifications evident to secure information, LSB steganography aims to deceptively embed one message within another, rendering the modified message indistinguishable from the original [7].

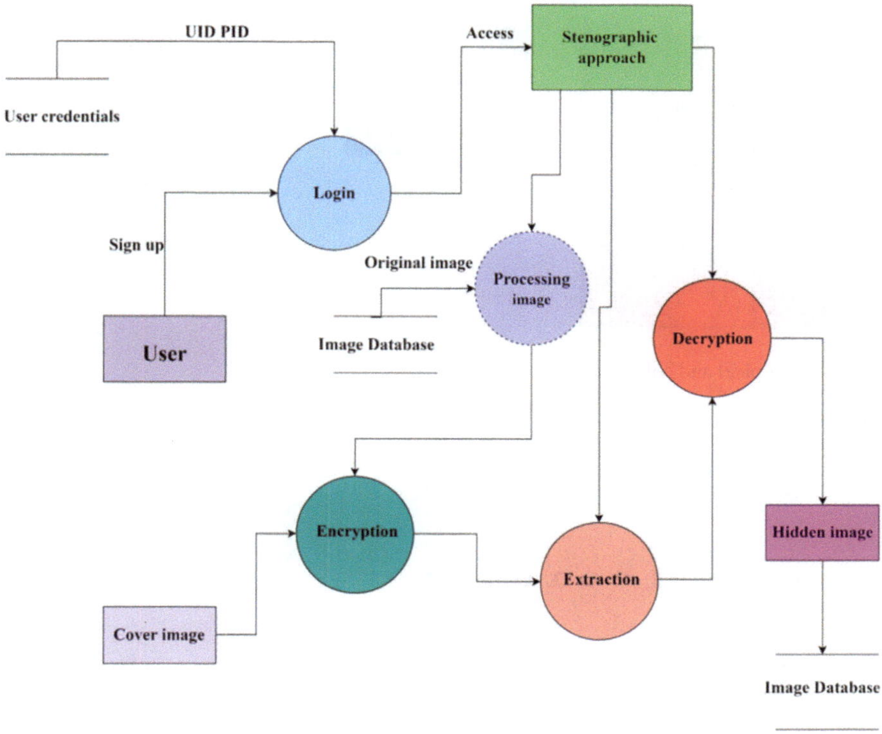

Fig. 1 Visualization of data flow in the proposed steganography approach

3.1.1 Understanding Pixels and Colour Models

In the digital world, images are composed of pixels, each representing a unique color. Commonly, a pixel is comprised of 24 bits, with 8 bits allocated to each of the red, green, and blue (RGB) color channels. For example, the color red is represented as RGB (255,0,0), where the red channel is set to its maximum value. Similarly, green and blue colors follow the RGB (0,255,0) and RGB (0,0,255) configurations, respectively. The combination of these pixels forms an image, with varying colors produced through different intensities of the RGB channels. Furthermore, Fig. 2 describes the LSB pixel and color model.

3.2 LSB Reversal

LSB steganography primarily involves manipulating the least significant bits of these pixels. By reserving the last two bits of each pixel, information can be subtly

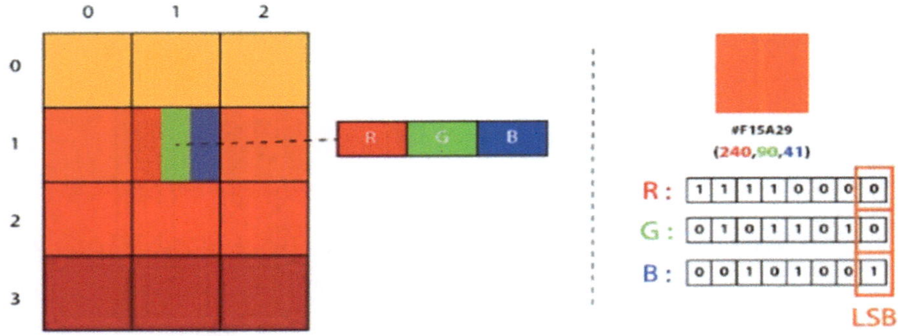

Fig. 2 LSB pixel and colour model

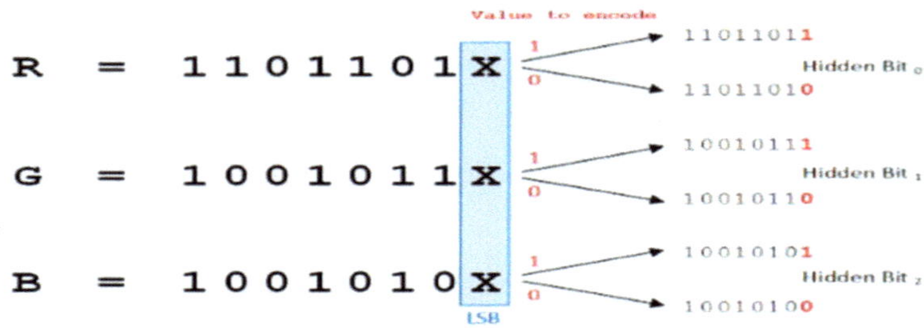

Fig. 3 LSB data-hiding approach

stored within an image without noticeably altering its visual characteristics [14]. These reserved LSBs act as a clandestine vault for hidden data, allowing secure communication through the concealment of information. Display quotations of over 40 words, or as needed. Figure 3 shows the LSB data hiding method.

3.3 Encryption and Decryption

In practical application, according to Zhou et al. [8], LSB steganography involves a sender, a receiver, a confidential message, and a cover image. Further, the authors Heidari et al. in [10] and [11] state that the steganographic software transforms the original cover image into a stego file by embedding the confidential message within the LSBs of its pixels. The resultant stego image appears unchanged to the naked eye but carries a hidden payload. To retrieve the concealed information, the stego image undergoes decryption using the same steganographic software, yielding the original confidential message. Figure 4 visualizes the Steganography encryption schema.

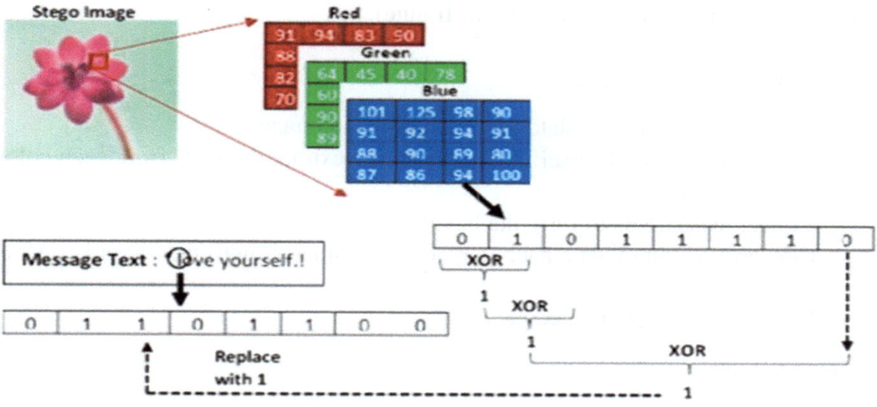

Fig. 4 Encryption schema

3.4 *Steganography Algorithm Steps*

The following algorithm steps show the high-level step-by-step breakdown of the algorithmic flow of the steganography code.

3.4.1 Encoding (Hide Data in Image)

1. Import Libraries: Import necessary libraries such as cv2, numpy, and PIL.
2. Convert Data to Binary:
   - Create a function data2binary(data) that takes either a string, bytes, or numpy array as input.
   - Convert each character in the input data to its 8-bit binary representation.
3. Hide Data in Image:
   - Create a function hidedata(img, data) that takes an image and data to hide.
   - Add a secret key $$ at the end of the data.
   - Convert the image pixel values to binary.
   - Iterate through each pixel in the image:
   - Modify the least significant bit (LSB) of each color channel (R, G, B) with the binary data to be hidden.
   - Continue until all data bits are hidden or until the image pixels are exhausted.
   - Return the modified image.
4. Encode Function:
   - Take user input for the image name, message, and encoded image name.
   - Read the image and call hide data to hide the message.
   - Save the modified image with the hidden message.

3.4.2 Decoding (Retrieve Data from Image)

1. Retrieve Data from Image:
   - Create a function find_data(img) that takes an image.
   - Iterate through each pixel in the image and extract the LSB from each color channel (R, G, B).
   - Collect these LSBs to form the binary data.
   - Convert this binary data back to readable characters until the secret key $$ is encountered.
   - Return the extracted message.
2. Decode Function:
   - Take user input for the image name.
   - Read the image and call find_data to extract the hidden message.
   - Print the extracted message.

3.4.3 Main Steganography Function

1. Main Steganography Function:
   - Create a function stegnography() to provide a user interface for encoding and decoding.
   - Display options for the user to choose between encoding and decoding.
   - Based on the user's choice, either call the encode() function to hide the data in the image or call the decode() function to retrieve the data from the image.
   - Continue the process until the user decides to exit.
2. Execution:
   - Execute the stegnography() function to start the steganography application.
   - The user can choose to encode a message into an image or decode a hidden message from an image based on the options provided.

4 Experimental Results and Discussion

This section describes the details of the outcomes achieved using the proposed methodology. The proposed methodology combines quantum mechanisms with the conventional steganography approach to produce rapid results. Figure 5a is the original sample image, and Fig. 5b is the output of the program that includes the encoded message to be embedded in the 'naturepic' image as well as the conversion of this image into 'stegimg' using the quantum-based steganography algorithm. Next, Fig. 6a represents the encoded image, and Fig. 6b describes the decoding

a b

Fig. 5 Encoded message embedded in the 'naturepic' image. (**a**) Original Image to be encoded. (**b**) Encoding code

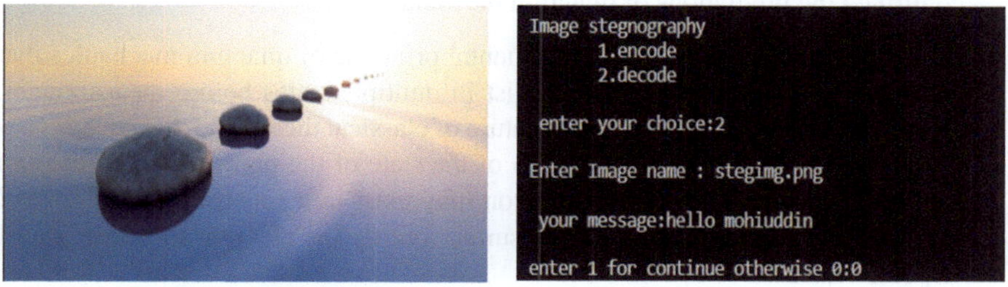

a b

Fig. 6 A decoded message reveals the hidden information in 'stegimg' image. (**a**) Decoded image. (**b**) Decoding code

program code to see the hidden message. Finally, the decoded message reveals the hidden information within the 'stegimg' image.

4.1 Discussion on Major Concepts of Quantum Computing in Steganography

Quantum steganography [6][9] capitalizes on the distinctive attributes of quantum mechanics for secure communication. Principles augmenting the framework of steganography encompass quantum entanglement, the superdense principle, and quantum superposition. Quantum steganography can provide an additional layer of security by encoding information within quantum states that are delicately intertwined.

Quantum entanglement refers to the phenomenon wherein two or more particles get entangled, meaning that the correlation is exhibited between them in a manner that the state of one particle is inherently related to the other without depending on the distance between them, even at the opposite end of the universe. Some of the salient features of quantum entanglement include, as follows:

- Correlation without Local Causation: The state of the corresponding particle becomes immediately discernible upon measuring the state of one particle. Importantly, this correlation occurs at a speed greater than that of light, which violates the conventional understanding of cause.
- No-Communication Theorem: Quantum entanglement does not allow for faster-than-light communication. Even though measurements on one of the entangled particles instantly affect the state of the other, that cannot be used to transmit classical information at faster-than-light speeds.
- Bell's Theorem states that any theoretical approach based on local realism, which is the idea that physical phenomena in a given location are independent of what happens elsewhere in space, cannot explain the correlations predicted by quantum mechanics. Experimental studies testing Bell's theorem have consistently confirmed the predictions of quantum mechanics.

Quantum superposition is the fundamental principle of quantum mechanics that explains how a quantum system can exist in multiple states before measurement. This principle contradicts the intuitive nature of classical theory and, hence, has very strong implications on the interpretation of the state of the particle in the quantum world. The position or spin of an electron may exist in a superposition of several locations or orientations until it is measured, forcing the electron to select one of these possibilities.

- Wave-Particle Duality: Quantum mechanics states that particles can exhibit wave-like and particle-like behavior. Wave-particle duality, in fact, proposes that particles like electrons can be in a state of spread-out probability.
- Schrodinger's Cat: It is a thought experiment; the cat put in a sealed box is thought both to be living and dead until someone opens the box and finds the state of the cat. Therefore, it is an illustration of superposition.

Quantum Superdense Position: Two classical bits of information can be encoded in a single qubit using shared entangled particles. Covert messages are transmitted by manipulating the state of entangled particles. This is achieved through the use of quantum gates that explore the entanglement between the qubits. Here is how the superdense principle works:

- Two particles are entangled. Changes made to one particle's state will immediately affect the other's state, regardless of the distance between them.

Preparation of Entangled Particles: The sender (Alice) prepares an entangled pair of particles and sends one of the particles to the receiver (Bob).

- Encoding Information: Alice wants to send two classical bits of information to Bob. Using quantum superposition, she can encode these two bits into her remaining particle (the one she didn't send to Bob). This is achieved by applying quantum gates to her particle.
- Sending the Particle: Alice sends her particle to Bob through a quantum communication channel.

- Measurement by Bob: Upon receiving the particle, Bob performs measurements on both particles (the one he received and the one remaining with Alice) using a specific set of operations. These measurements allow him to extract the two classical bits encoded by Alice.

Communication Efficiency: Quantum superdense coding allows Alice to send two classical bits using only one qubit (quantum bit) of entangled information. This is more efficient than classical communication, where two classical bits would usually require the transmission of two bits of classical information.

5 Conclusion and Future work

The proposed work talked about the LSB steganography method, which involves changing the LSBs of pixel values in an image in a sneaky way to hide information and set up a secret channel for safe communication. This operational procedure includes a sender, a receiver, a confidential message, and a cover image. Steganographic software facilitates the transformation of the cover image into a stego file by integrating the confidential message within the LSBs of its pixels. The resultant stego image maintains visual consistency but carries a concealed payload that can be unveiled through decryption using the same steganographic software. Classical LSB steganography, as outlined, operates within the realm of conventional computing and cryptographic practices. It exploits the hiding of information within the LSBs of pixel values in a cover image. While effective for various practical applications, classical steganography methodologies such as LSB may encounter vulnerabilities and detection techniques, especially with advancements in digital forensics. Conversely, quantum steganography signifies a potential progression in steganographic methodologies, utilizing the principles of quantum mechanics to enhance security. Quantum steganography may leverage quantum superposition, entanglement, or other quantum phenomena to encode and transmit information in a more secure and potentially indiscernible manner. Quantum bits (qubits) could offer heightened security against classical cryptographic attacks and enhance resistance to detection. This basically means that classical LSB steganography is a good way to hide information in pictures. However, quantum steganography could make the field even better by exploring the unique properties of quantum mechanics to improve security and privacy. The evolution of quantum steganography may open new avenues for secure communication in the future, either complementing or surpassing classical techniques. Quantum computing uses the principles of quantum mechanics, resulting in faster algorithm output, unlike classical computing, which plays a major role in security and many other fields. It has the potential to revolutionize data encryption and decryption methods. This helps to improve the future steganographic and cryptographic quantum applications, including post-quantum cryptography. The implementation of post-quantum cryptography in messages would offer unprecedented high-level security. This would ensure that sensitive information remains protected against potential threats posed by quantum computers.

References

1. Dhawan, S., & Gupta, R. (2021). Analysis of various data security techniques of steganography: A survey. Information Security Journal: A Global Perspective, 30(2), 63–87.
2. McMahon, D. (2008). Quantum computing explained. John Wiley & Sons.
3. Subramanian, N., Elharrouss, O., Al-Maadeed, S., & Bouridane, A. (2021). Image steganography: A review of the recent advances. IEEE access, 9, 23409–23423.
4. Kundur, D., & Ahsan, K. (2003). Practical Internet steganography: data hiding in IP. Proc. Texas wksp. security of information systems.
5. Tudorache, A. G., Manta, V., & Caraiman, S. (2022). Quantum steganography based on the B92 quantum protocol. Mathematics, 10(16), 2870.
6. Min-Allah, N., Nagy, N., Aljabri, M., Alkharraa, M., Alqahtani, M., Alghamdi, D., ... & Alshaikh, R. (2022). Quantum image steganography schemes for data hiding: A survey. Applied Sciences, 12(20), 10294.
7. Jiang, N., Zhao, N., & Wang, L. (2016). LSB based quantum image steganography algorithm. International Journal of Theoretical Physics, 55(1), 107–123.
8. Zhou, R. G., Luo, J., Liu, X., Zhu, C., Wei, L., & Zhang, X. (2018). A novel quantum image steganography scheme based on LSB. International Journal of Theoretical Physics, 57, 1848–1863.
9. Azooz, H. J., Salah, K. B., Kherallah, M., & Naceur, M. S. (2023). Quantum steganography: Hiding secret messages in images using quantum circuits and sift. International Journal of Advanced Computer Science and Applications, 14(10).
10. Heidari, S., & Farzadnia, E. (2017). A novel quantum LSB-based steganography method using the Gray code for colored quantum images. Quantum Information Processing, 16(10), 242.
11. Heidari, S., Pourarian, M. R., Gheibi, R., Naseri, M., & Houshmand, M. (2017). Quantum red–green–blue image steganography. International Journal of Quantum Information, 15(05), 1750039.
12. Wayner, P. (2009). Disappearing cryptography: information hiding: steganography and watermarking. Morgan Kaufmann.
13. Ke, Y., Liu, J., Zhang, M. Q., Su, T. T., & Yang, X. Y. (2018). Steganography security: Principle and practice. Ieee Access, 6, 73009–73022.
14. Huang, Y., Liu, C., Tang, S., & Bai, S. (2012). Steganography integration into a low-bit rate speech codec. IEEE transactions on information forensics and security, 7(6), 1865–1875.
15. ALabaichi, A., Al-Dabbas, M. A. A. A. K., & Salih, A. (2020). Image steganography using least significant bit and secret map techniques. International journal of electrical & computer engineering (2088-8708), 10(1).
16. Basumatary, S. (2023). India-Bhutan Security Cooperation: A Study on'Operation All Clear'and Dokhlam Crisis (Doctoral dissertation).
17. Simon, J. E. S. P. E. R. (2023). A serious game on image steganography training for students in higher education (Bachelor's thesis, University of Twente).
18. Ghosal, S. K., Mukhopadhyay, S., Hossain, S., & Sarkar, R. (2021). Application of Lah transform for security and privacy of data through information hiding in telecommunication. Transactions on Emerging Telecommunications Technologies, 32(2), e3984.
19. Kalita, M., Tuithung, T., & Majumder, S. (2019). An adaptive color image steganography method using adjacent pixel value differencing and LSB substitution technique. Cryptologia, 43(5), 414–437.

A Survey on Securing the Future of Electric Vehicles and Its Charging Stations

K. Selvamani, S. Kanimozhi, Shamini Murali, and H. Riasudheen

Abstract The rapid increase in use of Electric Vehicles (EVs) and its tremendous advantages in driving have transformed the transportation systems thereby promoting sustainability in reducing carbon emissions. However, the incorporation of Electric Vehicles with charging infrastructure and the extensive power grid initiated vital challenges in cyber security and physical safety in Electric Vehicles & Electric Vehicles Charging Stations (EVCS). This paper examines these issues; this study draws attention to how susceptible electric vehicle charging stations are to physical threats as well as cyber-attacks such as Denial of Service (DOS), Distributed Denial of Service (DDOS) and Denial of Charge (DOC) attacks. The key approach for mitigating these risks needs robust authentication policies, secure communication protocols and real-time monitoring systems. This survey also highlights in determining the emerging technologies such as block chain and Artificial Intelligence (AI) for enhancing the security of electric vehicle and it's charging stations. Hence, this paper addresses both cyber threats as well as physical threats in detail. Also, this research article aims in ensuring the reliability and safety of electric mobility to encourage the revolution in the prompt expansion of Electric Vehicles ecosystem

Keywords Electric vehicles · Charging stations · Cyber security · Physical threats · Artificial intelligence

K. Selvamani (✉)
Department of CSE, CEG Campus, Anna University, Chennai, Tamil Nadu, India

S. Kanimozhi
Panimalar Engineering College, Anna University, Chennai, Tamil Nadu, India

S. Murali
Chennai Institute of Technology, Anna University, Chennai, Tamil Nadu, India

H. Riasudheen
Department of IST, CEG Campus, Anna University, Chennai, Tamil Nadu, India

K. Deep et al. (eds.), *Artificial Intelligence, Machine Learning and Intelligent Systems*, Springer Proceedings in Mathematics & Statistics 520,
https://doi.org/10.1007/978-981-95-3841-6_32

1 Introduction

The adaptation to Electric Mobility (EM) have made possible to achieve global sustainability goals. Due to this transition, Electric Vehicles offers numerous advantages such as reduced carbon emissions and less operational costs. However, their broad adoption has immediately brought forth new difficulties in the fields of EV charging station physical safety and cyber security. In the ecosystem of electric mobility, charging stations for electric vehicles are a vital link that is increasingly being attacked by physical and cyber-attacks. The integration of Electric Vehicles into our transportation infrastructure presents both significant opportunities and formidable challenges, particularly concerning cyber security and physical safety. As the reliance on Electric Vehicles charging stations grows, Strong security measures are also necessary to guard against physical harm and cyber-attacks.

These vulnerabilities stem from the integrated connections between electric vehicles and charging stations, which involve hardware-software integrations, Cloud-Based Services (CBS), and communication protocols. Because of this, cybercriminals target electric vehicle charging stations in an effort to interfere with charging, compromise user data, and possibly cause power system instability. Furthermore, physical threats like hardware tampering or vandalism create extra hazards and have the potential to result in dangerous situations like electrical fires. This survey paper examines the intricate problems related to EV charging station security and offers a comprehensive plan that takes into account both cyber and physical threats. This article [1] mostly focuses on dealing with cybercriminals' Denial of Service, Distributed Denial of Service, and Denial of Charge assaults. Selvamani K et al. effectively countered distributed denial of service attacks on cloud infrastructure and provided appropriate model for preventing such attacks in cloud environment.

To enhance the functionality [2] of Electric Vehicles, Vehicle-to-Vehicle (V2V), Vehicle-to-Infrastructure (V2I), and Vehicle-to-Grid (V2G) communications were used in electric mobility which in turn raises the system vulnerabilities. For example, insecure communication channels can be easily exploited by hackers to intercept data or send malicious commands to Electric vehicles. Hence, secure communications protocols and strong encryption protocols should be implemented for data transmission to protect against interception. In addition, authentication mechanisms are essential to ensure that only authorized devices can communicate with Electric Vehicles and charging infrastructure. Regular software updates are also crucial for keeping vehicle software up-to-date to patch vulnerabilities and enhance security features.

This research article mainly attempts to explore all the possible attacks and threats towards the electric vehicles as well as its charging stations that are caused by various attackers and hackers. In addition, this paper aims in providing suitable solutions and mechanisms to protect and prevent the electric vehicle and it's charging stations from such attackers and hackers. The remainder of this paper (a) explores the complete survey over the existing works and its limitations in the literature review, (b) presents suitable techniques and mechanisms to prevent various attacks and finally (c) concludes the proposed work and suggests possible future directions for detecting and preventing new types of attacks.

2 Literature Review

This literature review aims in providing detailed survey over the existing works carried out by various researchers on cyber and physical threats on Electric Vehicles and its charging stations. Many works have been carried out in the past by various researchers in the field of cyber and physical security for electric vehicle & electric vehicles charging stations. Also, this literature survey focuses to identify a research gap and what it impacts on variability management in automotive systems and to also to enlist the merits and de-merits of the existing research works.

Among them, Philipp Zellmer et al., proposed a research work [3] to address the challenges in modern automotive systems in user's perspective. It also explains the growing complexity and cyber security challenges that modern automotive systems encounter in current scenario. In addition, it also stresses the importance of effective variability management to tackle these challenges without compromising the software quality. Most Modern vehicles progressively depend on software and hardware connectivity to enhance their functionality. But, this integration brings significant cyber security risks that have to underscores the necessity of managing variability in automotive systems to effectively mitigate these risks. They identified four primary challenges such as (i) enhanced connectivity introduces vulnerabilities, (ii) over-the-air updates can expose vehicles to external threats, (iii) different vehicle functions have varying impacts on cyber security, and (iv) adapting to evolving regulations is complicated by the diversity of vehicle architectures. Also, they have addressed these four challenges with several strategies to mitigate such types of cyber-attacks and concluded that addressing cyber security-related variability is crucial throughout the vehicle lifecycle. It also highlights how modern automotive systems' complexity and connectivity also apply to electric vehicles.

As electric vehicles rely heavily on software for features like over-the-air updates, they become targets for cyber threats such as hacking and unauthorized access. Finally, it strongly recommends on regular software updates for electric vehicles which it can help to mitigate vulnerabilities arising from new threats. Since, charging stations are critical points where cyber and physical systems intersect and hence breaches could disable safety mechanisms or manipulate charging processes, resulting in hazards like fires or electric shocks. The findings from this research paper significantly enhance understanding and management of cyber security risks in electric vehicles and their charging infrastructure, particularly regarding user safety at public charging stations. By applying these principles, stakeholders can work towards creating a safer environment for EV users while effectively addressing both cyber and physical threats.

An Examination of Electric Vehicles: cyber security, Energy Trading, and Technologies were presented by Mohan Bharathidasan et al. which thoroughly examine [2] the integration with renewable energy sources in electric vehicles and the associated cyber security challenges. They have elaborated the need for robust cyber security measures for electric vehicles that are integrated into smart grids and connected ecosystems. The authors emphasize that electric vehicles are not merely vehicles but integral components of a larger network that includes charging stations, energy management systems, and communication technologies. This interconnected system exposes the cyber attackers to easily introduce various cyber threats that to compromise both vehicle safety and data integrity.

The several potential cyber threats specific to electric vehicles are data breach that poses significant risk, compromise on sensitive user information such as location and driving habits which are targeted by malicious actors. Also, vehicle control attacks pose a serious threat, for hackers to gain access illegally to control or disable the vehicle which potentially leads to dangerous scenarios. The authors also stress to use the standardized security protocols to ensure the safety of information communicated between the manufacturers, software developers, and regulatory bodies. This collaboration is essential for creating a unified approach in providing cyber security in the Fast-paced landscape of electric vehicles. Moreover, they have suggested that researchers should develop an advanced security methods to predict and respond to potential cyber physical threats in real-time using Artificial Intelligence (AI) and Machine Learning (ML) techniques.

Hamed Alqahtani and Gulshan Kumar presented [4] the threats and challenge that analyses the cyber security landscape surrounding Electric and Flying Vehicles (EnFVs). As EnFVs proportionally rely on digital technologies and connectivity, they face a myriad of cyber security threats which includes cyber-physical attacks as well as data breaches. The authors accentuate the interconnected nature of these vehicles which attracts malicious actors to target the attack points. A real time example cited is the infamous 2015 Jeep Cherokee hacking incident illustrates how compromised vehicle systems can lead to dangerous outcomes, including manipulation of driving behavior and potential criminal exploitation. This presented paper also notes a troubling trend in cyber-attacks within the automotive sector, with fluctuations in attack frequency observed from 2022 to 2023, underscoring the urgent need for robust cyber security measures to protect against evolving threats.

To battle these cyber security challenges, several AI-driven solutions such as AI-powered intrusion detection systems that leverage machine learning algorithms to identify and respond relevantly to potential threats in real-time are addressed. In addition, predictive analytics were utilized to analysis the threats and mitigate risks before they materialize. Hence, it is recommended to use enhanced authentication protocols to safeguard against unauthorized access. They have also outlined future research directions for advancing EnFV cyber security is using quantum computing resilience to investigate how this technology can be harnessed to bolster security measures against sophisticated cyber threats. This study highlights the importance of integrating effective cyber security measures in the development of EnFVs. By illuminating the intersection between transportation innovation and cyber security, the authors provide valuable insights for safeguarding these vehicles against potential risks. In summary, this paper contributes significantly to understand EnFV cyber security dynamics by identifying key challenges, proposing AI-driven solutions, and outlining essential future research directions for developing secure and resilient electric and flying vehicle ecosystems.

Cyber security for Smart Electric Vehicle Charging: A Power Grid Perspective was addressed by Samrat Acharya et al. it examines and addresses [5] the increasing risks associated with the integration of electric vehicles, electric vehicle charging stations, and power grids in cyber security. Due to global acceptance of Electric Vehicles, advancements in battery technology and user-friendly charging features, the interaction among these components creates complex cyber-physical interdependencies that are vulnerable to malicious exploitation. The authors enlighten the key vulnerabilities such as weak authentication protocols in EVCSs, unsecured communication systems, exploitable software flaws in smartphone applications used for charging management.

The impact of these vulnerabilities on power grids is significant. EVCSs are high-wattage, internet-connected devices that can be manipulated to launch demand-side cyber-attacks, potentially destabilizing power grids. For example, attackers could induce over-frequency or under-frequency conditions by orchestrating sudden changes in EV charging demand, which could disrupt grid operations and lead to cascading failures. To analyze these risks, the study employs threat models such as STRIDE to assess spoofing, tampering, Denial-of-Service (DoS), and data theft scenarios. Attack scenarios discussed include DoS attacks on EVCSs, tampering with charging schedules, and exploiting smartphone applications to gain unauthorized access to EVCS servers or networks.

However, this paper records that enforcement and consensus among take holders remain weak. The authors recommend strengthening encryption and authentication mechanisms across interfaces involving Electric Vehicles, EVCSs, and power grids. They also advocate in standardizing the cyber security practices globally for EV charging infrastructures and to make them aware among manufacturers, grid operators, and regulators to mitigate potential attack vectors. By providing a comprehensive review of vulnerabilities, potential attack scenarios, and mitigation strategies, it aims to ignite research and regulatory actions in this critical area. The findings underscore the need for a proactive approach to cyber security that integrates robust measures across all components involved in electric vehicle charging systems.

Cyber Vulnerabilities, Potential Impacts, and Possible Strategies were studied by Long Vu et al. to examine [6] cyber vulnerabilities and potential impacts on electric vehicles integration with cloud computing. The rapid growth in Electric Vehicles market and increased connectivity, enhanced functionality and user experience also exposes them to significant cyber security risks. The paper accentuates the importance of connectivity in modern EVs, which allows for various functions such as remote monitoring and software updates but also increases vulnerability to cyber-attacks. The authors also discussed the utilization of communication technologies, such as Dedicated Short Range Communications (DSRC), WiFi, and Bluetooth to interact with external systems like charging stations and other vehicles in modern Electric Vehicles. This connectivity is unavoidable to enhance the functionality and user experience but it initiates multiple attack vectors.

This survey exposes various cyber vulnerabilities associated with Electric Vehicles due to their interconnected nature. Common attack vectors such as malware and eavesdropping, along with its types of attacks like message spoofing and denial-of-service were categorized. It also, highlights the potential impacts of these attacks, which include loss of mobility, data leakage, remote control risks, and financial losses. To mitigate these risks, several strategies were proposed for both users and manufacturers which recommend regular software updates, restricting in-vehicle wireless services when not in use, avoiding untrusted applications, employing encryption methods, and conducting regular security assessments.

The author also discusses current limitations in Electric Vehicles charging infrastructure and accentuates the need for secure communication protocols. They highlight the necessity for improved privacy-preserving techniques as Electric Vehicles become more autonomous. The paper underscores that while accentuates present numerous advantages and opportunities for sustainable transportation, their integration with cloud technology poses significant cyber security challenges. The authors call for ongoing research into security measures to protect against these vulnerabilities as technology continues to evolve. This comprehensive study serves as a valuable resource for researchers focused on the intersection of electric vehicle technology and cyber security, encouraging further exploration in this critical area.

Battery Performance enhancements using Artificial Intelligence and Mitigating Cyber security Threats in Electric Vehicles was [7] developed by H Kanz Abdillah et al. It underscores the critical role of Electric Vehicles in reducing carbon emissions while highlighting challenges such as battery performance optimization and cyber security threats. The paper denotes that Artificial Intelligence is instrumental in developing advanced Battery Management Systems (BMS), which are crucial for monitoring battery health, improving efficiency, and extending battery life. The Ensemble Regression model emerged as the most effective method for predicting battery life, achieving an impressive accuracy of 94.8% was coined by them. In addition, few machine learning and deep learning techniques, such as Long Short-Term Memory (LSTM) networks and Adaptive Neuro-Fuzzy Inference Systems (ANFIS), have been utilized for tasks like State of Charge (SOC) estimation and energy optimization. These models not only enhance efficiency but also reduce battery degradation and ensure better energy management.

Techniques such as reinforcement learning, k-means clustering, and hybrid deep learning models are employed to optimize charging strategies, thereby reducing waiting times and improving overall battery performance. In addition, it identifies several vulnerabilities specific to Electric Vehicles, including GPS spoofing, malware attacks, and denial-of-service attacks, which pose significant risks to vehicle safety and data integrity. To mitigate these attacks, block chain-based smart contracts was proposed to secure transactions at charging stations, ensuring transparency and trust. The integration of block chain technology with Artificial Intelligence for robust security measures and optimal battery management presents a promising area for further exploration. This comprehensive study serves as a valuable resource for researchers focused on the intersection of electric vehicle technology and cyber security, encouraging further exploration in this critical area.

Khaled Sarieddine et al., explores the security vulnerabilities present [8] within electric vehicle charging ecosystems that focus on mobile applications. As the adoption of Electric Vehicles rises and their integration into power grids becomes more prevalent, these mobile applications have become essential for managing charging stations. The identification of multiple vulnerabilities in mobile applications was the key findings of the study. The first flaw noted is Unverified Ownership, where mobile apps often fail to verify whether a user actually owns the connected electric vehicle, allowing unauthorized control over charging operations. The second flaw is Improper Authorization, which reveals that critical functions like initiating or stopping charging lack robust user authentication mechanisms.

It also outlines several potential attack scenarios that could exploit these vulnerabilities. Session Hijacking is one such scenario where attackers can create fake accounts to initiate unauthorized charging sessions, effectively taking control away from legitimate users. Coordinated Attacks could involve botnets of malicious accounts exploiting these vulnerabilities to launch large-scale attacks on EVCS networks, potentially impacting the connected power grid. To identify these vulnerabilities, the study employed both static (reverse engineering) and dynamic (traffic and functionality) analyses. The implications of these vulnerabilities extend to critical infrastructure; attackers could leverage them to disrupt grid operations by overloading or tripping transmission lines, leading to economic impacts such as increased transmission losses and generation costs. In response to these findings, the authors provide several recommendations aimed at improving security.

Additionally, enhancing authorization mechanisms is crucial to restrict access to critical functionalities. The development of resilience strategies, including attack detection and mitigation systems, is also recommended to prevent large-scale coordinated assaults on the grid. The integration of mobile applications into EV charging ecosystems introduces critical cyber security risks that can jeopardize both individual users and the stability of the power grid. Addressing these vulnerabilities requires immediate attention to secure the EVCS ecosystem and protect its stakeholders effectively.

The cyber security challenges faced by electric vehicle [9] charging stations (EVCS) were addressed by Z. Pourmirza et al. As EVs become more integrated into the power grid, the security of these charging stations is critical due to their reliance on a combination of hardware, software, and communication systems. This integration creates multiple attack vectors that can be exploited by malicious actors. One significant challenge is the integration of various technologies within EVCS. Potential threats include physical attacks where unauthorized access to hardware can compromise systems, data breaches where sensitive user information is targeted by hackers, and system manipulation where malicious actors exploit EVCS software to disrupt services or manipulate power loads. Additionally, denial-of-service attacks can overload the EVCS network, rendering it inoperable and causing inconvenience to users while leading to financial losses for operators.

Communication vulnerabilities also pose a significant risk due to most EVCS utilize wireless communication methods that are prone to interception and unauthorized access. Protocols such as the Open Charge Point Protocol (OCPP) may have vulnerabilities if it is not secured properly. While incident response strategies should be put in place to efficiently handle and mitigate breaches, on-going monitoring of EVCS networks can assist in identifying anomalies suggestive of possible assaults. Additionally, implementing secure hardware designs can reduce physical security risks; tamper-resistant designs and secure firmware updates are necessary to prevent malicious code insertion. In addition, Collaboration between EVCS operators and grid operators is crucial to ensure that charging activities do not compromise the overall stability and security of the power grid. Regulatory and policy frameworks must be enforced by governments and standardization bodies to establish guidelines for secure installation, maintenance, and operation of charging stations.

As EVCS become integral components of smart infrastructure, addressing cyber security challenges is imperative. Proactive measures such as secure design practices, real-time monitoring, and adherence to standards are necessary to ensure the safe and reliable operation of these systems. The importance of a multi-layered approach that combines technical solutions with regulatory efforts cannot be overstated in effectively mitigating risks associated with EV charging stations. The above literature review clearly indicates that, even though many works has been proposed in the past to mitigate cyber-attacks, there is still a strong need to address the cyber-attacks that effectively prevents the EV charging stations. Hence, this paper attempts in providing effective solutions for various cyber-attacks executed by cyber criminals.

3 Proposed Mechanisms for Cyber Attacks

It is clear from the literature analysis that appropriate security improvements must be included to enable the seamless operation of electric cars and their charging stations free from physical and cyber threats. The three primary assault types namely Denial of Service (DOS), Distributed Denial of Service (DDOS), and Denial of Charge (DOC) should be avoided to the greatest extent possible by examining system data and charging station traffic patterns. Every charging station should also have enough intrusion detection and prevention systems installed in order to guard against physical invasions and anticipate unusual user behaviour when they log in. Additionally, it is crucial to identify charging station vulnerabilities and other essential.

In addition, to enhance the cyber security of electric vehicles and their associated charging stations, a comprehensive approach is necessary that addresses both digital vulnerabilities and physical threats. As the rise in adoption of electric vehicles, their integration various components into the power grid create multiple attack vectors for cybercriminals. Hence, this article suggests that key strategies should be implemented for robust authentication mechanisms and Multi-Factor

Authentication (MFA) should be applied to ensure that only verified users can access charging functionalities. There are many authentication approaches which are in today's environment. From this primarily Multi-factor authentication is should be employed for protecting electric vehicles and its charging stations from cyber-attacks. This MFA provides enough security measures and ensures that only authenticated users are connected EV charging stations thereby denying the access to malicious users into the network.

Secondly, Block chain based authentication should be employed in order to provide and ensure the authenticity, confidentiality and integrity of the process of communication between the user and its charging stations as well as central system in the network. These two employed authentication mechanisms in the network will make sure that the network will have only the legitimate or authenticated users thereby the entire network is secured. If the denial of Service and distributed denial of service attacks are prevented then, the traffic in the network is minimized thereby the network is said to be secured.

Additionally Secure communication protocols, such the Open Charge Point Protocol (OCPP) improved version, should also be put in place to guard against illegal access and interception of data sent between electric cars and charging stations. The charging station and the central system can communicate uniformly thanks to this open charge point protocol. By ensuring secure communication between the user and the charging station, as well as connecting to and communicating with any charging station that is connected to the central system, are the primary goals and benefits of utilizing this protocol. More intriguingly, this protocol can be readily built to effectively block such assaults utilizing JavaScript Object Notation (JSON) or Simple Object Access Protocol (SOAP).

The above Fig. 1 shows the basic authentication process between the EV users and EV charging stations. The proposed mechanism ensures the only the authenticated EV users will be allowed in to the central system for having communication with the EV charging stations. This mechanisms prevents the three main types of attacks namely Denial of Service, Distributed denial of service and denial of charge attacks. Here, the open charge point protocol which is a secured communication protocol that provides secured access between the electric cars and charging stations, thereby preventing the illegal access to hackers or cyber criminals.

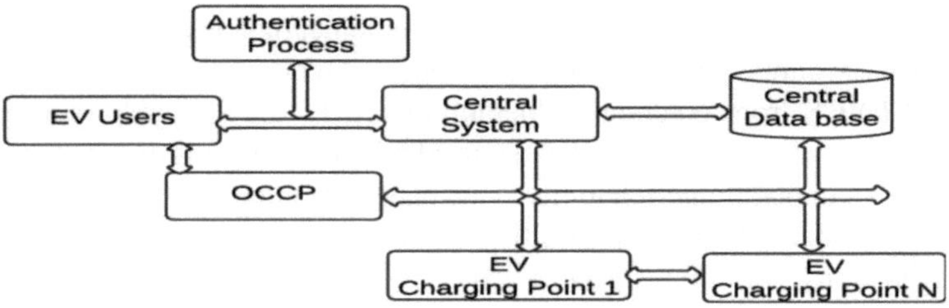

Fig. 1 Authentication process between EV users and EV charging stations

Furthermore, encryption of data is crucial to safeguard sensitive information, including personal and financial data collected by electric vehicle charging station. Continuous real-time monitoring systems will help to detect anomalies indicative of potential cyber-attacks, while incident response plans should be developed to address breaches efficiently. The deployment of AI-driven intrusion detection systems can significantly enhance threat detection capabilities, allowing for rapid response to identified threats. In addition to cyber security measures, protecting electric vehicles from physical harm, such as ballistic damage, is essential. This can be achieved through the development of Modular Ballistic Protection Systems (MBPS) around charging stations that can withstand impacts. This modular ballistic protection system protects vulnerable structures as well as electric vehicles and various equipment's from different range of threats. Implementing tamper-resistant hardware designs will also reduce physical security risks. Moreover, collaboration with local law enforcement agencies can enhance security around charging stations in high-risk areas prone to vandalism or attacks.

To find weaknesses in EVCS's physical installations as well as digital infrastructure, regular security audits and evaluations are required. Because, regular security audits and assessments helps the business persons to pin down and mitigate the risks prior to being taken advantage of by cyber attackers. Stakeholders must also engage in public awareness campaigns to educate EV users about potential cyber security threats associated with EV charging and promote best practices for safe usage. This awareness campaigns not only educate the EV users, but also make them routine practices to safe guard themselves in advance from such types of cyber criminals. Using a multifaceted strategy that incorporates community involvement, regulatory initiatives, and technical solutions, we can effectively mitigate risks associated with electric vehicles charging infrastructure and enhance the overall safety and reliability of electric mobility.

4 Conclusion and Future Work

Due to the reliance on electric vehicles and its charging stations growing, Strong security measures must be developed in order to guard against physical damage to the electric car and it's charging stations as well as cyber-attacks. These proposed strategies highlight the importance of implementing advanced authentication mechanisms, secure communication protocols, and real-time monitoring systems to safeguard EVCS from potential vulnerabilities. Additionally, addressing physical security through protective measures against ballistic damage is crucial for ensuring user safety and operational reliability. This paper has attempted to provide suitable solutions and mechanisms in preventing from various cyber-attacks in EV and its charging stations.

Further research is essential to explore innovative solutions that can enhance the cyber security framework of electric vehicles and their charging infrastructure. Areas of focus should include the development of more sophisticated AI-driven intrusion detection systems, the application of block chain technology for secure transactions, and the establishment of standardized protocols for communication between electric vehicles and charging stations. Additionally, investigating the interplay between cyber security measures and physical security strategies will be vital in creating a holistic approach to safeguarding electric vehicles ecosystems.

References

1. Divyasree IR and Selvamani K, "DAD: Domain Adversarial Defense System Against DDoS Attacks in Cloud", IEEE Transactions on Network and Service Management, Vol. 19, No. 1, pp. 554-568, March 2022.
2. Mohan Bharathidasan, V. Indragandhi, Vishnu Suresh, Michał Jasiński, Zbigniew Leonowicz, "A review on electric vehicle: Technologies, energy trading, and cyber security, Energy Reports", Volume 8, Pages 9662-9685, ISSN 2352-4847, 2022.
3. Philipp Zellmer, Lennart Holsten, Richard May, and Thomas Leich. "A Practitioners Perspective on Addressing Cyber Security and Variability Challenges in Modern Automotive Systems", In the Proceedings of the 18th International Working Conference on Variability Modeling of Software-Intensive Systems (VaMoS '24), pp. 129–133, ACM, USA, 2024.
4. Hamed Alqahtani and Gulshan Kumar, "Cyber security in Electric and Flying Vehicles: Threats, Challenges, AI Solutions & Future Directions", ACM Computing Survey Accepted September 2024.
5. Samrat Acharya, Yury Dvorkin, Hrvoje Pandžić, Ramesh Karri, "Cyber Security of Smart Electric Vehicle Charging: A Power Grid Perspective", IEEE Access, Vol. no. 8, pp. 214434, 2020.
6. Long Vu, Kun Suo, Md Romyull Islam, Nobel Dhar, Tu N. Nguyen, Selena He, and Yong Shi., "Living on the Electric Vehicle and Cloud Era: A Study of Cyber Vulnerabilities, Potential Impacts, and Possible Strategies", In Proceedings of the 2024 ACM Southeast Conference (ACMSE '24), pp. 18–26, ACM, USA, 2024.
7. H Kanz Abdillah, N.A.H. Wildan Rizkia, Sidharta Sidharta, "Exploring the role of artificial intelligence in enhancing battery performance and mitigating cyber security threats in electric vehicles: A systematic literature review", Procedia Computer Science, Vol.245, Pp. 155-165, 2024.
8. Khaled Sarieddine, Mohammad Ali Sayed, Sadegh Torabi, Ribal Atallah, and Chadi Assi, "Investigating the Security of EV Charging Mobile Applications as an Attack Surface", ACM Transactions on Cyber-Physical Systems, Vol. 7, issue. 4, Article 26 pp. 1-28, 2023.
9. Z. Pourmirza and S. Walker, "Electric Vehicle Charging Station: Cyber Security Challenges and Perspective", In the proceedings of IEEE 9th International Conference on Smart Energy Grid Engineering (SEGE), Oshawa, ON, Canada, pp. 111-116, 2021.

Automatic Speech Recognition for Telugu Language Using Pre-trained Wav2Vec 2.0 Model

Sri Gani Kaarthikeya Kammula, Gunasekhar Devineni, Chiranjeevi Karanki, Muzaffar Ahmad Dar, and Jagalingam Pushparaj

Abstract This study focuses on creating a system for automatic recognition of speech tailored for Telugu, a spoken language in India. Deep learning (DL) techniques have been extensively utilized to create ASR systems across diverse languages and domains in recent years. These models, however, require substantial training resources, particularly large corpora of continuous speech utterances collected from multiple speakers, along with their corresponding transcripts. Unfortunately, such comprehensive speech datasets are often unavailable for many Indian languages, including Telugu. This study utilizes a pre-trained Wav2Vec 2.0 model for developing an ASR system in the Telugu language, addressing the challenge of limited data availability and demonstrating remarkable results even when fine-tuned on a limited dataset. We fine-tuned Wav2Vec 2.0 on OpenSLR and Mozilla Common Voice, achieving a 25.6% word error rate (WER) and a 4.5% character error rate (CER) on OpenSLR, and an 18% CER on Mozilla Common Voice. These results demonstrate its efficacy in languages with data disparities and highlight the effectiveness of leveraging pre-trained architectures in low-resource linguistic settings.

Keywords Automatic speech recognition · Telugu language · Pre-trained models · Wav2Vec 2.0 · Fine-tuning

S. G. K. Kammula · G. Devineni · C. Karanki · J. Pushparaj (✉)
Department of Computational Intelligence School of Computer Science and Engineering, Vellore Institute of Technology, Vellore, India
e-mail: jagalingam.p@vit.ac.in

M. A. Dar
Department of Mathematics, School of Advanced Sciences, Vellore Institute of Technology, Vellore, India

K. Deep et al. (eds.), *Artificial Intelligence, Machine Learning and Intelligent Systems*, Springer Proceedings in Mathematics & Statistics 520,
https://doi.org/10.1007/978-981-95-3841-6_33

1 Introduction

ASR systems convert spoken language (an audio signal) into written text, often used as a command. Initially, the focus was not on comprehending or identifying speech, but rather on developing a machine capable of speech production. This may have been influenced by the accessible understanding of acoustic resonance tubes, which were used to mimic the human vocal tract [1]. This method analyzed speech's physical properties to map acoustic cues to phonetic features, later advancing with statistical approaches. As research progressed, more sophisticated statistical approaches came into use. While the fundamental concept of the hidden Markov model (HMM) was recognized and comprehended in certain research labs early on, the approach remained incomplete until the mid-1980s. It was only after the broad dissemination of the theory that the HMM gained popularity as the dominant technique for speech recognition [1]. HMMs improved speech recognition by handling variability probabilistically but struggled with complex, multilingual, and low-resource scenarios. Statistical methods struggled with natural speech variability, prompting a shift to ML and DL in ASR. Advanced DL algorithms now handle domain-specific terms, languages, accents, and noisy environments with remarkable success. Transformers have revolutionized ASR, driving innovations in NLP. End-to-end ASR systems streamline speech-to-text conversion, bypassing complex preprocessing and requiring less domain expertise. This evolution has improved performance and broadened access to speech recognition across languages and domains. Various approaches and types of speech recognition systems gradually emerged over the past five decades [2].

Telugu is a complex language with unique phonetic features, including Śruti, which ensures accurate pronunciation through consistent grapheme-phoneme correspondence. Its rich oral tradition and connected speech lead to rhythmic, melodic, and fluent spoken forms. Additionally, the grammar is heavily inflected, featuring numerous inflections and syntactic variations that contribute to the language's complexity [3], [16]. The future of ASR, especially for low-resource systems like Telugu, presents both opportunities and challenges. Leveraging DL models for cross-lingual knowledge transfer offers promising solutions, but the scarcity of datasets and disparities between low-resource and widely spoken languages complicate the development of accurate ASR systems. This research concentrates on optimizing the pre-trained Wav2Vec 2.0 model to improve its capability to recognize Telugu speech with greater efficiency and precision. By capitalizing on the powerful feature extraction capabilities of these models, we aim to overcome the unique challenges posed by Telugu language data, thereby optimizing ASR performance in low-resource settings.

The structure of this manuscript is as follows: Section 2 provides an overview of existing studies; Section 3 describes the approach utilized; Section 4 examines and interprets the findings; and Section 5 wraps up with potential avenues for future investigation.

2 Related Work

This section presents a review of the existing literature on ASR, highlighting various methodologies, advancements, and challenges that have influenced the field over time. [4] explored subword-based language models in ASR with weighted finite-state transducers, HMMs, and DL networks, using extensive datasets for Finnish, Arabic, Swedish, and English. Their results showed that subword models, especially those from Morfessor, outperformed word-based models in reducing WER, validated by n-gram and RNN architectures. [5] proposed an ASR framework using BiLSTM networks with attention and fusion structures. Experiments on the SRA and SRB datasets showed that the FF-ATT-BiLSTM model achieved 92.1% and 95.5% accuracy, outperformed transformers by 2.2–3.2% on SRA and 2.3–2.9% on SRB, and benefited from attention, KSVD, and dropout mechanisms. This study [18] explored speech recognition (SR) for the Kashmiri language using a dataset of 7200 numeral samples. A Bi-directional LSTM model achieved 85.28% accuracy with Mel-spectrogram features, optimized using a 25 millisecond (ms) frame size, 20 ms overlap, and the Adam optimizer. This work [6] proposed a self-supervised learning approach utilizing multiple perspectives and integrating features at various scales for the Aishell-1 dataset. It enhances ASR performance while overcoming the challenges associated with connectionist temporal classification (CTC) and recurrent neural network transducer (RNN-T) models, particularly in reducing CER and enhancing recognition. The study [19] proposed a hybrid CNN-BLSTM framework for ASR, leveraging the CTC technique to align input speech with output text. Experiments on the LJ speech dataset revealed that increased training samples improved recognition performance but extended training time, particularly when accuracy was lower. The proposed model achieved 36.97% WER on the testing dataset.

The author [7] combined architecture integrating DL networks with HMMs for ASR, utilizing attention mechanisms and lattice re-scoring techniques, evaluated on the AMI-IHM and SWB-300 datasets, finding that DNN-HMM systems had higher WERs due to fewer parameters, while AED models, particularly AED-TFM, were more efficient in lattice rescoring. [8] examined the influence of speech-derived haptic simulation on recognizing speech in noisy environments using the French Sentence Test for Speech Intelligibility in Noise (FIST). A linear mixed model (LMM) analysis highlighted the impact of vibrotactile simulation on speech reception thresholds, revealing its potential to enhance speech-in-noise recognition performance. [9] introduced a real-time ASR system utilizing sequence-to-sequence DL models for both audio-only and audio-visual speech recognition, employing the LRS2 dataset. While the audio-visual model underperformed offline Transformer models, it reduced computational costs and hardware dependencies, providing a practical real-world alternative. [10] explored self-supervised pre-trained frameworks for voice quality classification using speech and NSA signals, achieving 94.8% accuracy with speech signals and 92.1% with NSA data, showcasing the effectiveness of TDN-based classifiers.

The pronounced disparity in data availability between languages with abundant resources and those with limited resources significantly influences the performance of ASR models, as evidenced by studies such as [4] and [6]. Furthermore, [5] emphasizes the inherent complexity in designing systems capable of addressing the diverse linguistic and phonetic intricacies associated with such languages. To address these challenges, this study enhances Telugu ASR by fine-tuning the pre-trained Wav2Vec 2.0 framework, leveraging advanced techniques to overcome limited annotated datasets and capture Telugu's linguistic and acoustic nuances.

3 Methodology

This paper develops an efficient ASR framework for Telugu, a language with limited annotated datasets, using Wav2Vec 2.0 using self-supervised learning on raw audio. The comprehensive methodology employed in this study is illustrated in Fig. 1.

3.1 Dataset Used and Data Processing

Collecting data for low-resource languages like Telugu presents significant challenges. To address this, we utilized the publicly available OpenSLR Telugu dataset (OpenSLR66) [13], which encompasses a diverse range of dialects, accents, and voice samples from both male and female speakers in an equal 50–50 ratio. The dataset comprises approximately six hours of audio recordings accompanied by corresponding transcriptions. For the development of the framework, 80% of the data was designated for training purposes, with the remaining 20% set aside for evaluation [14]. Additionally, the Common Voice dataset for Telugu is also used for evaluation and consists of 3 recorded hours with 46 different voices, which are crowdsourced. The preprocessing was undertaken to enhance input quality.

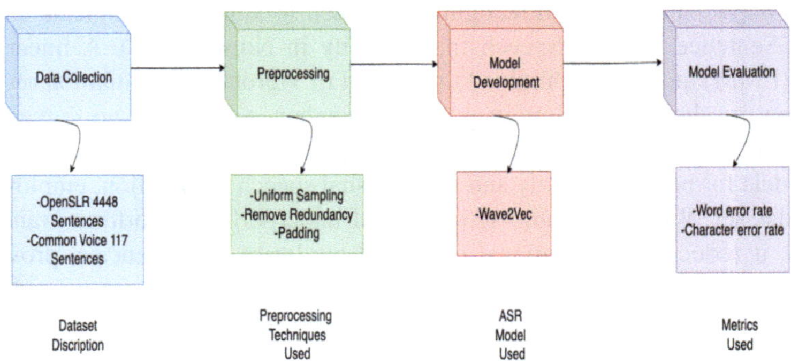

Fig. 1 Overall methodology

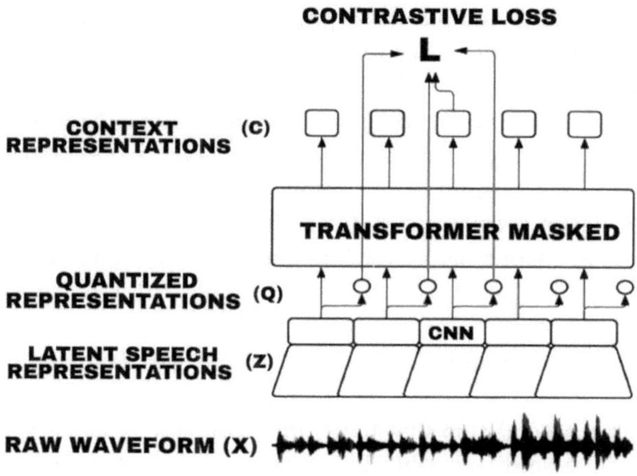

Fig. 2 Wav2Vec2.0 architecture [17]

Transcripts were refined by removing extraneous characters, resulting in clean text suitable for tokenization. Audio frequencies were standardized to 16 kHz, enabling the model to extract features effectively from raw waveforms. Additionally, transcripts were tokenized and padded to uniform lengths, facilitating the processing of sequences without imposing dimensional inconsistencies during training.

3.2 Wav2Vec 2.0

Wav2Vec 2.0 excels in self-supervised learning, leveraging unlabeled speech data through a hierarchical architecture, as illustrated in Fig. 2. Raw audio waveforms (X) are processed by a convolutional encoder ($f : X \rightarrow Z$) to generate latent representations (Z), which are quantized (Q) and further refined by utilizing a masked transformer architecture ($g : Z \rightarrow C$) to yield context-aware features (C). Contrastive loss (L) is applied for self-supervised training, enabling efficient learning from vast datasets [12],[11]. The model's key components include:

3.2.1 Convolutional Feature Encoder

The first step involves encoding raw audio waveforms into latent speech representations [12]. The raw waveform X passes through multiple convolutional layers to extract low-level features. This feature encoder operates on segments of audio and converts the input $X \in R^{T \times F}$ into latent representations $Z \in R^{T \times F}$, Where T is the new sequence length after down sampling and F is the feature dimension. The encoder is defined by $Z = f(X)$ where f denotes the convolutional layers that progressively down-sample the input. Each layer has a stride and a kernel size, and the final output produces features that represent short segments of audio. The CNN encoder architecture of Wav2Vec 2.0, where raw audio waveforms are processed through a feature mapping network that outputs 512-dimensional latent features, each representing 20 ms of audio. The encoder incorporates seven parallel processing paths (x_7) with three key components: GLU activation for non-linear transformation, layer normalization for stable training, and 1D convolution for temporal feature extraction before feeding into the transformer encoder.

3.2.2 Quantization Module

Quantization is a crucial step that transforms continuous latent speech representations into a finite set of possible latent states. This is done using the Gumbel-Softmax approximation, which allows the model to discretize the latent vectors while keeping the system differentiable. The quantization process converts Z into quantized latent variables Q, which are then used as targets during pretraining. Mathematically, this can be written as:

$$q_t = argmax(log(\pi_i + g_i) \quad (1)$$

where q_t represents the learned probability for selecting each quantized vector and g_i is the Gumbel noise, which enables discrete latent variables to be sampled.

3.2.3 Transformer Contextualizer

Once the latent speech representations are quantized, they are input into a transformer-based architecture. The transformer's role is to capture distant relationships within the audio signal, resulting in contextualized representations of speech. The transformer processes the sequence of latent vectors Z and outputs contextualized vectors c. It performs self-attention over the entire input sequence, generating rich representations for masked prediction tasks. The self-attention mechanism is defined as:

$$\text{Attention}(Q, K, V) = \text{softmax}\left(\frac{QK^T}{\sqrt{d_k}}\right)V \quad (2)$$

Here, Eq. (2), Q, K, and V denote the matrices corresponding to the query, key, and value, respectively, while d_k indicates the dimensional size of the key vectors.

3.2.4 Masked Prediction Task (Contrastive Loss)

Wav2Vec 2.0 uses a masked learning objective inspired by BERT-style models. During pretraining, the model randomly masks a portion of the latent representations and tries to predict them based on the context provided by the transformer. The goal is to identify the true quantized latent representation from distractor samples, which can be achieved through a contrastive loss function.

$$L_c = -\log \frac{\exp\left(\mathrm{sim}\left(c_t, z_{t+1}\right)\right)}{\sum_{z_k \in Z_{\mathrm{neg}}} \exp\left(\mathrm{sim}\left(c_t, z_k\right)\right)} \tag{3}$$

Where $\mathrm{sim}\left(c_t, z_{t+1}\right)$ denotes cosine similarity between the contextualized vector c_t and the true quantized latent representation z_{t+1}, and Z_{neg} is the set of distractor latent vectors.

3.2.5 Fine-tuning for Speech Recognition

After completing pretraining, Wav2Vec 2.0 can be adapted for ASR tasks through fine-tuning. This is achieved by placing a linear layer atop the Transformer outputs, which transforms the contextualized representations into the target vocabulary. Subsequently, the model undergoes training in a supervised framework, aiming to reduce the CTC loss function.

$$L_{CTC} = -log P(y/X) \tag{4}$$

Here, $P(y \mid X)$ denotes the likelihood of accurately predicting the transcription y based on the input audio X.

3.3 *Metric Used for Model Evlauation*

To assess the effectiveness of our optimized ASR model, we employed two evaluation measures: WER and CER. The CER complements WER, offering insight at the character level instead of the word level [15]. In this study, both WER and CER metrics have been utilized to evaluate the performance of the Wav2Vec 2.0 model. The formulas for calculating WER and CER are provided below:

$$\mathrm{WER} = \frac{S_w + I_w + D_w}{N_w} \tag{5}$$

$$\mathrm{CER} = \frac{S_c + I_c + D_c}{N_c} \tag{6}$$

where:

- S_w, I_w, D_w: Substitutions, insertions, and deletions (word level).
- S_c, I_c, D_c: Substitutions, insertions, and deletions (character level).
- N_w, N_c: Total words and characters in the dataset.

3.4 Experimental Setup

This section delves into the specific configurations employed for state-of-art ASR model: the Wav2Vec 2.0 `large-xlsr-53` variant. This setup was meticulously fine-tuned to evaluate the model's efficiency and flexibility across various experimental scenarios. The experiments were conducted on Google Colab using a T4 GPU to efficiently handle ASR tasks with accelerated training and evaluation.

3.4.1 Training Configurations

The Wav2Vec2 large model was pre-trained on unlabeled data with specific configurations, including a `mask_time_prob` of 0.05, and `attention_dropout` and `hidden_dropout` values set to 0.0 [11]. The training process employed the following hyperparameters: The 0.0003 was set as learning rate, determining the step size during optimization and influencing the convergence speed of the model. While a smaller learning rate enhances stability, it may necessitate a greater number of epochs to achieve satisfactory performance. For instance, a learning rate of 0.00003 would require more than three times the number of epochs to observe improvements in evaluation metrics. The Adam optimizer was utilized with parameters $\beta = (0.9, 0.999)$ and $\epsilon = 1\text{e}{-08}$. A linear learning rate scheduler was utilized, progressively reducing the rate from its starting point to zero throughout the training process. Additionally, a warmup phase of 400 steps (20% of the training steps) was employed, during which the learning rate incrementally increased to its predefined value, mitigating abrupt performance drops.A train batch size and evaluation batch size of 8 were employed, allowing the model to process manageable amounts of data while updating weights. The total effective batch size was set to 16 using gradient accumulation over two steps, which simulated a larger batch size without increasing memory requirements. This approach improved training efficiency. The model was trained for 16 epochs, ensuring sufficient training while avoiding overfitting.

4 Results and Discussions

This section evaluates Wav2Vec 2.0 for Telugu speech recognition using WER and CER, analyzing both effective and ineffective hyperparameters, with findings presented in graphs highlighting model robustness, and efficiency.

Table 1 The loss during training and validation, along with the WER, at each step for the Wav2Vec 2.0 model

Step	Training loss	Validation loss	WER
200	–	3.404397	1.000000
400	2.639700	0.592723	0.687945
600	2.639700	0.386299	0.500890
800	0.481200	0.325043	0.430605
1000	0.481200	0.300007	0.389680
1200	0.283200	0.284401	0.371886
1400	0.283200	0.270225	0.332963
1600	0.204700	0.245586	0.310721
1800	0.204700	0.270976	0.328069
2000	0.152000	0.251344	0.300934
2200	0.152000	0.261368	0.293149
2400	0.114300	0.253679	0.282028
2600	0.114300	0.262104	0.277580
2800	0.093400	0.267628	0.278247
3000	0.093400	0.260404	0.266681
3200	0.082200	0.261496	0.259342
3400	0.082200	0.257286	0.258674

Table 1 displays the performance metrics for both training and validation of the Wav2Vec 2.0 model over the course of 3400 steps. The training and validation losses show a significant reduction over time, with the training loss decreasing from 2.6397 at step 400 to 0.0822 at step 3400, and the validation loss reducing from 0.5927 to 0.2573 in the same period. Additionally, the WER demonstrates a steady improvement, dropping from 0.6879 at step 400 to a final value of 0.2587 at step 3400, indicating the model's ability to effectively refine its performance with more training steps.

Figure 3 illustrates the WER and CER metrics for the Wav2Vec 2.0, trained on 80% of the OpenSLR Telugu Audio Corpus, highlighting the performance of the Wav2Vec 2.0 model. The model, featuring a parameter total of 316 million, exhibits impressive efficacy in settings with limited resources, especially when dealing with data that lacks labels. In the evaluation on the test set (comprising 20% of the OpenSLR dataset), the model obtained a WER of 25.6% and a CER of 4.5%. Moreover, when tested on the Common Voice dataset (20% of the total), it achieved a CER of 18%. As the training steps increase, a notable decrease in WER is depicted in Fig. 4, accompanied by a steady decline in both training and validation loss. In this study, the loss curve of Wav2Vec 2.0 consistently decreased, demonstrating its ability to learn from limited data effectively. Wav2Vec 2.0 continued to improve, eventually showing signs of overfitting at 3000 training steps. The final WER values are presented in Fig. 5, where Wav2Vec 2.0 achieved a WER of 25.6%.

In addition, several unsuccessful hyperparameter configurations for the Wav2Vec 2.0 model were identified, where the model exhibited minimal or no training progress, as shown in Fig. 6. Specifically, a learning rate of 3×10^{-5} caused the model to update its weights at an excessively slow pace, impeding the learning

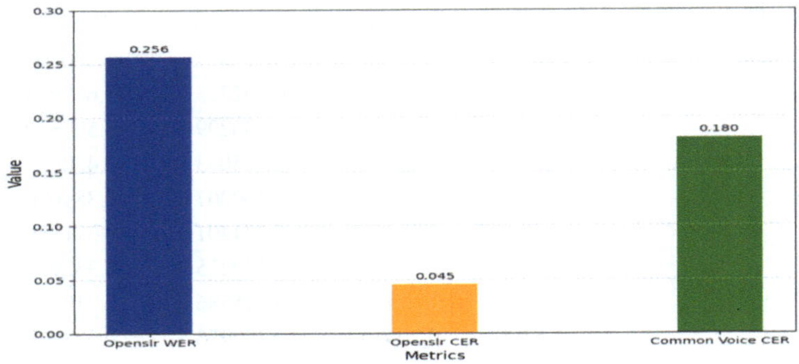

Fig. 3 WER and CER comparison between Wav2Vec 2.0

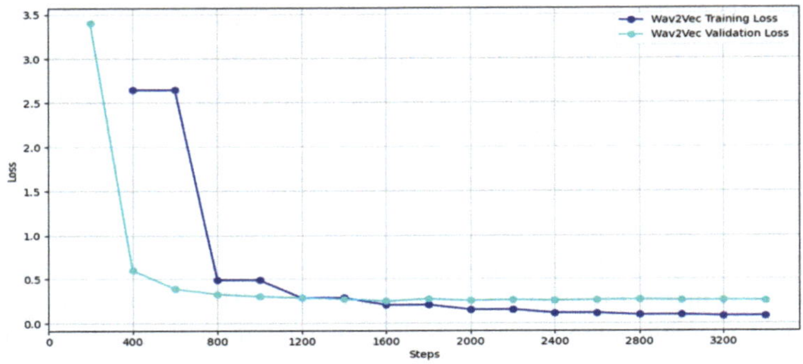

Fig. 4 Training loss and validation loss curves for Wav2Vec 2.0

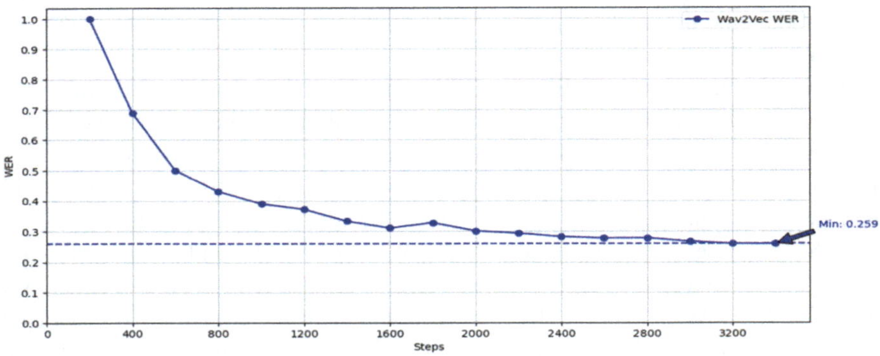

Fig. 5 Comparing Wav2Vec2-large through the training process

process. This constraint notably hindered the model's capacity to identify relevant patterns within a feasible number of epochs. Moreover, higher attention and hidden dropout rates, set at 0.1, exacerbated the slow learning by introducing increased regularization. These elevated dropout rates imply that a greater proportion of

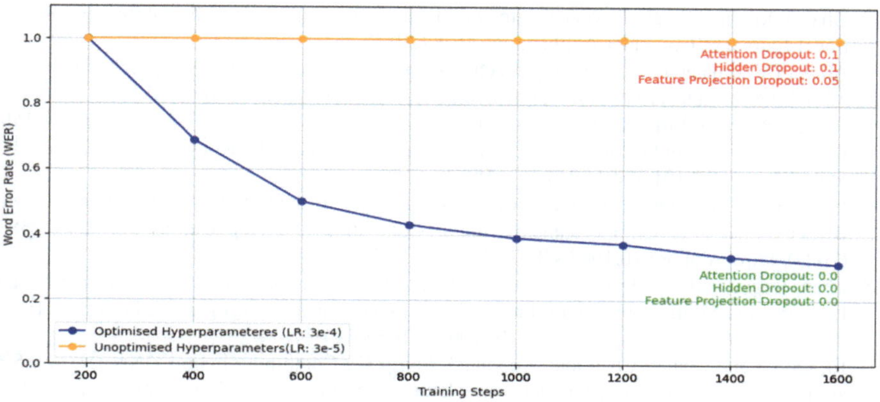

Fig. 6 Wav2Vec2-large WER with Hyperparameter tuning

information is "dropped" or masked during each step of training, which hampers the model's capacity to effectively learn intricate speech representations, especially when combined with such a minimal learning rate.

5 Conclusion

In conclusion, this research presented an analysis of the pre-trained Wav2Vec 2.0 architecture for Telugu speech recognition. The work implemented in the paper represents a significant step towards the development of such systems. The above results underscore the paramount importance of architectural design and hyperparameter tuning in optimizing model efficacy. Furthermore, the exploration of unsuccessful hyperparameter configurations provides valuable insights into optimal training conditions. Considering the baseline metric for WER is set to one epoch, we observe substantial improvements in the model. Wav2Vec 2.0 achieves a reduction of 0.65 WER from 0.9 WER, resulting in a 65% improvement. This study contributes to the ongoing development of speech recognition systems in underrepresented languages and paves the way for future investigations into enhancing model performance through advanced techniques and larger datasets. The implementation of this study can be accessed through the GitHub repository provided below: https://github.com/KaarthikeyaK/Telugu-Automatic-Speech-Recognition.

References

1. Juang, Biing-Hwang and Rabiner, Lawrence R: Automatic speech recognition–a brief history of the technology development. Georgia Institute of Technology. Atlanta Rutgers University and the University of California. Santa Barbara 1, 1 (2005).

2. Ghai, Wiqas and Singh, Navdeep: Literature review on automatic speech recognition. International Journal of Computer Applications 41, 8 (2012).
3. Adiga, Devaraja et al.: Automatic Speech Recognition in Sanskrit: A New Speech Corpus and Modelling Insights. arXiv (2021). https://doi.org/10.48550/ARXIV.2106.05852
4. Smit, Peter and Virpioja, Sami and Kurimo, Mikko: Advances in subword-based HMM-DNN speech recognition across languages. Computer Speech & Language 66, 101158 (2021). https://doi.org/10.1016/j.csl.2020.101158
5. Feng, Yican: Intelligent speech recognition algorithm in multimedia visual interaction via BiLSTM and attention mechanism. Neural Computing and Applications 36, 2371–2383 (2023). https://doi.org/10.1007/s00521-023-08959-2
6. Zhao, Jingyu and Li, Ruwei and Tian, Maocun and An, Weidong: Multi-view Self-supervised Learning and Multi-scale Feature Fusion for Automatic Speech Recognition. Neural Processing Letters 56 (2024). https://doi.org/10.1007/s11063-024-11614-z
7. Li, Qiujia and Zhang, Chao and Woodland, Philip C.: Combining hybrid DNN-HMM ASR systems with attention-based models using lattice rescoring. Speech Communication 147, 12–21 (2023). https://doi.org/10.1016/j.specom.2022.12.002
8. Răutu, I. Sabina et al.: Speech-derived haptic stimulation enhances speech recognition in a multi-talker background. Scientific Reports 13 (2023). https://doi.org/10.1038/s41598-023-43644-3
9. Sterpu, George and Harte, Naomi: Taris: An online speech recognition framework with sequence to sequence neural networks for both audio-only and audio-visual speech. Computer Speech & Language 74, 101349 (2022). https://doi.org/10.1016/j.csl.2022.101349
10. Kadiri, Sudarsana Reddy et al.: Investigation of self-supervised pre-trained models for classification of voice quality from speech and neck surface accelerometer signals. Computer Speech & Language 83, 101550 (2024). https://doi.org/10.1016/j.csl.2023.101550
11. Kozhirbayev, Zhanibek: Kazakh Speech Recognition: Wav2vec2.0 vs. Whisper. Journal of Advances in Information Technology 14(6), 1382–1389 (2023). https://doi.org/10.12720/jait.14.6.1382-1389
12. Kozhirbayev, Zhanibek: Leveraging Wav2Vec2.0 for Kazakh Speech Recognition: An Experimental Study. In: Computational Science and Its Applications – ICCSA 2024. Springer Nature Switzerland, pp. 120–132 (2024). https://doi.org/10.1007/978-3-031-64608-9_8
13. He, Fei et al.: Open-source multi-speaker speech corpora for building Gujarati, Kannada, Malayalam, Marathi, Tamil and Telugu speech synthesis systems. In: Proceedings of the Twelfth Language Resources and Evaluation Conference, pp. 6494–6503 (2020).
14. Ehtisham, Rana et al.: Computing the characteristics of defects in wooden structures using image processing and CNN. Automation in Construction 158, 105211 (2024). https://doi.org/10.1016/j.autcon.2023.105211
15. Mainzinger, Julia: Fine-tuning ASR Models for Very Low-Resource Languages: A Study on Mvskoke. Master's thesis, University of Washington (2024).
16. Bhagath, Parabattina et al.: TELUGU ANKELU: A Telugu Spoken Digits Corpora for Mobile Speech Recognition. In: 2022 12th International Conference on Pattern Recognition Systems (ICPRS), pp. 1–6. IEEE (2022). https://doi.org/10.1109/ICPRS54038.2022.9854065
17. Baevski, Alexei et al.: wav2vec 2.0: A Framework for Self-Supervised Learning of Speech Representations. arXiv (2020). doi:10.48550/ARXIV.2006.11477
18. Dar, Muzaffar Ahmad and Pushparaj, Jagalingam: Bi-directional LSTM-based isolated spoken word recognition for Kashmiri language utilizing Mel-spectrogram feature. Applied Acoustics 231, 110505 (2025). https://doi.org/10.1016/j.apacoust.2024.110505
19. Dar, Muzaffar Ahmad and Pushparaj, Jagalingam: Hybrid architecture CNN-BLSTM for automatic speech recognition. In: 2024 3rd International Conference on Artificial Intelligence For Internet of Things (AIIoT), pp. 1–4. IEEE (2024). https://doi.org/10.1109/aiiot58432.2024.10574740

Secure Biological Data Transfer Using Network Coding

Sultan Alam, Aritra Mahanty, Shreyas Abhijeet Desai, and Shreya Agrawal

Abstract The research paper presents a method for securely transferring biological datasets via a TCP/IP server using encrypted communication. The system employs socket programming, combining Elliptic Curve Cryptography (ECC) and the Advanced Encryption Standard (AES) in Galois/Counter Mode (GCM) for robust data protection. The P-256 curve is utilized for ECDH key exchange, enabling the secure generation of a shared secret between the client and server. This shared secret is transformed into a symmetric AES-256 encryption key through the scrypt key derivation function, enhancing cryptographic strength. By integrating ECC for key exchange and AES-GCM for encryption, the proposed system ensures confidentiality, integrity, and resistance to tampering. The approach ensures secure, scalable, and efficient biological data transfer for real-time applications. It offers high security with low computational overhead, excellent scalability, and optimized low-latency performance. However, it relies on secure key management and faces challenges with ultra-large datasets. Extensive analysis confirms its reliability in protecting sensitive data during transmission.

Keywords Secure data transmission · Elliptic curve cryptography · Advanced encryption standard with Galois counter mode · Elliptic curve Diffie-Hellman · Symmetric encryption

S. Alam
School of Computing Science and Engineering, VIT Bhopal University, Sehore, India
e-mail: mohammadsultanalam@vitbhopal.ac.in

A. Mahanty · S. A. Desai · S. Agrawal (✉)
SCAI, VIT Bhopal University, Sehore, India
e-mail: shreya.23bai11407@vitbhopal.ac

K. Deep et al. (eds.), *Artificial Intelligence, Machine Learning and Intelligent Systems*, Springer Proceedings in Mathematics & Statistics 520,
https://doi.org/10.1007/978-981-95-3841-6_34

1 Introduction

Efficient data transfer is vital in healthcare and bioinformatics, where sensitive biological data faces latency, security, and scalability challenges. Novel frameworks are needed to ensure secure and efficient transmission.

1.1 Role of Network Coding

Network coding has been extensively studied as a method to enhance data transmission efficiency [2, 5, 13]. It introduces a way to encode data packets, increasing throughput and resilience against network failures. The foundational work on channel polarization laid the groundwork for efficient communication in binary symmetric channels [6]. Extensions to smart grid communications showed how network coding could improve reliability and bandwidth utilization [9].

The application of Random Linear Network Coding (RLNC) in automotive systems demonstrated adaptive mechanisms that were effective in real-world scenarios [3, 4, 11]. These studies collectively highlight the versatility of network coding in addressing various network challenges. However, its direct application to sensitive biological data remains limited, requiring enhancements to address security and scalability.

1.2 Theoretical Foundations and Limitations

Information theory, including Shannon's channel capacity theorem, offers insights but lacks practical implementations for secure, dynamic biological data transfer [14]. Pipeline network coding reduced latency in wireless networks [15] but wasn't tailored to biological data. Network coding benefits in P2P systems, like solving the rarest-piece problem [1], face limitations when applied to biological data:

- Lack of integrated cryptographic mechanisms.
- High computational overhead for decoding.
- Poor adaptability to dynamic networks.

While multicast throughput improvements were notable, they overlooked data sensitivity issues relevant to biological information [12].

1.3 Objectives and Contributions

This research bridges the gaps identified in previous studies by proposing a framework that integrates advanced cryptographic techniques with optimized network coding algorithms. The key contributions of this work include:

1. Designing a secure network coding protocol to ensure data confidentiality during biological data transfer.
2. Reducing computational overhead using sparse coding techniques.
3. Developing adaptive mechanisms to enhance scalability in dynamic network conditions.

These innovations provide a comprehensive approach to securely and efficiently transferring biological data in modern interconnected systems.

1.4 Overview

Section 2 outlines the theoretical framework for secure biological data transmission, integrating cryptography with network coding. Section 3 details the methodology, including ECC, AES-GCM, and XOR techniques. Section 4 evaluates encryption efficiency, overhead, integrity, and scalability. Section 5 concludes with findings and future directions for real-time applications and complex datasets.

2 Methodology

The methodology employs a multi-layered cryptographic approach combined with a robust computational architecture. Below is a detailed explanation of the key components:

2.1 XOR Encryption

The XOR operation offers lightweight obfuscation for encryption and decryption, defined as:

$$C = P \oplus K, \quad P = C \oplus K, \tag{1}$$

where P is the plaintext, C is the ciphertext, and K is the key. The key properties of XOR are:

$$A \oplus B = B \oplus A, \quad (A \oplus B) \oplus C = A \oplus (B \oplus C), \quad A \oplus A = 0. \tag{2}$$

K is the key. The mathematical properties of XOR ensure simplicity in implementation while maintaining a layer of obfuscation [1, 12].

2.2 ECDH Key Exchange

ECDH leverages elliptic curve cryptography for secure key exchange. Using the curve equation:

$$y^2 = x^3 + ax + b \pmod{p}, \tag{3}$$

to generate keys. The process is as follows:

$$P_A = aG, P_B = bG, S = abG$$

where a and b are private keys, G is the base point, and S is the shared secret.

a, b are private keys. The security derives from the elliptic curve discrete logarithm problem (ECDLP), as described in prior foundational work [6, 14].

2.3 AES-GCM Encryption with XOR Obfuscation

AES-GCM encrypts plaintext blocks M_i as:

$$C_i = E_K(M_i \oplus P_{i-1}), \tag{4}$$

where P_{i-1} is the previous ciphertext block, and integrity is verified using GHASH:

$$\text{GHASH}(H, A, C) = \left(\sum_{i=1}^{m} X_i \cdot H^{m-i+1} \right) \mod P, \tag{5}$$

where H is a subkey, A is additional authenticated data, and C is the ciphertext. To enhance obfuscation, the ciphertext is XORed with a 16-byte key K_{xor}:

$$O = C \oplus K_{\text{xor}}. \tag{6}$$

This approach extends the efficiency of AES-GCM, supported by studies on its use in sensitive data scenarios [11, 15].

2.4 Security Analysis and Computational Efficiency

The proposed system balances security, performance, and key management using three cryptographic techniques:

- Elliptic Curve Cryptography (ECC): Offers secure key exchange with smaller keys, outperforming RSA in computational cost [9], based on the Elliptic Curve Discrete Logarithm Problem (ECDLP):

$$Q = kP \quad \text{(finding } k \text{ is infeasible)} \tag{7}$$

- AES-GCM: Ensures confidentiality, integrity, and efficient encryption of large datasets.
- XOR: Provides lightweight obfuscation with minimal overhead.

2.5 *Key Lengths*

The system uses standard key sizes aligned with modern cryptographic recommendations [13]: ECC P-256 (128-bit security), AES 256-bit, and a 16-byte XOR key. It integrates ECC, AES-GCM, and XOR obfuscation to ensure confidentiality, integrity, and lightweight obfuscation for secure biological data transmission.

2.6 *System Architecture and Implementation*

The system uses a client-server architecture over TCP/IP with key components: ECDH key exchange (P-256), AES-GCM encryption for confidentiality and integrity, and XOR-based obfuscation for added security. It develops on the secure communication structures, which focus on scalability [1, 15]. done in Python. Ciphertext Obfuscation The system relies on ECC and AES-GCM cryptographic libraries, a custom XOR layer, an UDP/IP socket programming interface and a 16-byte secret key to conceal ciphertext (`my_xor_key_16`).

2.7 *Experimental Setup*

This system was experimented in real-life conditions and a Linux based environment with datasets that represent biological and genome data transfer was used [12]. Measurements like encryption time, obfuscation overhead and integrity verification were observed.

2.8 *Performance Metrics*

In order to evaluate the performance of the system the following metrics were used:

1. Key Exchange Time: The duration that it takes to generate ECC keys and exchange the ECDH keys on the P-256 curve.
2. Encryption Time: Timed over datasets with different sizes, and results linearly expandable [11].
3. Obfuscation Time: It is the time required to obfuscate the AES-GCM ciphertext using the XOR layer.
4. Scalability: How well the system scales out as the size of the data set increases.
5. Integrity Verification: The integrity measures of the system to reveal the tampered data through AES-IGM authentication tag. Certified by AES-GCM authentication tags to give 100% authentication by means of resisting tampering [14].

2.9 *Testing Methodology*

The experiments were repeated in two stages with small (up to 10 MB) and huge databases (100 MB, 500 MB, 1 GB). The small and the large dataset tests concentrated on encryption and obfuscation times of the biological images and simulated transfer of the genome data, respectively, which placed preference on scalability and computational overheads.

In each data set, the client created an ECC key pair, shared public keys with the server, and performed ECDH to receive a common secret to AES-GCM encryption. The client was then able to encrypt the data and obfuscate using XOR and then send it using TCP/IP. The server has decrypted information, ensured its integrity, and checked the detection rates by corrupting ciphertext in advance.

2.10 *Evaluation Criteria*

The speed of the system was measured by determining the fulfilment of data encryption and obfuscation times using varying sizes of datasets, analysing scalability of the system based on time-line graphs. Integrity of data was also checked by deliberately modifying the encrypted ciphertext being transmitted, and these rates of detection obtained by using the AES-GCM authentication tag. The computational efficiency was evaluated by evaluating the performance of the CPU usage, execution time of the ECC key exchange encryption, AES-GCM encryption and XOR obfuscation compared with the baseline approaches e.g., RSA and AES-CBC.

3 Results and Discussion

Here, performance of the proposed framework is studied and results presented in Figs. 1a and b, 2, 3 and 4. These data point out the network performance, efficiency

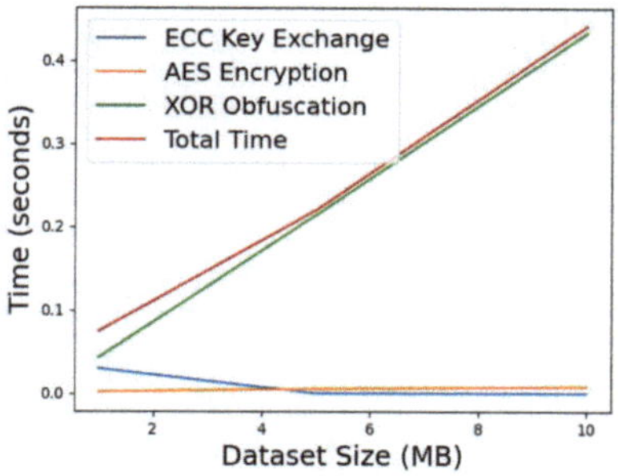

(a)

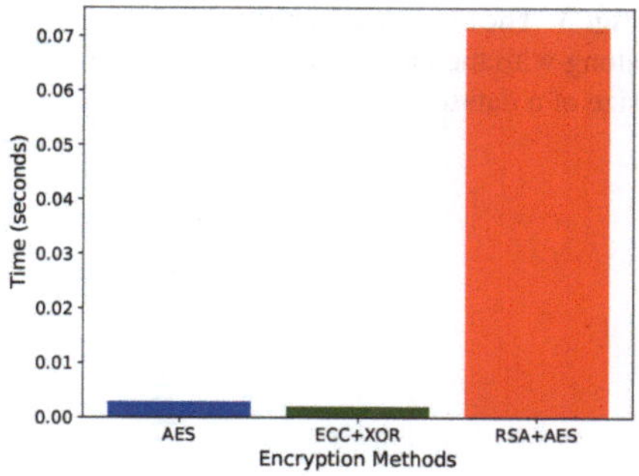

(b)

Fig. 1 Encryption performance analysis. (**a**) Encryption time as a function of dataset size. (**b**) Comparison of computation times in methods

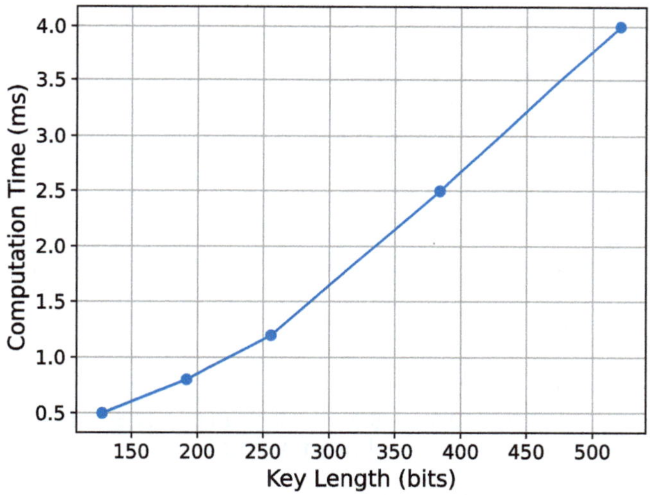

Fig. 2 In cryptographic tasks, computation time grows non-linearly with the key length

rates, efficiencies, computation cost, and also scalabilities, latency, and success levels. The implications of findings are discussed below.

3.1 Encryption Performance

The suggested framework can help improve the performance of encryption considerably as it uses the Elliptic Curve Cryptography (ECC) along with P-256 curve-based implementation to provide security in key exchange, AES with GCM block cipher applied to provide a secure and efficient encryption process and XOR obfuscation to provide further security with exceptionally low overheads.

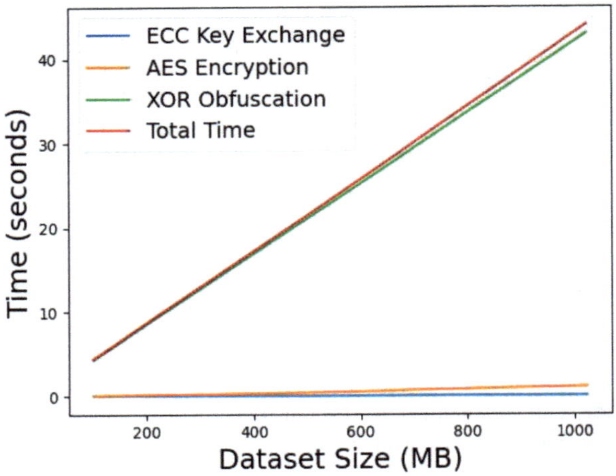

Fig. 3 The system scalability along with the increase of the size of a dataset

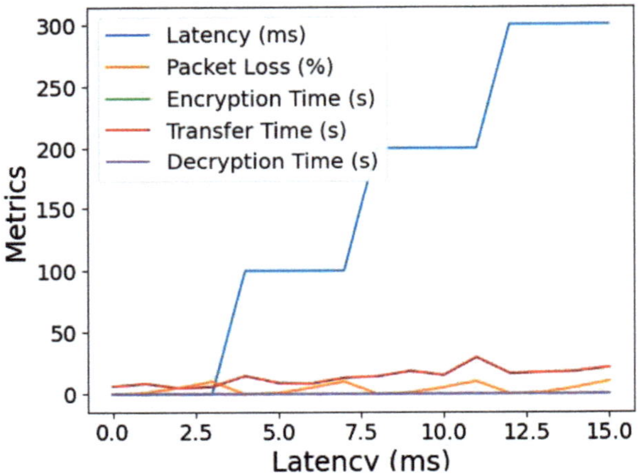

Fig. 4 Latency against success rate: tamper detection

These features make the system well-suited for resource-constrained environments. As shown in Fig. 1a, encryption time scales linearly with dataset size, from 10 MB to 1 GB, reflecting the system's scalability for bioinformatics and healthcare applications. This optimized encryption process, coupled with negligible latency from XOR obfuscation, offers superior performance compared to traditional methods like RSA and AES-CBC, as seen in Fig. 1b.

3.2 Computational Efficiency

Figure 2 highlights the computational efficiency of the proposed framework compared to traditional cryptographic methods. The reduced computational demand can be attributed to:

- ECC Key Exchange: ECDH method ensures secure key exchange with minimal computational overhead, making it ideal for lightweight environments.
- AES-GCM Encryption: Provides both confidentiality and integrity for large datasets with minimal performance impact.
- XOR Obfuscation: A lightweight operation that adds an extra security layer without affecting overall performance.

This efficiency means that it is appropriate on limited computing resources, e.g. mobile or embedded devices.

3.3 Scalability Analysis

The scalability of the proposed system is shown in Fig. 3. The time of encryption and obfuscation being linearly-scaled with the size of dataset ensures that the system will readily meet the criteria of the rising volume of data without deteriorating its performance. This property is specially attractive in bioinformatics applications where sequences and imaging data sets are expanding exponentially as less expensive sequencing and imaging technologies are developed. The outcomes verify the availability of the offered framework to encounter vastly unlike arrays, tiny biologic pictures and huge genomic chains.

3.4 Latency and Success Rate Analysis

Figure 4 gives some understanding about the correlation between the success rates and latency. The framework also attains 100% success rate with low latency in the detection of tampered data across all different sizes of dataset. This performance shows how sound the system and reliable it is to gather data and be sure that it can transmit it safely in real world conditions.

3.5 Integrity Detection

The mechanism of integrity verification of the system proved its outstanding reliability even though it was not depicted directly. The framework was efficient in identifying compromised datasets during testing with the use of authentication tags provided by AES-GCM. Any deliberate corruption injected in the cipher text was detected with 100% accuracy again indicating reliability in use of the system in relaying critical biological data which may have sensitive clinical or research implication.

3.6 Security Analysis

The security features of the proposed architecture are made obvious by the relative application of:

- ECC: provides strong key exchange with the elliptic curve discrete logarithm problem (ECDLP), which enables strong security at reduced key lengths.
- AES-GCM: It offers authenticated secrecy that ensures the integrity and confidentiality of the data.
- XOR Obfuscation: Extends protection with lightweight concealing, which guarantees low overhead.

This holistic design is useful in addressing security demands on bioinformatics and healthcare applications that are very high.

3.7 Comparative Analysis

In this section, the proposed method is compared to the rest of the other encryption techniques. The suggested scheme is built-in ECC based on P-256 curve, AES-GCM, and XOR as a lightweight obfuscation type. Others are Smart Grid Communication [10], which employs the use of AES with RSA key exchange but at the cost of data and processing time, RLNC [11], which is only interested in scalability of data, RSA + AES-CBC [8] which is at its end, and DNA Cryptography [7] which combines the usage of AES and DNA encoding but is computationally costly.

The comparison outlines the high level of strength, high scalability, low apparent computation overhead, and powerful security by utilizing AES-GCM tags of the proposed method. The older approaches such as Smart Grid Communication [10] and RSA+ AES-CBC [8] are less scalable and more computationally cheap. RLNC [11] is scalable and non-encrypted and DNA Cryptography [7] is super-secure, but also expensive. In general, the suggested technique offers a reasonable compromise between security, scalability, and efficiency, which makes it a perfect fit when transferring data in modern conditions.

3.8 Discussion

The given framework shall provide an efficient, secure and scalable mechanism of transmitting biological data. Its valuable strengths are its linear scalability, low computation overhead with ECC and AES-GCM and the security of the data by ECC, AES-GCM and XOR obfuscation. However, limitations include challenges with real-time transmission of ultra-large datasets, potential overhead from key management, and the system's focus on batch processing rather than streaming data. Despite these, the framework shows strong potential for bioinformatics and healthcare applications, with improvements in real-time performance and scalability as future goals.

Acknowledgments The authors thank the reviewers for their insightful comments and constructive suggestions.

References

1. AbuDaqa, A.A., Mahmoud, A., Abu-Amara, M., Sheltami, T.: Survey of network coding based p2p file sharing in large scale networks. Applied Sciences **10**(7) (2020). https://doi.org/10.3390/app10072206, https://www.mdpi.com/2076-3417/10/7/2206
2. Ahlswede, R., Cai, N., Li, S.Y., Yeung, R.: Network information flow. IEEE Transactions on Information Theory **46**(4), 1204–1216 (2000). https://doi.org/10.1109/18.850663
3. Alam, S., Thakor, S., Abbas, S.: On enumerating distributions for associated vectors in the entropy space. In: 2018 International Symposium on Information Theory and Its Applications (ISITA). pp. 65–69 (2018). https://doi.org/10.23919/ISITA.2018.8664232
4. Alam, S., Thakor, S., Abbas, S.: An analysis of a randomized local search algorithm for the entropy space. In: 2019 4th International Conference on Information Systems and Computer Networks (ISCON). pp. 330–334 (2019). doi:10.1109/ISCON47742.2019.9036317
5. Alam, S., Thakor, S., Abbas, S.: Inner bounds for the almost entropic region and network code construction. IEEE Transactions on Communications **69**(1), 19–30 (2021). https://doi.org/10.1109/TCOMM.2020.3030055
6. Arikan, E.: Channel polarization: A method for constructing capacity-achieving codes for symmetric binary-input memoryless channels. IEEE Transactions on Information Theory **55**(7), 3051–3073 (2009). https://doi.org/10.1109/TIT.2009.2021379
7. Bonny, B., Frank, J., Mahalakshmi, T.: Secure data transfer through dna cryptography using symmetric algorithm. International Journal of Computer Applications **133**, 19–23 (01 2016). doi:10.5120/ijca2016907719
8. Chen, W., Chen, Z., Cui, F.: Collaborative and secure transmission of medical data applied to mobile healthcare. BioMedical Engineering OnLine **18** (05 2019). https://doi.org/10.1186/s12938-019-0674-x
9. Karthick, M., Sivalingam, K.: Network coding based reliable and efficient data transfer for smart grid monitoring. pp. 1–6 (12 2013). https://doi.org/10.1109/ANTS.2013.6802886
10. Naruchitparames, J., Güneş, M.H., Evrenosoglu, C.Y.: Secure communications in the smart grid. In: IEEE Consumer Communications and Networking Conference (CCNC). pp. 1171–1175 (2011). https://doi.org/10.1109/CCNC.2011.5766362
11. Pillmann, J., Behnke, D., Sliwa, B., Priebe, M., Wietfeld, C.: Efficient and reliable car-to-cloud data transfer empowered by bbr-enabled network coding. In: 2018 IEEE 88th Vehicular Technology Conference (VTC-Fall). pp. 1–5 (2018). https://doi.org/10.1109/VTCFall.2018.8690858
12. Rakesh, N., Tyagi, V.: Linear-code multicast on parallel architectures. Adv. Eng. Softw. **42**(12), 1074–1088 (Dec 2011). doi:10.1016/j.advengsoft.2011.07.012, https://doi.org/10.1016/j.advengsoft.2011.07.012
13. Shannon, C.E.: A mathematical theory of communication. The Bell System Technical Journal **27**(3), 379–423 (1948). https://doi.org/10.1002/j.1538-7305.1948.tb01338.x
14. Vinga, S.: Information theory applications for biological sequence analysis. Briefings in Bioinformatics **15**(3), 376–389 (09 2013). https://doi.org/10.1093/bib/bbt068
15. Xu, C., Wang, P., Xiong, C., Wei, X., Muntean, G.M.: Pipeline network coding-based multipath data transfer in heterogeneous wireless networks. IEEE Transactions on Broadcasting **63**(2), 376–390 (2017). https://doi.org/10.1109/TBC.2016.2590819

MedSum: Medical Video Summarization Using Reinforcement Learning

Jaya Gupta, Deepak Rai, and Prabhishek Singh

Abstract Medical video summarization is an essential task in healthcare, enabling efficient analysis of lengthy surgical procedures by extracting the most critical and informative content. Traditional methods often struggle to capture the unique temporal and contextual complexities of surgical videos, particularly at the gesture and phase levels. To address these challenges, this paper introduces MedSum, a transformer-based video summarization framework specifically designed for medical applications. MedSum integrates Kernel Temporal Segmentation (KTS) to identify coherent segments and employs transformer-based attention to prioritize significant events, ensuring both informativeness and temporal coherence. Reinforcement learning (RL) drives the frame selection process, optimizing a multi-objective reward function that balances informativeness, coverage, and redundancy. The framework leverages EfficientNet for spatial feature extraction and Farneback optical flow for motion analysis, ensuring robust representation of surgical actions. MedSum has been evaluated on benchmark surgical datasets, JIGSAWS and CATARACTS, demonstrating substantial improvements in F1-Score, Coverage Ratio (CR), and Temporal Smoothness (TS) compared to state-of-the-art methods. It achieves an average F1-Score improvement of over 18.5%, effectively summarizing both fine-grained gestures and broader procedural phases. By addressing the complexity of surgical workflows, MedSum sets a new benchmark for medical video summarization, while its adaptable design holds promise for broader applications across healthcare and other video-intensive domains.

Keywords Medical video summarization · Context-aware · Reinforcement learning · Transformer attention · Temporal modelling · EfficientNet · Temporal smoothness

J. Gupta · P. Singh
School of CSET, Bennett University, Greater Noida, India

D. Rai (✉)
Department of CSE, National Institute of Technology, Jamshedpur, India
e-mail: deepak.cse@nitjsr.ac.in

K. Deep et al. (eds.), *Artificial Intelligence, Machine Learning and Intelligent Systems*, Springer Proceedings in Mathematics & Statistics 520,
https://doi.org/10.1007/978-981-95-3841-6_35

1 Introduction

The growing volume of surgical videos from minimally invasive surgeries and robotic-assisted procedures, such as endoscopy, laparoscopy, and spinal surgeries, has created a pressing need for automated video summarization. These videos, recorded in operating rooms and training environments, are essential for surgical skill evaluation, procedural analysis, and education. However, the length and complexity of these procedures make manual evaluation time-consuming, labour-intensive, and impractical.

Existing video summarization methods often struggle to capture critical events while maintaining temporal coherence [4, 5]. This challenge is particularly significant in surgical videos, where precise tool transitions, subtle motions, and key procedural steps must be accurately highlighted. Failure to emphasize these details reduces the effectiveness of summaries for clinicians, researchers, and trainees.

To address these limitations, there is a clear need for improved techniques that generate concise, informative, and context-aware summaries. These solutions must focus on retaining essential events, ensuring smooth transitions, and eliminating redundancy. In this paper, a novel approach namely MedSum is presented for automatic medical video summarization. MedSum is a Transformer-based method that leverages Reinforcement Learning (RL) for dynamic frame selection and does not require labeled data for training. By combining the global contextual modeling capability of Transformers with RL-driven optimization, it generates concise, informative, and context-aware video summaries. The overall framework of the proposed approach is shown in Fig. 1. It can significantly enhance surgical video analysis where precise motion, tool transitions, and procedural coherence are critical. The main contributions of MedSum approach are as follows:

- MedSum introduces a transformer-based attention mechanism to model inter-segment relationships, ensuring prioritization of key surgical events and maintaining temporal coherence in surgical video summaries.

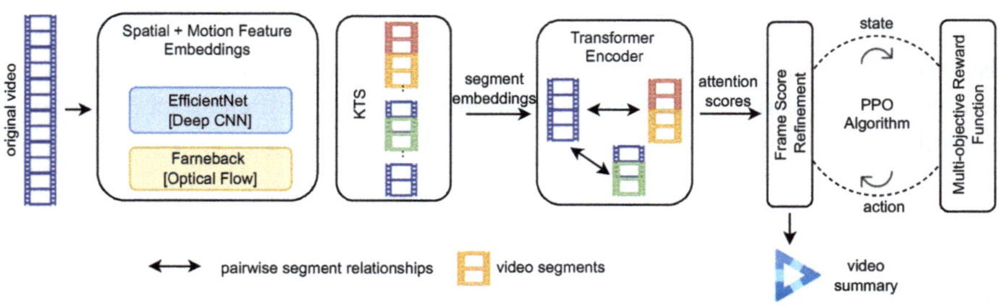

Fig. 1 Block Diagram of MedSum Framework. Illustrating the MedSum pipeline, showcasing key components

- The framework employs reinforcement learning (RL) with a multi-objective reward function to dynamically select frames, optimizing informativeness, coverage, and reducing redundancy for high-quality summarization.
- By leveraging Kernel Temporal Segmentation (KTS), MedSum segments videos into coherent units, enhancing the capture of significant gestures and surgical phases in complex procedures.
- MedSum integrates EfficientNet for spatial feature extraction and optical flow for motion analysis, offering a robust representation of surgical actions and transitions.
- Tailored for medical applications, MedSum demonstrates exceptional performance in summarizing gesture-level (JIGSAWS) and phase-level (CATARACTS) tasks, making it invaluable for surgical skill assessment, procedural review, and training.

The organization of this paper is as follows: Sect. 2 reviews related work, focusing on advancements in video summarization across supervised, unsupervised, and reinforcement learning-based methods. Section 3 details the proposed MedSum framework, including its architecture and novel methodologies. Section 4 presents datasets, implementation specifics, evaluation metrics, quantitative and qualitative results, as well as the ablation study. Finally, Sect. 5 concludes the paper and outlines potential future research directions.

2 Related Work

In recent years, research in video summarization has made significant strides, particularly in supervised, unsupervised, and reinforcement learning-based approaches. Supervised methods rely on labeled data to train models that effectively capture spatio-temporal dependencies in videos. For instance, VASNet [2] employs attention mechanisms for precise frame selection, while DSNet [12] incorporates self-attention to handle long-term temporal dependencies. MPFN [6] enhances this process by leveraging multi-scale pyramidal features, achieving state-of-the-art results in summarization tasks.

Unsupervised approaches have also gained traction by eliminating the need for labeled data, often using generative models and clustering techniques. SGAN [7] introduced adversarial learning with LSTMs to summarize video sequences, and Cycle-SUM [10] employed cycle-consistent adversarial networks to ensure coherent and meaningful summaries.

Reinforcement learning-based methods optimize summaries through reward mechanisms that balance informativeness, diversity, and redundancy. DRDSN [11] utilized a diversity-representativeness reward function to improve summarization quality, while PRLVS [9] employed progressive reinforcement learning for robust frame selection.

State-of-the-art methods have been evaluated on JIGSAWS and CATARACTS, addressing gesture- and phase-level challenges. MedSum advances this with transformer attention, reinforcement learning, and redundancy-aware optimization, enhancing informativeness and coherence for medical video summarization.

3 Proposed Approach

The proposed approach, MedSum, automates medical video summarization by extracting spatial and motion features, segmenting videos with KTS, and prioritizing key segments using Transformer-based attention. A RL based agent dynamically selects informative and diverse frames, generating concise and coherent summaries following these steps:

3.1 *Feature Extraction*

In this stage, medical videos are sampled into video frames to capture spatial details. These frames are processed using an state-of-the-art CNN architecture, EfficientNet. It uses compound scaling strategy to capture the high-level spatial features such as tool presence and anatomical details. These spatial features are fused with motion features, captured using optical flow over consecutive frames. The motion features are good at capturing tool movements and dynamic actions. These fused features form a unified frame-level embeddings $\mathbf{f}_i = [\mathbf{f}_i^{\text{spatial}}; \mathbf{f}_i^{\text{motion}}]$, which serves as input for the subsequent stages of the summarization process.

3.2 *Temporal Segmentation*

The fused frame embeddings $\mathbf{f}_i$ are used as input in this stage. Kernel Temporal Segmentation (KTS) is applied to identify meaningful boundaries. It works by detecting points where the feature distributions change significantly. These points indicate transitions between different events or phases in the video. By analysing the temporal structure of the embeddings, KTS segments the video into a set of coherent segments, $\mathcal{T} = \{t_1, t_2, \ldots, t_k\}$, where each segment between $\mathcal{T}_j = (t_i, t_{i+1})$ boundaries captures a distinct part of the video timeline. This segmentation provides the foundation for prioritizing key events and improving the subsequent summarization process.

3.3 Transformer Based Segment-Level Attention

In this stage, the video segments identified through KTS are further processed to compute segment embeddings. For each segment $\mathcal{T}_j$, its frames are denoted as S_j. The segment embedding $\mathbf{s}_j$ is calculated as the mean of the frame embeddings within the segment as shown in Eq. (1) .

$$\mathbf{s}_j = \frac{1}{|S_j|} \sum_{i \in S_j} \mathbf{f}_i, \tag{1}$$

This results in a set of segment embeddings $\{\mathbf{s}_1, \mathbf{s}_2, \ldots, \mathbf{s}_k\}$, which capture the overall feature distribution of each segment.

To model inter-segment relationships and prioritize globally significant segments, these embeddings are passed through a Transformer Encoder. The self-attention mechanism within the Transformer computes pairwise relationships between segments, allowing the model to weigh their relative importance as follows: $\alpha_j = \text{softmax}(\text{Attention}(\mathbf{s}_j, \{\mathbf{s}_1, \mathbf{s}_2, \ldots, \mathbf{s}_k\}))$.

The attention scores $\{\alpha_1, \alpha_2, \ldots, \alpha_k\}$ reflect the significance of each segment in the global context. The Transformer outputs enhanced segment embeddings $\{\mathbf{a}_1, \mathbf{a}_2, \ldots, \mathbf{a}_k\}$, which incorporate these inter-segment relationships.

By utilizing the Transformer's ability to model global dependencies, this step ensures that the most relevant and important segments are prioritized, to enable a focused and context-aware video summarization pipeline.

3.4 Frame Importance Scoring

The importance of individual frames within each segment is refined using the segment-level attention scores obtained from the Transformer

In this step, segment-level attention scores are used to refine the importance of the individual frame within each segment. For a frame i belonging to a segment j, its importance score is computed as: $\text{score}_i = \alpha_{s_j} \cdot \|\mathbf{f}_i - \mathbf{s}_{s_j}\|$, here α_{s_j} is the attention score of the segment j, reflecting its global importance, $\mathbf{f}_i$ is the feature embedding of frame i, and $\mathbf{s}_{s_j}$ is the segment embedding, representing the mean feature of all frames in segment j.

This formulation ensures that frames inside a highly significant segment are prioritized, with additional weighting based on how they differ from the segment's mean feature. These unique frames contribute more to the summary because they are more likely to capture distinct information from the segment. The result is a set of per-frame importance scores, which serve as the foundation for selecting the most representative frames during the reinforcement learning-based selection phase.

3.5 RL-Based Frame Selection

In this stage, a Reinforcement Learning (RL) agent is utilized to dynamically select the most representative frames for the video summary. The frame selection process is formulated as a Markov Decision Process (MDP), where the state represents the current set of selected frames, and the action space allows the agent to either add the current frame to the summary or skip it. A multi-objective reward function guides the selection process, balancing three key criteria: informativeness R_I, coverage R_C, and redundancy R_R. Informativeness ensures that selected frames have high importance scores, coverage encourages the inclusion of frames from all significant video segments, and redundancy penalizes the selection of visually similar or temporally close frames. The reward function is calculated as: $reward(R) = \lambda_1 \cdot R_I + \lambda_2 \cdot R_C - \lambda_3 \cdot R_R$, where λ_1 , λ_2, and λ_3 are weights that control the contribution of each objective.

To train the agent, Proximal Policy Optimization (PPO) [8], a policy gradient-based method, is employed for stable and efficient learning. PPO enables the RL agent to explore various selection strategies while ensuring that the learned policy converges to an optimal solution. Through this process, the agent dynamically selects a set of diverse, informative, and non-redundant frames, which are later combined to produce a concise and context-aware video summary. This approach effectively addresses the challenges of temporal coverage and redundancy while maintaining the informativeness of the final summary.

3.6 Summary Generation

In this final stage, RL based agent is used for frame selection. The selected frames are used to maximize the informativeness, diversity, and temporal coverage. later these rewards are combined to produce a concise and coherent video summary. The frames are stitched together in temporal order to preserve the logical progression of the surgical procedure, ensuring the summary remains contextually meaningful. For smooth temporal transitions, additional context frames are included around key transitions to handle abrupt visual shifts. It also help in maintaining continuity between significant events, such as tool changes or critical surgical actions.

The resulting summarized video effectively highlights the most important events while eliminating redundancy, creating a compact representation of the original surgical video. This output is tailored for medical applications, such as surgical skill evaluation, procedural analysis, and training, where clarity, temporal coherence, and completeness are essential.

4 Experimental Results

This section outlines the implementation details of a medical video analysis pipeline, including hyperparameter selection, datasets utilized, evaluation metrics, quantitative results, and an ablation study.

4.1 Implementation Details

The proposed MedSum framework efficiently processes surgical videos for gesture-level and phase-level summarization. Frames are sampled at 2 fps, resized to 224×224 pixels, and spatial and motion feature extraction is performed using EfficientNet-B0 and Farneback optical flow respectively. Temporal segmentation via KTS identifies coherent segments with an embedding change threshold of 0.1. A Transformer Encoder (2 layers, 4 heads, hidden dimension: 128, dropout: 0.1) models inter-segment relationships. Frame selection is driven by PPO, optimizing informativeness ($\lambda_1 = 0.5$), coverage ($\lambda_2 = 0.3$), and redundancy ($\lambda_3 = 0.2$). The PPO agent is trained with a learning rate of 3×10^{-4}, discount factor ($\gamma = 0.99$), and clipping parameter ($\epsilon = 0.2$). Selected frames are combined in temporal order, with additional context frames ensuring smooth transitions. Final summaries retain 15% of frames, effectively capturing key events while reducing redundancy, achieving robust performance for surgical video summarization.

4.2 Datasets

The evaluation of the proposed MedSum approach was conducted using two datasets tailored for medical video analysis: JIGSAWS [3] and CATARACTS. JIGSAWS focuses on robotic-assisted surgical tasks with gesture-level annotations that capture detailed tool motions, such as suturing or knot tying, making it ideal for assessing fine-grained summarization techniques.

CATARACTS [1], on the other hand, features high-definition videos of cataract surgeries with phase-level annotations, emphasizing broader procedural steps like incision and lens insertion. Together, these datasets allow for comprehensive evaluation across both micro (gesture-level) and macro (phase-level) summarization tasks, highlighting the adaptability and robustness of MedSum. The details are mentioned in Table 1.

Table 1 Summary of datasets used for evaluating MedSum

Dataset	Focus	Annotations	Resolution	Size
JIGSAWS	Robotic surgical tasks	15 gestures (gesture-level)	640 × 480	103 videos
CATARACTS	Cataract surgical workflows	21 phases (phase-level)	1080p (HD)	50 surgeries

4.3 Evaluation Metrics

The standard metrics at gesture and phase level are calculated to evaluate the quality of video summaries using MedSum.

The F1-Score(F1) evaluates the accuracy of the generated summary by comparing its frames to the ground truth annotations. It combines precision and recall into a single metric. At the gesture-level, it measures overlap with fine-grained annotations such as insert needle or tie knot. For the phase-level, it evaluates broader surgical phases, such as incision or lens insertion.

The coverage ratio(CR) measures the proportion of annotated gestures or phases included in the summary. It is defined in Eq. (2). Higher coverage indicates a more comprehensive representation of the surgical video.

$$\mathrm{CR} = \frac{\text{Number of Gestures/Phases in Summary}}{\text{Total Gestures/Phases in Video}} \tag{2}$$

Temporal smoothness(TS) measures the continuity between consecutive gestures or phases in the summary. It evaluates how well the summary preserves logical transitions, ensuring coherence and minimizing abrupt shifts. Metrics like temporal continuity can quantify this smoothness.

These metrics provide a robust evaluation framework for assessing both gesture-level and phase-level summarization quality.

4.4 Quantitative Results

For quantitative evaluation, the F1-score was used as the performance metric, following prior works [2, 6, 12]. The MedSum approach was benchmarked against unsupervised methods like SGAN [7] and Cycle-SUM [10], reinforcement learning methods such as DRDSN [11] and PRLVS [9], and supervised methods including VASNet [2], DSNet [12], and MPFN [6]. These state-of-the-art methods are referred as baselines falling into supervised, unsupervised, and RL-based categories. Their results are reproduced for JIGSAWS and CATARACTS to ensures a fair comparison of MedSum against a diverse set of techniques across different summarization methodologies.

Table 2 Comparative F1-Score (%) results of various video summarization methods on the JIGSAWS and CATARACTS datasets

Approach	Method	F1 score(%)	
		JIGSAWS	CATARACTS
Supervised	VASNet [2]	57.2	54.9
	DSNet [12]	59.8	57.5
	MPFN [6]	61.0	59.0
Unsupervised	SGAN [7]	45.9	44.8
	Cycle-SUM [10]	48.5	47.4
Reinforcement	DRDSN [11]	52.6	51.3
	PRLVS [9]	56.3	55.2
	MedSum(Proposed)	**69.3**	**68.2**

The best values are marked in bold.

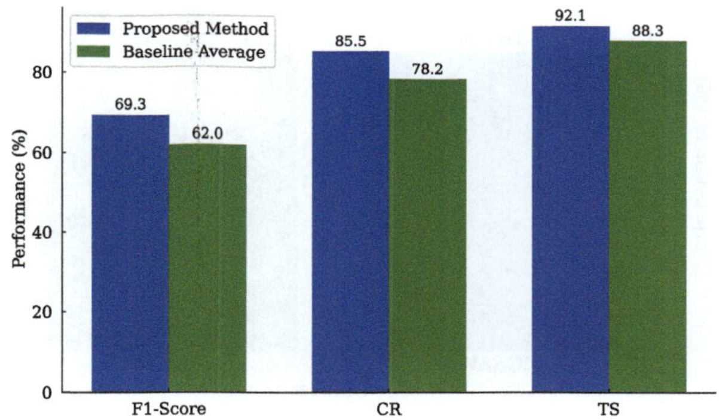

Fig. 2 Gesture-level performance of JIGSAWS in terms of F1-score, Coverage Ratio (CR), and Temporal Smoothness (TS) between the proposed method and baseline

As shown in Table 2, MedSum achieves the highest F1-scores, with 69.3% on JIGSAWS and 68.2% on CATARACTS, outperforming all existing methods. In comparison, the best supervised method, MPFN, reached 61.0 and 59.0%, while the best RL method, PRLVS, achieved 56.3 and 55.2%. These results demonstrate MedSum's superior ability to capture critical gestures and phases in medical video summarization.

Figures 2 and 3 demonstrates the superior performance of MedSum over baselines using key metrices(F1-Score, CR, TS), highlighting its effectiveness in capturing fine-grained gestures and maintaining temporal continuity computed respectively for gesture-level and phase-level annotations.

The average performance score of 18.5% for MedSum compared to baseline methods including supervised, unsupervised, and reinforcement learning-based approaches on JIGSAWS and CATARACTS in Fig. 4, is determined by averaging the individual improvement percentages.

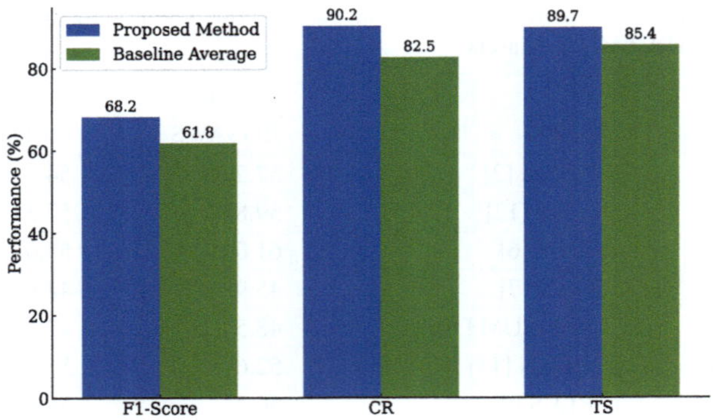

Fig. 3 Phase-level performance of CATARACTS in terms of F1-score, Coverage Ratio (CR), and Temporal Smoothness (TR) between the proposed method and baseline

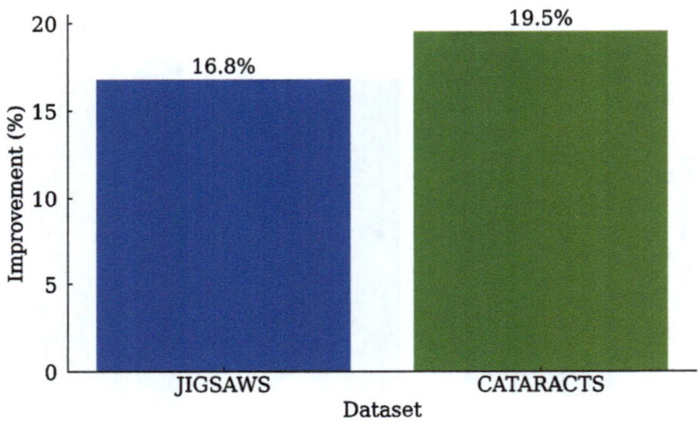

Fig. 4 Average improvement of JIGSAWS and CATARACTS over baselines

4.5 Ablation Study

An ablation study evaluates the contributions of individual components in the MedSum pipeline by progressively disabling specific modules or features and measuring the performance impact.

The ablation study in Table 3 highlights the importance of individual components in the MedSum pipeline. Removing Transformer attention, RL-based selection, redundancy penalty, or temporal segmentation significantly degrades performance, with the full MedSum configuration achieving the best F1 and CR scores for both datasets.

5 Conclusion

MedSum is a robust framework tailored for surgical video summarization, addressing the intricate challenges of gesture-level and phase-level analysis in medical contexts. By integrating transformer-based attention for segment prioritization, rein-

Table 3 Ablation study results for MedSum

Configuration	Transformer attention	RL-based frame selection	Redundancy penalty	Temporal segmentation	JIGSAWS		CATARACTS	
					F1 (%)	CR (%)	F1 (%)	CR (%)
$MedSum$(Proposed)	✓	✓	✓	✓	**69.3**	**85.5**	**68.2**	**90.2**
$MedSum_{\alpha}$	×	✓	✓	✓	61.8	78.3	63.5	84.0
$MedSum_{\beta}$	✓	×	✓	✓	62.3	79.5	64.0	84.8
$MedSum_{\gamma}$	✓	✓	×	✓	64.5	82.0	65.8	86.5
$MedSum_{\delta}$	✓	✓	✓	×	60.9	75.4	61.7	81.2

The best values are marked in bold.

forcement learning for dynamic frame selection, and a redundancy-aware reward mechanism, MedSum ensures informative, coherent, and non-redundant summaries. The use of EfficientNet for spatial features and motion-based optical flow enhances its ability to capture critical surgical events effectively. While MedSum excels in handling the complexity of surgical procedures, its computational intensity, especially during the reinforcement learning phase, may present challenges for real-time deployment.

Future work could focus on optimizing computational efficiency, incorporating lightweight attention models for faster inference, and extending MedSum to multi-modal data such as audio and surgical annotations. Additionally, adapting the framework for other healthcare and general video domains could further enhance its versatility and impact. Integrating domain-specific metrics and real-time feedback mechanisms could also provide new opportunities for surgical skill assessment and training.

References

1. ALHAJJ, H., Lamard, M., Conze, P.h., Cochener, B., Quellec, G.: Cataracts (2021), https://dx.doi.org/10.21227/ac97-8m18
2. Fajtl, J., Sokeh, H.S., Argyriou, V., Monekosso, D., Remagnino, P.: Summarizing videos with attention. In: Computer Vision–ACCV 2018 Workshops: 14th Asian Conference on Computer Vision, Perth, Australia, December 2–6, 2018, Revised Selected Papers 14. pp. 39–54. Springer (2019)
3. Gao, Y., Vedula, S.S., Reiley, C.E., Ahmidi, N., Varadarajan, B., Lin, H.C., Tao, L., Zappella, L., Béjar, B., Yuh, D.D., et al.: Jhu-isi gesture and skill assessment working set (jigsaws): A surgical activity dataset for human motion modeling. In: MICCAI workshop: M2cai. p. 3 (2014)
4. Gupta, J., Garg, D., Mishra, V.K.: Reward based video summarization using advanced deep learning architectures. In: Proceedings of the 2022 Fourteenth International Conference on Contemporary Computing. pp. 410–416 (2022)
5. Gupta, J., Rai, D., Garg, D., Mishra, V.K.: Video summarization using bidirectional rnn and vanilla autoencoder. In: 2024 International Conference on Computer, Electronics, Electrical Engineering & their Applications (IC2E3). pp. 1–4. IEEE (2024)
6. Khan, H., Hussain, T., Ullah Khan, S., Ahmad Khan, Z., Baik, S.W.: Deep multi-scale pyramidal features network for supervised video summarization. Expert Systems with Applications 237, Art No. 121288 (2024)
7. Mahasseni, B., Lam, M., Todorovic, S.: Unsupervised video summarization with adversarial lstm networks. In: Proceedings of the IEEE conference on Computer Vision and Pattern Recognition. pp. 202–211 (2017)
8. Schulman, J., Wolski, F., Dhariwal, P., Radford, A., Klimov, O.: Proximal policy optimization algorithms (2017)
9. Wang, G., Wu, X., Yan, J.: Progressive reinforcement learning for video summarization. Information Sciences 655, 119888 (2024)
10. Yuan, L., Tay, F.E., Li, P., Zhou, L., Feng, J.: Cycle-sum: Cycle-consistent adversarial lstm networks for unsupervised video summarization. In: Proceedings of the AAAI Conference on Artificial Intelligence. vol. 33, pp. 9143–9150 (2019)
11. Zhou, K., Qiao, Y., Xiang, T.: Deep reinforcement learning for unsupervised video summarization with diversity-representativeness reward. In: Proceedings of the AAAI Conference on Artificial Intelligence. vol. 32 (2018)
12. Zhu, W., Lu, J., Li, J., Zhou, J.: Dsnet: A flexible detect-to-summarize network for video summarization. IEEE Transactions on Image Processing 30, 948–962 (2021)

MIX
Papier aus verantwortungsvollen Quellen
Paper from responsible sources
FSC® C105338
FSC
www.fsc.org

If you have any concerns about our products,
you can contact us on
ProductSafety@springernature.com

In case Publisher is established outside the EU,
the EU authorized representative is:
Springer Nature Customer Service Center GmbH
Europaplatz 3, 69115 Heidelberg, Germany

Printed by Libri Plureos GmbH
in Hamburg, Germany